MW01169390

Trends and impacts of foreign investment in developing country agriculture

Evidence from case studies

Food and Agriculture Organization of the United Nations
Rome, 2013

Reprint June 2013

The designations employed and the presentation of material in this information product do not imply the expression of any opinion whatsoever on the part of the Food and Agriculture Organization of the United Nations (FAO) concerning the legal or development status of any country, territory, city or area or of its authorities, or concerning the delimitation of its frontiers or boundaries. The mention of specific companies or products of manufacturers, whether or not these have been patented, does not imply that these have been endorsed or recommended by FAO in preference to others of a similar nature that are not mentioned.

The views expressed in this information product are those of the author(s) and do not necessarily reflect the views or policies of FAO.

ISBN 978-92-5-107401-5 (print)
E-ISBN 978-92-5-107786-3 (PDF)

© FAO 2012

FAO encourages the use, reproduction and dissemination of material in this information product. Except where otherwise indicated, material may be copied, downloaded and printed for private study, research and teaching purposes, or for use in non-commercial products or services, provided that appropriate acknowledgement of FAO as the source and copyright holder is given and that FAO's endorsement of users' views, products or services is not implied in any way.

All requests for translation and adaptation rights, and for resale and other commercial use rights should be made via www.fao.org/contact-us/licence-request or addressed to copyright@fao.org.

FAO information products are available on the FAO website (www.fao.org/publications) and can be purchased through publications-sales@fao.org.

Acknowledgements

This book was produced by the Team on International Investment in Agriculture, Standards and Partnerships for Sustainable Trade of FAO's Trade and Market Division[1].

The work was co-funded by the Government of Japan through a Trust Fund project to the Food and Agriculture Organization of the United Nations "Support to study on appropriate policy measures to increase investments in agriculture and to stimulate food production".

This publication was edited by, in alphabetical order, Pedro Arias, David Hallam, Suffyan Koroma and Pascal Liu. The editors are very grateful to Massimo Iafrate for the processing of data and preparation of graphs, to Rita Ashton, Daniela Piergentili and Ettore Vecchione for the formatting and layout of the book, and to Jesper Karlsson for his assistance in the proofreading.

They would like to thank colleagues from other divisions of FAO for their support and/or inputs, in particular Mafa Chipeta, Sarah Lowder, Paul Mathieu, Masahiro Miyazako, Jakob Skoet, Saifullah Syed and Diana Tempelman.

They are grateful to the International Institute for Environment and Development (IIED), in particular Lorenzo Cotula, for the coordination of the case studies on Ghana, Mali and Zambia. Special thanks go to the following national researchers and consultants who undertook the original field research and wrote the draft reports:

- Fison Mujenja for the Zambia case study
- Alice K. Gowa for the Uganda case study
- Waleerat Suphannachart and Nipawan Thirawat for the Thailand case study
- Bede Lyimo for the Tanzania case study
- Adama Ekberg Coulibaly for the Senegal case study
- Moussa Djiré, Amadou Kéita and Alfousseyni Diawara for the Mali case study
- John Bugri and Adama Ekberg Coulibaly for the Ghana case study
- Saing Chan Hang, Hem Socheth, Ouch Chandarany, Phann Dalis and Pon Dorina, Cambodia Development Resource Institute, for the Cambodia case study
- Jose Rente Nascimento for the Brazil case study

[1] The team members are Pedro Arias, Maria Arnal, Gianluca Gondolini, David Hallam, Massimo Iafrate, Jesper Karlsson, Suffyan Koroma, Pascal Liu, Daniela Piergentili and Manitra Rakotoarisoa

Acronyms

ABRAF	Brazilian Association of Plantation Forest Producers
ADB	African Development Bank
ADB	Asian Development Bank
ADM	Archer Daniels Midland Company
ADRA	Adventist Development and Relief Agency
AGOA	African Growth Opportunity Act
AGRA	Alliance for a Green Revolution in Africa
ANRM	Assemblée Nationale de la République du Mali (Malian National Assembly)
APEC	Asia Pacific Economic Cooperation
APEP	USAID's Agricultural Productivity Enhancement Program
APEX	Brazilian Agency for Export and Investment Promotion
APIX	Agence de Promotion des Investissements du Sénégal
ASDP	Agriculture Sector Development Programme
ASEAN	Association of South East Asian Nations
BCB	Brazilian Central Bank
BDS	Business Development Services
BEST	Business Environment Strengthening for Tanzania
BFS	Brazilian Forest Service
BIT	Bilateral Investment Treaties
BNDES	Brazilian Development Bank
BOI	Board of Investment
BOT	Bank of Thailand
BPF	Business and Property Formalization
CAADP	Comprehensive Africa Agriculture Development Programme
CC	Coordinating Committee to implement the PROMECIF
CCG	Codes of Corporate Governance
CCROs	Customary Certificates of Rights of Occupancy
CCSR	Codes of Corporate Social Responsibility
CDC	Commonwealth Development Corporation
CDC	Council for the Development of Cambodia
CDF	Code Domanial et Foncier (Mali) (Property and Land Law)
CDRI	Cambodian Development Resource Institute
CDTF	Cotton Development Trust Fund
CEA	Cambodia Economic Association
CEDAC	Centre d'Étude et de Développement Agricole Cambodgien
CERFLOR	Brazilian Forest Certification Scheme

Acronyms

CFA	Communauté Financière Africaine (African Finance Community)
CGI	Global Competitiveness Index
CIB	Cambodia Investment Board
CIBRAZEM	Brazilian Storage Company
CICOL	Civil Society Coalition on Land
CILSS	Comité Inter - Etats de Lutte Contre la Sécheresse dans le Sahel (Permanent InterState Committee for Drought Control in the Sahel)
COMESA	Common Market for Eastern and Southern Africa
CPI	Corruption Perceptions Index
CROs	Certificates of Rights of Occupancy
CSO	Central Statistical Office
CSR	Corporate Social Responsibility
CT	Collectivité Territoriale (Local Authority)
CTSP	Children to School Support Programme
DANIDA	Danish International Development Agency
DEP	Department of Export Promotion
DFI	Development Finance Institution
DFiD	Department for International Development
DGID	Direction Générale des Impôts et des Domaines
DISEZ	Dakar Integrated Special Economic Zone
DN	Domaine National
DNCN	Direction Nationale de la Conservation de la Nature (Malian Nature Conservation Department)
DNEF	Direction Nationale des Eaux et Forêts (Malian Water and Forestry Department)
DNH	Direction National de l'Hydraulique (Malian Hydraulics Department)
DNP	Direction Nationale de la Pêche (Malian Fisheries Department)
DNPIA	Direction Nationale des Productions et des Industries Animales (Malian Livestock Production and Industry Department)
DNSV	Direction Nationale des Services Vétérinaires (Veterinary Services Department)
DRA	Direction Régionale de l'Agriculture (Regional Agriculture Department)
DRC	Democratic Republic of the Congo
DRPIA	Direction Régionale des Productions et des Industries Animales (Regional Livestock Production and Industry Department)
DSIP	Development Strategy and Investment Plan
DTAs	Double Taxation Agreements

EAC	East African Community
EC	Executive Committee
ECOWAP	Economic Community of West Africa States Agricultural Policy
ECOWAS	Economic Community of West Africa States
EFE	Zone Franche d'Exportation
EGF	Etats Généraux du Foncier (Malian Land Tenure Congress)
EIA	Environmental Impact Assessment
ELC	Economic Land Concession
EMBRAPA	Brazilian Agricultural Research Corporation
EMBRATER	Brazilian Enterprise of Technical Assistance and Rural Extension
EPA	Environmental Protection Agency
EPZ	Export Processing Zone
ERR	Expected Rate of Return
ESIA	Environmental and Social Impact Assessment
ESMP	Environmental and Social Management Plan
EU	European Union
FAO	Food and Agriculture Organization of the United Nations
FAQ	Fair Average Quality
FASDEP	Food and Agriculture Sector Development Policy
FBG	Farmer Business Group
FBOs	Farm Base Organizations
FCC	Fair Competition Commission
FDI	Foreign Direct Investment
FFA	Free Fatty Acids
FGD	Focus Group Discussion
FIAS	Foreign Investment Advisory Service
FINAME	Special Agency for Industrial Financing, a branch of BNDES
FIPA	Foreign Investment Protection Agreement
FISP	Farmer Input Support Programme
FSC	Forest Stewardship Council
FSP	Fertilizer Support Programme
FTA	Free Trade Agreement
FVL	Forest Vocation Lands
GCF	Gross Capital Formation
GDP	Gross Domestic Product
GDS	Grands Domaines du Sénégal
GEPC	Ghana Export Promotion Council
GERSDA	Groupe d'Etude et de Recherche en Sociologie et Droit Appliqué (Research and Study Group on Sociology and Applied Law)
GFC	Ghana Forestry Commission

GFZB	Ghana Free Zones Board
GHC	Ghanaian Cedi
GIF	Ghana Investment Fund
GIPC	Ghana Investment Promotion Centre
GLOBALGAP	Global Good Agricultural Practices
GMO	Genetically Modified Organism
GNP	Growth National Product
GOANA	Grande Offensive Agricole pour la Nourriture et l'Abondance
GoB	Government of Brazil
GOPDC	Ghana Oil Palm Development Companies
GPS	Geographical Positioning System
GRZ	Government of the Republic of Zambia
GSE	Ghana Stock Exchange
GTC	Green Trade Company
GTZ	German Technical Cooperation
GVC	Global Value Chains
HACCP	Hazard Analysis Critical Control Point
HDI	Human Development Index
HIV/AIDS	Human Immunodeficiency Virus/Acquired Immunodeficiency Syndrome
HS	Harmonized System
IBRD	International Bank for Reconstruction and Development. Also WB
ICSID	International Centre for the Settlement of Investment Disputes
ICT	Information and Communication Technology
IDB	Inter-American Development Bank
IDEA	USAID's Investment in Developing Export Agriculture
IFAD	International Fund for Agricultural Development
IFC	International Finance Corporation
IFPRI	International Food Policy Research Institute
IICA	Inter-American Institute for Cooperation on Agriculture
IIA	International Investment Agreement
IIED	International Institute for Environment and Development
ILO	International Labour Organization
IMF	International Monetary Fund
IOSCO	International Organization of Securities Commissions
IPCC	Intergovernmental Panel on Climate Change
IPPA	Investment Protection and Protocol Agreement
ISO	International Organization for Standardization
ITFC	Integrated Tamale Fruit Company
IUCN	International Union for the Conservation of Nature

JV	Joint Venture
Kascol	Kaleya Smallholders Company Ltd (Kascol)
KHR	Cambodian riel
LAC	Latin America and the Caribbean
LDC	Least Developed Country
LGA	Local Government Authority
LICUS	Lower Income Country Under Stress
LOA	Loi d'Orientation Agricole (Agricultural Framework Law)
M&A	Merger and Acquisitions
M&E	Monitoring and Evaluation
MAAIF	Ministry of Agriculture, Animal Industry and Fisheries
MAFF	Ministry of Agriculture, Forestry and Fisheries
MAFSC	Ministry of Agriculture, Food Security and Cooperatives
MCC	Millennium Challenge Corporation
MDC	Mpongwe Development Company
MDGs	Millennium Development Goals
MDRE	Ministère du Développement Rural et de l'Environnement (Malian Ministry for Rural Development and the Environment)
MEDIZON	Ministère Délégué auprès du Premier Ministre, chargé du Dévéloppement Intégré de la Zone de l'Office du Niger (Secretary of State, attached to the Prime Minister's office, responsible for the Integrated Development of the Office du Niger Area
MEF	Ministry of Economy and Finance
METASIP	Medium Term Agriculture Sector Investment Plan
MFC	Ministère des Finances et du Commerce (Malian Ministry of Finance and Trade)
MFEZ	Multi-facility Economic Zone
MFPED	Ministry of Finance, Planning and Economic Development
MiDA	Millennium Development Authority
MIGA	Multilateral Investment Guarantee Agency
MLAFU	Ministère du Logement, des Affaires Foncières et de l'Urbanisme (Malian Ministry of Housing, Land and Town Planning)
MLHHSD	Ministry of Lands, Housing and Human Settlements Development
MMA	Ministry of Environment of Brazil
MMEE	Ministère des Mines, de l'Énergie et de l'Eau (Ministry of Mines, Energy and Water)
MNEs	Multinational Enterprises
MoE	Ministry of Environment
MoFA	Ministry of Food and Agriculture
MoP	Ministry of Planning
MoU	Memorandum of Understanding
MoWRaM	Ministry of Water Resources and Meteorology
MSME	Micro, Small and Medium Enterprise

Mt	million metric tonnes
NAADS	National Agricultural Advisory Service
NAMBOARD	National Agricultural Marketing Board
NAROS	National Agricultural Research Organization
NEPAD	New Economic Partnership for Africa Development
NESBD	National Economic and Social Development Board
nFVL	Non-forest Vocation Lands
NGFSRP	Northern Ghana Food Security Resilience Project
NGO	Non-Governmental Organization
NIC	Newly Industrialized Countries
NSDP	National Strategic Development Plan
NSTDA	National Science and Technology Development Agency
OAE	Office of Agricultural Economics
OAPI	African Organization for Intellectual Property
ODS	Ozone Depleting Substances
OECD	Organization for Economic Co-operation and Development
OMOA	Organic Mango Outgrowers Association
ON	Office du Niger
ONFH	Observatoire National du Foncier et de l'Habitat (National Land and Housing Observatory)
PAP	Project-affected people
PEAP	Poverty Eradication Action Plan
PICS	Productivity and Investment Climate Surveys
PIV	Périmètres Irrigués Villageois (Village irrigation schemes)
PKC	Palm Kernel Cake
PKO	Palm Kernel Oil
PMA	Plan for the Modernization of Agriculture
PMO-RALG	Prime Minister's Office, Regional Administration and Local Government
PND	National Development Plan
POME	Palm Oil Mill Effluent
PPP	Public Private Partnership Policy
PPRA	Public Procurement Regulatory Authority
PPTA	Project Preparatory Technical Assistance
P-RM	Présidence de la République du Mali (Presidency of the Republic of Mali)
PROMECIF	Forestry Investment Business Climate Improvement Process
PRONAF	National Programme for Family Agriculture Strengthening
PRSC	Parastatal Sector Reform Commission
PRSP	Poverty Reduction Strategy Paper
PSM	Projet Sucrier de Markala (Markala Sugar Project)
PSOM	Dutch Government Programme

PTF	Privatization Trust Fund
PTS	Pesticides and Toxic Substances
R&D	Research and Development
RAI	Responsible Agricultural Investment (Principles)
REDD	Reducing Emissions from Deforestation and Forest Degradation in developing countries
ROI	Return on Investment
RPKO	Refined Palm Kernel Oil
SAGCOT	Southern Agricultural Growth Corridor of Tanzania
SAGF	Rabo Sustainable Agriculture Guarantee Fund
SDC	Swiss Development Cooperation
SDDZON	Schéma Directeur de Développement de la Zone de l'Office du Niger (Development Master Plan for the Office du Niger Area)
SEDIZON	Secrétariat d'Etat auprès du Premier Ministre, chargé du Dévéloppement Intégré de la Zone de l'Office du Niger (Secretary of State, attached to the Prime Minister's office, responsible for the Integrated Development of the Office du Niger Area)
SEAFMD	South East Asia Food and Mouth Disease Regional Coordination
SEKAB	Swedish Ethanol Chemistry AB
SEZ	Special Economic Zone
SIDA	Swedish International Development Agency
SIF	Forestry Research Society
SLC	Social Land Concession
SLPIA	Service Local des Productions et des Industries Animales (Local Livestock Production and Industry Service)
SME	Small and Medium Enterprise
SOCAS	Société de Conserves Alimentaires du Sénégal
SOSUMAR	Société Sucrière de Markala (Markala Sugar Company)
SPILL	Strategic Plan for Implementation of Land Laws
SSA	Sub-Saharan Africa
SSNIT	Social Security and National Insurance Trust
TAFTA	Thailand-Australia Free Trade Agreement
TECHNOSERVE	Technology in the Service of Mankind
TIC	Tanzania Investment Centre
TNC	Transnational Corporation
TRIMs	WTO Agreement on Trade-related Investment Measures
UBR	Uganda Business Register
UBS	Uganda Bureau of Statistics
UCDA	Uganda Coffee Development Authority
UEMOA	Union Économique et Monétaire Ouest Africaine
UEPB	Uganda Export Promotion Board
UFEA	Uganda Flower Exporters Association

UFPA	Uganda Fish Processors Association
UFV	Federal University of Viçosa
UIA	Uganda Investment Authority
UK	United Kingdom
UN	United Nations
UNCTAD	United Nations Conference on Trade and Development
UNDP	United Nations Development Programme
URA	Uganda Revenue Authority
USA	United States of America
USAID	United States Agency for International Development
USDA	United States Department of Agriculture
US$	United States dollar
VAT	Value Added Tax
WARDA	West African Rice Development Association
WB	The World Bank. Also IBRD.
WCGA	Western Cotton Growing Area
WEF	World Economic Forum
WFP	World Food Programme
WFS	World Food Summit
WIPO	World Intellectual Property Organization
WIR	World Investment Report
WTO	World Trade Organization
ZDA	Zambia Development Agency
ZMK	Zambian Kwacha
ZEMA	Zambia Environmental Management Agency
ZSC	Zambia Sugar Company

Foreword

Large-scale international investments in developing country agriculture, especially acquisitions of agricultural land, continue to raise international concern. Certainly, complex and controversial issues – economic, political, institutional, legal and ethical – are raised in relation to food security, poverty reduction, rural development, technology and access to land and water resources. Yet at the same time, some developing countries are making strenuous efforts to attract foreign investment into their agricultural sectors. They see an important role for such investments in filling the gap left by dwindling official development assistance and the limitations of their own domestic budgetary resources, creating employment and incomes and promoting technology transfer. More investment is certainly needed – more than US$80 billion per year according to FAO analysis. But can foreign direct investment be compatible with the needs of local stakeholders as well as those of the international investor? And can these investments yield more general development benefits?

Analyzing the impacts of foreign direct investment in developing country agriculture and even understanding its extent and nature has been hampered by the weakness of the available information and the lack of comprehensive statistical data. Much discussion of the phenomenon has been based on media stories but these are potentially misleading unless very carefully triangulated. This lack of reliable detailed information means serious analysis has tended to rely on case studies. This book collects together case studies undertaken by FAO in nine different countries. These add to the increasing volume of evidence from similar case studies undertaken by other international organizations.

It is important that any international investment should bring development benefits to the receiving country in terms of technology transfer, employment creation, upstream and downstream linkages and so on if these investments are to be "win-win" rather than "neo-colonialism". These beneficial flows are not automatic: care must be taken in the formulation of investment contracts and selection of business model. Appropriate legislative and policy frameworks need to be in place. The case studies in this book describe the extent, nature and impacts of international investments and examine the effectiveness of policy and legal frameworks. Obviously, generalizations are difficult both on the impacts of foreign investments and on the best regulatory approaches but the studies provide a wealth of insights which should be valuable to host country governments and investors alike. Their findings shed light on a number of issues including the extent to which forms of investment other than land acquisition – such as contract farming, out-grower schemes and other joint ventures - are more likely to yield development benefits to host countries. They highlight the importance of stronger governance in the host country and provide some indications of the priority areas of focus for international efforts to formulate guiding principles for responsible agricultural investments.

David Hallam
Director
Trade and Markets Division

Table of contents

Page

Page

Page

PART 4: BUSINESS MODELS FOR AGRICULTURAL INVESTMENT - IMPACTS ON LOCAL DEVELOPMENT **157**

Page

Page

Page

Page

Page

PART 5: SYNTHESIS OF FINDINGS 321

PART 6: CONCLUSIONS AND RECOMMENDATIONS 333

List of figures

Page

PART 2: OVERVIEW OF TRENDS IN FOREIGN DIRECT INVESTMENT (FDI)

PART 3: POLICIES FOR ATTRACTING FDI AND IMPACTS ON NATIONAL ECONOMIC DEVELOPMENT

BRAZIL: IMPROVING THE BUSINESS CLIMATE FOR FDI

Page

Page

PART 4: BUSINESS MODELS FOR AGRICULTURAL INVESTMENT - IMPACTS ON LOCAL DEVELOPMENT

CAMBODIA: LOCAL IMPACTS OF SELECTED FOREIGN AGRICULTURAL INVESTMENTS

GHANA: PRIVATE INVESTMENT FLOWS AND BUSINESS MODELS IN GHANAIAN AGRICULTURE

SENEGAL: ASSESSING THE NATURE, EXTENT AND IMPACT OF FDI IN SENEGAL'S AGRICULTURE

List of tables

Page

UGANDA: ANALYSIS OF PRIVATE INVESTMENT IN THE COFFEE, FLOWERS AND FISH SECTORS OF UGANDA

PART 4: BUSINESS MODELS FOR AGRICULTURAL INVESTMENT - IMPACTS ON LOCAL DEVELOPMENT

CAMBODIA: LOCAL IMPACTS OF SELECTED FOREIGN AGRICULTURAL INVESTMENTS

Page

ZAMBIA: INVESTMENT IN AGRICULTURAL LAND AND INCLUSIVE BUSINESS MODELS

List of boxes

Page

List of charts

Page

PART ONE

INTRODUCTION

1. Global context and issues

After several decades of under-investment in the agricultural sector in developing countries, the late 2000s witnessed a surge in foreign direct investment (FDI) in primary agricultural production. The reasons for this surge are diverse and complex, but the main drivers can be linked to the steep rise in commodity prices in 2007-2008 and the realization that demand for finite natural resources is set to continue increasing significantly in the next four decades. The spike in food prices prompted countries that are heavily dependent on food imports to invest in other countries where land and other natural resources (in particular water) are abundant with a view to securing supply. They view the ownership of production and the possibility to export the harvest back home as a more reliable strategy for food security than depending on international markets. In addition, high energy prices triggered international investment in the production of feedstock crops for biofuels. Beyond causes that are linked to the current situation of markets, other drivers indicate that the trend is likely to continue in the longer term. These drivers include expectations of rising prices, population growth, growing consumption rates and market demand for food, biofuels, raw materials and carbon sequestration.

Expectations of rising prices for land and other natural resources have given rise to financial speculation. In turn, speculation on land and other natural resources has been fuelled by the poor market performance of more traditional asset classes such as equity and bonds in the wake of the financial crisis that started in 2007. According to a survey of 25 large investment firms prepared for the OECD (2010), investment in farmland and agricultural infrastructure offers the following attractions as an emerging asset class: strong long-term macroeconomic fundamentals; attractive historical returns on land investment; a mix of current income and capital appreciation; uncorrelated returns with the equities market and a strong hedge against inflation.

While foreign investment in agriculture is not a completely new trend, the current situation differs from more traditional forms of international investment in the agro-food sector, which primarily aimed to provide a better access to markets or cheaper labour. Through the new investment forms, investors seek to gain access to natural resources, in particular land and water. Another feature is that the new forms of investment involve acquisition of land and actual production rather than looser forms of association with local producers. The new investors emphasize production of basic foods, including animal feed, for export back to the investing country rather than tropical crops for wider commercial export (Hallam 2011). According to the OECD survey (2010), 83 percent of the farmland acquired or leased on a long-term basis by survey respondents was dedicated to the production of major row crops (soft oilseeds, corn, wheat and feed grains), with 13 percent being invested in livestock production (typically grazing of beef cattle, dairy, sheep and swine) and 4 percent of farmland dedicated to permanent crops such as sugar cane and viticulture, agricultural infrastructure and set-asides.

2. Assessing the extent of foreign investment in agriculture

Although there is ample evidence of increasing investment in developing country agriculture, it is difficult to quantify the current phenomenon due to the lack of reliable data. For 2007 and 2008, comparable data on total FDI to all sectors are only available for 27 countries. For these countries, average annual inward FDI flows in the two years were estimated at US$922.4 billion (UNCTAD 2011). Of this total, FDI into agriculture (including hunting, forestry and fisheries) represented only 0.4 percent. A larger share, 5.6 percent, went to the food, beverages and tobacco sectors, primarily in high-income countries.

Chapter prepared by Pascal Liu, Trade and Markets Division, FAO

Trends over time in FDI are difficult to monitor because the number of countries for which data are available varies from year to year. Looking at agriculture alone, comparable data are available for 44 countries. FDI to these countries more than doubled between 2005-06 and 2007-08. However, the majority of these flows went to upper-middle and high-income countries (Lowder and Carisma 2011). These figures probably underestimate actual flows of foreign investment in agriculture, because data are missing for many countries. Furthermore, investments made by large private institutional investors, such as mutual funds, banks, pension funds, hedge funds and private equity funds are not included in estimates of FDI. A broad, though not comprehensive, recent survey of agricultural investment funds in several developing regions (excluding East Asia and the Pacific) found that such funds have increased in number and value (Miller et al., 2010). The second section of this report provides more estimates of FDI flows in the agricultural sector of selected developing regions and countries.

While foreign capital is invested in a wide array of agricultural assets, international debates and research have recently focused on foreign investments for the control of agricultural land on a large scale. This focus can be partly explained by the multifunctional characteristic of land. Beyond its economic value, land also has social, cultural and religious values in many countries. Large-scale land acquisition raises complex issues across various dimensions: legal, economic, social, environmental, ethical and cultural. Studies show that foreign investment in land takes place through purchase or long-term leases. Long-term lease of agricultural land is a more frequent arrangement than purchase in the case of foreign investment, partly due to the fact that several countries have regulations prohibiting the sale of land to foreigners. However, the economic and social implications tend to be similar as for outright sale since lease contracts are generally for a long period (typically 50 years and sometimes up to 99 years). In some cases of purchase, a local counterpart to the foreign investor is involved.

Several organizations have tried to estimate the area of land that has been the object of large-scale transactions in recent years using different sources. The non-governmental organization GRAIN has operated an online database of land acquisition mainly based on media reports www.farmlandgrab.org 2011). Estimates that are solely based on the collection of media reports may be misleading, as a substantial share of the announced projects does not materialize in an actual transaction for various reasons (including decision by the investor not to proceed). Systematic inventories of land deals based on official government records, crosschecked with third-party sources are likely to produce more reliable estimates. The figures gathered through these national inventories are usually lower than those based on media reports. In Mozambique, for example, media sources arrived at more than 10 million hectares acquired between 2008 and 2010, whereas a national inventory for 2004–2009 calculated a figure closer to 2.7 million hectares (Cotula and Polack 2012). The average size of individual transactions is also smaller than that suggested by media reports. The World Bank estimates that an area of 46.6 million hectares was acquired between October 2008 and August 2009 (Deininger and Beyerlee 2011).

The Land Matrix, a partnership between the Centre for Development and Environment (CDE) at the University of Bern, the Centre de coopération Internationale en Recherche Agronomique pour le Développement (CIRAD), the German Institute of Global and Area Studies (GIGA), the German Agency for International Cooperation (GIZ) and the International Land Coalition (ILC), systematically collates and seeks to verify information on large-scale land acquisitions. The data collected by the Partnership originate from media reports, international and non-governmental organizations and academics. The Partnership has collected reports for 1 217 agricultural land deals in developing countries accounting for over 83 million hectares of land over the period 2000-2012 (Anseeuw et al 2012)[1].

[1] The main sections of the database are now publicly available (http://www.landportal.info/landmatrix).

However, it estimated that the area concerned by transactions that it judged as "reliable" (i.e. cross-checked with other sources) accounted for only 39.3 percent of this area (32.7 million hectares)[2].

The difference between estimates primarily derives from differences in the methods used for calculation. There are differences in the considered time periods (some surveys cover a whole decade, others only a couple of recent years), in the type of investments that is included (for example some surveys do not record transactions for establishing a tree plantation), in the status of the project (some databases include projects announced by the media while other only include approved transactions) and in the minimum area for the transaction to be recorded (for example, the Land Matrix only records deals that cover 200 hectares or more).

While it is clear that some figures highlighted by the media are overestimated, there is also evidence that not all land transactions are reported. Investors may have various reasons for not reporting a deal, including commercial confidentiality and fear for their corporate image. Similarly, some governments may be reluctant to publicize a transaction for a variety of reasons. Consequently, the transactions that are not reported may somewhat offset those that are announced but do not materialize. Finally, it should be noted that even when agreements are signed and the transaction takes place, the share of land that is cultivated in reality is often much less than what was announced by the investor.

In terms of destination of FDI, Africa is the most targeted region: the Land Matrix estimates that 754 land deals covering 56.2 million hectares are located in Africa, compared with 17.7 million hectares in Asia, and 7 million hectares in Latin America. Reported land deals in Africa concern an area equivalent to 4.8 percent of Africa's total agricultural area, or the territory of Zimbabwe (Anseeuw et al 2012). The majority of reported acquisitions are concentrated in a few countries. A large number of countries (84) are reported to be targeted by foreign investors, but only 11 of them concentrate 70 percent of the reported targeted area. Among those 11 countries, 7 are African, namely Sudan, Ethiopia, Mozambique, United Republic of Tanzania, Madagascar, Zambia and Democratic Republic of the Congo. In South-East Asia, the Philippines, Indonesia and Lao People's Democratic Republic are particularly affected.

In conclusion, even though the real scale of foreign investment in agricultural land may be smaller than what the media suggest, the available evidence shows that it is important.

The fact that most of the debates on large-scale land acquisitions has focused on foreign investments is easy to understand. Foreign investments raise a number of delicate issues related to national sovereignty and independence which are all the more sensitive in view of the colonial history of many countries. In addition, foreign investments in land can be large-scale with many involving more than 10,000 hectares and some more than 500 000 hectares (Hallam 2011). Investments by foreign firms tend to cover a larger area than those made by domestic companies. For example, in the Office du Niger area in Mali, no foreign investor acquired less than 500 hectares, while local investors acquired much more modest areas.

Nevertheless, the international attention given to foreign investment should not conceal the fact that in most countries domestic investors acquire more agricultural land than foreign ones. The World Bank (2011) estimates that domestic investors were responsible for 80 percent of the land transactions in the surveyed developing countries. Even though the average area covered by the transactions was smaller than that of foreign investments, domestic investors still accounted for 60 percent of the total acquired area. Case studies have shown the critical role of national elites in land acquisition. Nationals accounted for the following percentages of the area acquired in the following countries: 97

[2] Even the "cross-checked figures" should be treated with caution due to the lack of reliability of alternative sources in some cases.

percent in Nigeria, 70 percent in Cambodia, 53 percent in Mozambique and around 50 percent in Sudan and Ethiopia. In some cases, though, domestic companies act as an entry point for foreign investors, facilitating their access to agricultural assets (Burnod et al. 2011).

3. Origins of agricultural FDI

A variety of actors from both the private and public sectors are involved in this new investment trend. Private sector actors include investment funds, pension funds, hedge funds, agricultural and agro-industrial companies, and in some cases, energy companies. Public sector actors include governments, sovereign wealth funds and other state-owned companies. Increasingly, governments prefer to support investment by their home companies rather than investing directly into agricultural land in developing countries. This results partly from a strategy of risk reduction, including financial risks and risks to their reputation in the wake of negative media coverage. This support can take the form of public private partnerships whereby the government provides or guarantees loans and provides tax rebates, technical assistance or other means of assistance. A recent survey suggests that investments made by public-private partnerships accounted for some 600 000 hectares in 2012 (Anseeuw et al 2012).

In terms of geographical origin, recently-published data from the Land Matrix indicate that investment originates from three groups of countries: emerging economies in East Asia and South America; Gulf countries; and countries from North America and Europe (Anseeuw et al 2012). International media have highlighted the role played by Middle Eastern and East Asian countries, in particular China. However, the World Bank finds that it is only in Sudan that Middle Eastern countries account for a majority of foreign investment in agriculture (Deininger et Byerlee 2011). As for China, Cotula and Polack (2012) suggest that it is a key investor in Southeast Asia but has a less important contribution to investment in agricultural land in Africa. There is evidence that companies from Southeast Asia have been investing significantly in African agriculture. Southeast Asia has become both a destination for and a source of foreign agricultural investment. South America is in a similar situation. Although North American and European investors have attracted less media attention, there is evidence that they account for a significant share of foreign investment in developing country agriculture. According to a survey done for the OECD (2010), most investment funds investing in farmland across the world are based in Europe and North America. Schoneveld (2011) argues that European firms account for 40 percent of all land acquired in Africa, while North American companies account for 13 percent. In particular, European and North American firms dominate investments for the production of biofuels in Africa.

4. Patterns of FDI flows

There is a strong tendency towards intra-regional investment in Asia and South America, as local firms seek to replicate the success in their home country by investing across the national borders. In Africa, South African companies have been successfully investing in other countries of the continent. In some cases they channel investment from companies based in another continent into other African countries, such as Mozambique, United Republic of Tanzania or Zambia, taking advantage of their expertise in African agriculture (Cotula and Polack 2012). Partnerships are important for investors, as they can contribute to reducing the costs of complex local administration, and for legal reasons in some contexts. For example in 12 percent of the cases collected by the Land Matrix Project, foreign investors had built partnerships with domestic companies. Foreign investors also often act in partnership with each other. Investors from the United States, United Kingdom and South Africa have formed such partnerships in about a third of the deals in which there are involved (Anseeuw et al. 2012).

As for inter-regional investment, a particular pattern of bilateral investment flows has emerged following established cultural, political and business ties and geographical restrictions on investment funds. Gulf countries have favoured investments in Sudan and other, mainly African, OIC member states, for example. China has favoured Southeast Asia and, in Africa, Zambia, Angola and Mozambique (von Braun and Meinzen-Dick 2009).

5. Implications for food security

Various studies suggest that investors are targeting countries with weak land tenure security, although they seek countries that, at the same time, offer relatively high levels of investor protection (Anseeuw et al. 2011, Deininger and Byerlee 2011). The data from the Land Matrix reveal a tendency for investors to focus on the poorest countries, and those that are also less involved in world food exchanges. Investors are targeting countries that are among the poorest, are poorly integrated into the world economy, have a high incidence of hunger, and weak land institutions. Some 66 percent of the deals reported in the Land Matrix were in countries with high prevalence of hunger.

The implications for food security are even more significant when one considers the type of land that is being acquired. In most cases these are good quality, fertile lands with irrigation. Investors have a tendency to target land with high yield gaps, good accessibility and considerable population densities. Spatial analysis of land deals reveal that they tend to target cropland where the yield gap is relatively large, and where additional inputs (water, fertilizers, seeds, infrastructure and know-how) may create greater yields. For example, land acquisitions in Mali and Senegal are heavily concentrated in the irrigable areas of the Ségou Region and the Senegal River valley, respectively (Cotula and Polack 2012). Accessibility is another criterion for choice of target area: the majority of deals may be less than three hours away from the next city. The lands targeted by investors are located near roads and markets. More than 60 percent of all land deals target areas with population densities of more than 25 persons per km2 (Anseeuw et al 2012). Approximately 45 percent of the land deals included in the Land Matrix database concern cropland or crop-vegetation mosaics. Intensive competition for cropland with local communities is therefore likely. Even where national indicators may suggest large reserves of suitable land, transactions are often found within cultivated areas and farmland. This finding questions the assumption that investments are mostly focused on non-utilized land and serve to bring it into production. It has important implications for food security, especially if the crop is destined for exportation.

In addition to the direct risks in terms of reduced food availability at the local level, there are other risks associated with large-scale land acquisition, especially in countries where local land rights are not clearly defined and governance is weak. These risks include the displacement of local smallholders, the loss of grazing land for pastoralists, the loss of income for local communities, and in general, negative impacts on livelihoods due to reduced access to resources, which may lead to social fragmentation. For example, while rural communities often derive incomes from the collection of timber and non-wood products in forests, forested areas are highly affected by land acquisitions. Some 24 percent of the land deals surveyed by the Land Matrix Project are located in forested areas, representing 31 percent of the total area of land acquisitions.

These negative effects may generate conflict. The risk of adverse environmental impacts is important too. All these risks have been highlighted by a wide range of institutions including farmer organizations, research institutes, regional farm groups, governments, the media, development agencies, non-governmental organizations and multilateral organizations. They have rightly generated much concern and international debates. To some extent, this focus on large-scale land acquisition and its risks has tended to overshadow the fact that developing countries have a considerable need for more investment in their agricultural sector. The question of agricultural investment is much broader than land acquisition and many investment projects do not involve the transfer of control over land.

6. Urgent need for agricultural investment in developing countries

Agricultural investment is the most important and most effective strategy for poverty reduction in rural areas, where the majority of the world's poorest people live (World Bank 2008). Investing in agriculture reduces poverty and hunger through multiple pathways. Farmers invest to enhance their productivity and incomes. From society's point of view, this in turn generates demand for other rural goods and services and creates employment and incomes for the people who provide them -- often the landless rural poor. These benefits ripple from the village to the broader economy. Agricultural investment is also key to eradicating hunger through all of the dimensions of food and nutrition security. Agricultural investment by farmers or the public sector that increases productivity at the farm level can also increase the availability of food on the market and help keep consumer prices low, making food more accessible to rural and urban consumers (Alston et al. 2000). Lower priced staple foods enable consumers to supplement their diets with a more diverse array of foods, such as vegetables, fruit, eggs, and milk, which improves the utilization of nutrients in the diet (Bouis, Graham and Welch 2000). Finally, agricultural investments can also reduce the vulnerability of food supplies to shocks, promoting stability in consumption.

However, low investment in the agricultural sector of most developing countries over the past 30 years has resulted in low productivity and stagnant production. The recent food crisis has exposed these weaknesses, as agricultural production was slow to respond to rising prices. Yet, the agricultural sector faces a considerable challenge over the next four decades. World agriculture must feed a projected population of 9 billion people by 2050, some 2.5 billion more than today, and most of the growth in population will occur in countries where hunger and natural resource degradation are already rife. Crop and livestock production systems must become more intensive to meet growing demand but they must also become more sustainable (FAO 2011, Save and Grow). Sustainable intensive production systems are capital-intensive; they require more physical, human, intellectual and social capital in order to sustain and rebuild the natural capital embodied in land and water resources. Additional investments of at least US$83 billion annually are needed in agriculture to meet targets for reducing poverty and the numbers of malnourished (Schmidhuber, Bruinsma and Boedeker 2009). Doing so in a sustainable manner that preserves natural resources and is conducive to long-term development will require even more funds. Increased investment by the public sector in developing countries will be necessary, which implies a reversal of the declining trend observed over the past decades. The share of public spending on agriculture in developing countries has fallen to around 7 percent, and even less in Africa (Hallam 2011). Investment is stagnant or falling in regions where hunger is most widespread (FAO 2012). Higher and more volatile food prices have reawakened policymakers to the importance of agriculture, and they have responded by increasing commitments to supporting the sector. This renewed attention to agriculture offers an opportunity to prepare for these challenges. Public investment by governments plays an essential role in creating the necessary conditions and enabling environment in which farmers can thrive, and in catalyzing and channelling private investment towards socially beneficial outcomes. The public sector also provides public goods which benefit society but for which private incentives are lacking. However, public-sector investments alone will not be sufficient. An increase in investment by the private sector is needed, in particular a rise in the investments made by farmers themselves, who account for the bulk of investment in agriculture. A recent study shows that farmers are by far the largest investors in agriculture (Lowder, Carisma and Skoet 2012). Annual investment in on-farm agricultural capital stock exceeds government investment by more than 3 to 1 and other resource flows by a much larger margin. On-farm investments are more than twice as important as all other sources of investment combined. Particular attention must be paid to ensuring that smallholders, many of whom are women, are able to invest on their farms and benefit from other public and private investment. This requires the existence of an enabling investment climate and the provision of public goods such as research and extension, market institutions and infrastructure, training and education, and risk management tools.

However, in spite of the new priority given to agriculture, many developing countries have limited financial capacity to fill the investment gap. Commercial bank lending to agriculture is less than 10 percent in sub-Saharan Africa, while microfinance loans are usually too small and not suited to capital formation in agriculture (Da Silva and Mhlanga 2009). It is unlikely that the solution will come from international donors either, as the share of official development assistance going to agriculture has fallen from around 10 percent to 5 percent (Hallam 2011). Recent summits of the G8 and G20 have made strong commitments to supporting increased investment in developing country agriculture for food security. This is a positive development. Nevertheless, in view of the unfolding economic crisis in the major industrialized nations and the slowing of growth in large emerging economies, international aid is unlikely to increase sufficiently to meet the investment needs in the short and medium terms.

Given the limitations of alternative sources, foreign direct investment could make a contribution to bridging the investment gap in developing countries' agriculture. The available data show that agricultural FDI is very small compared with domestic agricultural investment. Further, the agricultural sector still accounts for a very small percentage of total FDI inflows in most developing countries. A review of case studies on sub-Saharan Africa suggests that less than 5 percent of FDI goes to agriculture (Gerlach and Liu 2010). There is a potential for growth if more investments can be directed to the sector. While FDI cannot be expected to become the main source of capital, it can potentially generate various types of benefits for the agricultural sector of the host country such as employment creation, technology transfer and better access to capital and markets. However, these benefits cannot be expected to arise automatically and the risks discussed above are real. Consequently, the challenge for policy makers, development agencies and local communities is to maximize the benefits of foreign agricultural investment while minimizing its risks. This requires the capacity to orient foreign investments towards the right type of projects. Whether this objective can be met will depend on a large number of factors, among which the legal and institutional framework in place in the host country and the local context are critical.

7. The development potential of inclusive business model

In view of the risks associated with large-scale acquisition of land and a number of prominent project failures, there have been calls for the promotion of alternative business models that would involve the local community more actively. Arguably, inclusive business models that involve smallholders in production and/or other related activities have the potential to minimize the risks and maximize the benefits of agricultural investment. In 2009, FAO, the International Fund for Agricultural Development (IFAD) and the Swiss Development Cooperation (SDC) contracted the International Institute for Environment and Development (IIED) to prepare a conceptual paper on inclusive business models for investment in agricultural land aimed at raising productivity and promoting agricultural production for the market. IIED reviewed relevant literature and its own stock of field research and knowledge to identify key issues related to various business models for investment in agriculture, and the land tenure implications of such models. The study found that among the different business models reviewed, no single model was the best possible option for smallholders in all circumstances. The adequacy of a model was found to depend closely on the local context and to be contingent on tenure, policy, culture, history and biophysical and demographic factors. None of the models could be described as a holistic solution to rural development. In addition, the study suggests that the practical arrangements of the project may be more important than the category of model (Vermeulen and Cotula 2010). As a result, there is a need for a deeper understanding of inclusive business models through the detailed analysis of concrete experiences in the field.

8. Objectives, scope and methodology

Although there has been much debate about the potential benefits and risks of international investment, there is no systematic evidence on the actual impacts on the host country. In particular, there is a lack of detailed and reliable data. Also, there is a need for more evidence on the workings and impacts of inclusive business models through the detailed analysis of projects implemented in the field. In order to acquire an in-depth understanding of potential benefits, constraints and costs of foreign investment in agriculture and of the business models that are more conducive to development, FAO's Trade and Market Division (EST) has undertaken research on the impacts of international agricultural investment. The research aims to provide better knowledge on the trends and impacts of foreign direct investment on host communities and countries, to gather evidence on inclusive business models, to identify good practices and to develop guidance for host governments. To this end, FAO designed and directed case studies in selected developing countries. The studies were conducted in partnership with research institutions (the International Institute for Environment and Development (IIED) for Ghana, Mali and Zambia; the *Cambodia Development Resource Institute* (CDRI) for Cambodia) or through the direct recruitment of local researchers and consultants.

The studies covered three developing regions where foreign investment in primary agricultural production has tended to concentrate in the past six years, namely Africa, Asia and Latin America. Among these regions, the studies give particular emphasis to Africa, as it is arguably the region where the problems raised by large-scale land acquisition are the most urgent. More specifically, the African studies presented in this publication focus on Sub-Saharan Africa, as North Africa was already covered to some extent by the analyses undertaken by FAO's Regional Office for the Near East in 2009-2010 (Tanyeri-Abur and Hag Elamin 2011).

The studies examined the trends in agricultural FDI and its economic, social and environmental impacts in host countries. They reviewed the recent trends and current situation of large scale agricultural investments and land acquisitions in the selected countries, with special attention to various types of business models, distinguishing those with and without land acquisition. They analysed the factors determining the impacts and their relative significance. Two types of case studies were undertaken. The first type focused on national policies to attract FDI in agriculture and their impacts on national economic development. These studies covered Brazil, United Republic of Tanzania, Thailand and Uganda. The second type also reviewed the national policy framework, but then went on to examine the business models of selected agricultural investments in five developing countries and assess their economic, environmental and social impacts at the local and, when possible, national levels. This group of studies covered Cambodia, Ghana, Mali, Senegal and Zambia. Although the main subject of the studies was foreign investment, a few relevant large-scale agricultural investment projects by domestic investors were also examined.

More specifically, these studies analysed the drivers and the main actors (national and international) in each country, as well as the institutional process and national governance context framing the process of decision resulting in investments and land allocations (or the absence of land acquisitions, where relevant). They examined the specific policy measures that had an impact on the investment project, the economic inclusion of local smallholder farmers in the business model of the large investment projects and the participation of women where relevant. Where possible, the research investigated the contextual situation prior to the investments concerning land tenure patterns (land ownership, use and control), human capital situation with respect to education, training, extension and vocational education and the employment opportunities available (farm and non-farm as well as the working conditions by sex/age). It analysed the design and implementation of different business models in each country, including land-based and non-land investments; the process that led to the choice of a particular model; the policy measures (incentives, support, constraints) that influenced the process; and the success factors, the constraints encountered and the solutions adopted to overcome them. The studies also analysed the actual economic, social and environmental impacts of the business models studied. In particular, they assessed the effects on smallholder farmers and local communities within a gender and equity perspective such as income generation, improvement in welfare, employment/working conditions on and off farm, value addition, knowledge diffusion/spillovers, transfer of technology, skills development, forward and backward linkages, improvement in access to markets/capacity to trade and involvement of institutions such as farmers organizations.

Finally, the studies identified best practices and lessons learnt in terms of policy measures that are conducive to successful investment projects where the host country, the local community and the investor all benefit from the investment.

9. Contents

This publication examines the trends and impacts of FDI in developing country agriculture, in particular through the presentation of the main findings of the case studies. After the introduction, the second part provides an overview of the global trends in foreign agricultural investment in developing regions using various sources of statistical data. Part three presents case studies on policies to attract FDI in agriculture and their impacts on national economic development in selected countries in Africa, Asia and Latin America. The fourth part examines the business models that were used in selected agricultural investments in five developing countries. It assesses their economic, environmental and social impacts at the local level and how they are influenced by national policies. The fifth part draws a synthesis of the studies' findings. Finally, part six offers conclusions and recommendations.

References

Anseeuw, W., Boche, M., Breu, T., Giger, M., Lay, J., Messerli, P. & Nolte, K. 2012. *The State of Large-Scale Land Acquisitions in the 'Global South' Analytical Report based on the Land Matrix Database*. By The Land Matrix Partnership (ILC, CDE, CIRAD, GIGA, GIZ).

Burnod, P. et al. 2011. *From international land deals to local informal agreements: regulations of and local reactions to agricultural investments in Madagascar*. Paper presented at the International Conference on Global Land Grabbing, April 2011, UK.

Cotula, L. & Polack, E. 2012. *The global land rush: what the evidence reveals about scale and geography*. IIED Briefing. April 2012. London, UK.

Da Silva, C., & Mhlanga, N. (2009). *Models for investment in the agricultural sector*. Paper presented at the FAO Expert Meeting on Foreign Investment in Developing Country Agriculture, 30–31 July 2009, Rome. Rome, Italy: Food and Agriculture Organization of the United Nations.

Deininger, K & Byerlee, D. 2011. *Rising Global Interest in Farmland - can it yield sustainable and equitable benefits?* The World Bank. Washington D.C.

FAO. 2011. *Save and grow: A policy-maker's guide to the sustainable intensification of smallholder crop production.* Rome.

FAO. 2012 (forthcoming). *The State of Food and Agriculture 2012; Investing in agriculture for a better future.* Rome.

Gerlach, A. & Liu, P. (2010). *Resource-seeking foreign direct investments in Africa: A review of country case studies.* Trade policy research working paper. Rome: Food and Agriculture Organization of the United Nations.

GRAIN. 2011. www.farmlandgrab.org

Hallam, D. 2011. *International investment in developing country agriculture—issues and challenges*. Food Security journal, 3 (Suppl 1):S91–S98.

Lowder & Carisma. 2011. *Financial resources flows to agriculture: A review of data on government spending, official development assistance and foreign direct investment*; FAO-ESA Working Paper No 11-19; December 2011. Rome, Italy: Food and Agriculture Organization of the United Nations

Lowder, S., Carisma, B. & Skoet, J. 2012 (forthcoming). *Who invests in agriculture and how much? An empirical review of the relative size of various investments in agriculture in low- and middle-income countries*. Agricultural Development Economics Division, Working Paper No. 12-XX, Rome, FAO.

Miller, C., Richter, S., McNellis, P. & Mhlanga, N. 2010 *Agricultural investment funds for developing countries*. Rome: Food and Agriculture Organization of the United Nations.

OECD. 2010. *Private Financial Sector Investment in Farmland and Agricultural Infrastructure*. Working Party on Agricultural Policies and Markets Report. OECD, Paris.

Schmidhuber, J., Bruinsma, J., & Boedeker, G. (2009). *Capital requirements for agriculture in developing countries to 2050*. Paper presented at the FAO Expert Meeting on How to Feed the World in 2050, 24–26 June 2009, Rome.

Schoneveld, G.C. 2011. *The Anatomy of Large-Scale Farmland Acquisitions in Sub-Saharan Africa*. Working Paper 85. CIFOR, Bogor. See: www.cifor.org/nc/online-library/browse/view-publication/publication/3732

Tanyeri-Abur, A. & Hag Elamin, N. 2011. *International Investments in Agriculture in the Near East: Evidence from Egypt, Morocco and Sudan*. Edited by Aysen Tanyeri-Abur and Nasredin Hag Elamin, FAO Regional Office for the Near East, Cairo, Egypt.

UNCTAD. 2011. *World Investment Report* 2011.

Vermeulen, S., & Cotula, L. (2010). *Making the most of agricultural investment: A survey of business models that provide opportunities for smallholders*. Rome: Food and Agriculture Organization of the United Nations. London, UK: International Institute for Economic Development. Rome. International Fund for Agricultural Development. Bern. Swiss Development Cooperation.

Von Braun, J. & Meinzen-Dick, R. (2009). *"Land grabbing" by foreign investors in developing countries: Risks and opportunities*. IFPRI Policy Brief 13.

World Bank. 2008. *World Development Report* 2008. Washington, D.C., United States.

PART TWO

OVERVIEW OF TRENDS IN FOREIGN DIRECT INVESTMENT (FDI)

1. Introduction[1]

Foreign Direct Investment (FDI) has contributed significantly to growth and development in many developing countries over the last three decades, although, the benefits have not been evenly distributed. The countries that have benefited the most are those (e.g. Brazil, Malaysia, Republic of Korea, etc.) in which the conditions for harnessing inflows of foreign capital were in place and the opportunities and risks associated with current and future market developments were clearly understood by both investors and host country policy makers. These include – political stability, investments-friendly regulatory and policy frameworks, skilled or easy-to-train manpower, market size or proximity to large markets with minimal trade and physical barriers, etc. However, several developing countries have seen FDI's contribution to growth (in terms of GDP) at very high rates even without the development-friendly conditions in place. In such countries, (e.g. Nigeria, Zambia, etc.) this has been mainly due to the very high returns on investments from mainly extractive industries although the development benefits are still indeterminate.

In many developing countries, FDI in the agricultural sector has been mostly concentrated in the up-stream sub-sectors – food processing, beverages and related allied sectors. However, in many developing countries, the ongoing food and financial crises have witnessed a surge in investments in large tracks of land to grow and export food and biofuel crops to investor countries.

This recent upswing in domestic private and foreign investments in agricultural industries has come about as a result of several factors. First, as the expanding populations of emerging nations experience rapid economic growth, individual incomes have increased and they are spending more on food. Further, their tastes are shifting to a richer diet including more meat, fish and milk products. In order to satisfy demand, these countries have to import some of these food items thereby creating opportunities for both domestic and foreign investors to invest in agricultural industries in developing host countries. Because of policies limiting land use for agriculture in many developed countries, some of this investment is now happening across emerging nations--South-South investment. Another factor is the increase in biofuel initiatives around the world, particularly in Brazil, the United States, and the European Union. These have resulted in an increase in investment in developing countries in crops such as sugarcane, cereals and oilseed. In addition, countries such as Saudi Arabia, Republic of Korea and United Arab Emirates; all with limited arable land and/or insufficient water for irrigation, are buying large plots of land in soil rich developing countries in order to counteract export restrictions. Finally, speculation and portfolio diversification have also been noted as key factors.

Using data from UNCTAD[2], FAO[3] and fDi Markets databases[4], this chapter examines broad trends in FDI (inward flows and stocks) and where possible their general tendencies in the agricultural sectors of developing regions (Africa, Asia and Latin America and the Caribbean) and the nine countries whose agricultural sector investment structures, profiles, incentives, business models, etc., are evaluated in the ensuing chapters. The countries are: Brazil, Cambodia, Ghana, Mali, Senegal, United Republic of Tanzania, Thailand, Uganda and Zambia.

[1] Chapter prepared by Suffyan Koroma and Massimo Iafrate, Trade and Markets Division, FAO.

[2] FDI data from UNCTAD were obtained from: http://unctadstat.unctad.org/ReportFolders/reportFolders.aspx

[3] FAO data on agricultural capital stocks were obtained from: http://faostat3.fao.org/home/index.html#DOWNLOAD

[4] The fDi Markets Database is available at: http://www.fdimarkets.com/

2. FDI's contribution to growth

Although FDI has made significant contribution to growth in many developing countries, for a good number of them, the development effects are yet to be realized. However, considerable efforts are needed to collect and maintain data and databases on FDI flows in a coherent and consistent manner to enable analysis of its long-term development effects.

FDI has been shown to play an important role in promoting economic growth, raising a country's technological level, and creating new employment in developing countries (Borenzstein, De Gregorio, and Lee. 1998)[5]. It has also been shown that FDI works as a means of integrating developing countries into the global market place and increasing the capital available for investment, thus leading to increased economic growth needed to reduce poverty and raise living standards. At the same time many countries have understood the role played by FDI and they have taken steps to remove investment barriers. For example, during the 1990s, 1000 FDI law and regulations were amended of which 94 percent were amended principally to attract FDI (UNCTAD, 2010)[6]. In an effort to attract FDI, many countries have implemented incentives including tax exemption, government pledges, tariff reduction on equipment and machinery imports, subsidy, etc. These are dealt with in greater detail in the country case studies.

It is worth pointing out, at the outset, that data on investment flows and stocks are often not collected in a consistent manner and suffers from several shortcomings including country coverage, inconsistent sectoral classification and categorization, etc. In this regard, it was observed by (Lowder and Carisma, 2011)[7] that the long-term aggregate growth in FDI is more due to the expansion of countries reported with data than an overall trend movement. However, in this analysis, country level data are utilized as much as possible which might overcome the problem noted above.

Figure 1 depicts the long-term trend (1980-2010) of FDI flows for Africa, Asia and Latin America. FDI flows to each of these three regions over the last two decades starting from the 1990s have been growing at an average annual rate of 15.3 percent in Africa, 14.3 percent in Latin America and 16.8 percent in Asia.

Figure 2 presents the trends in FDI for the African countries under consideration with the behaviour of the trends exhibiting identical

FIGURE 1
Trends in FDI flows to Africa, Asia and Latin America, 1980-2012

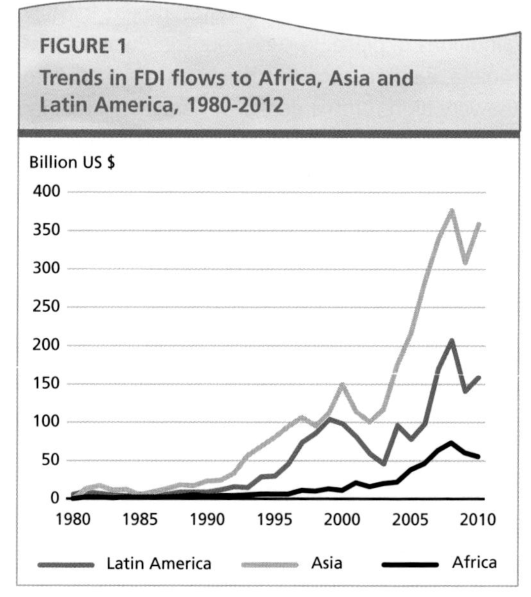

Source: Adapted from UNCTAD, 2009

[5] Borenzstein, Eduardo; Jose De Gregorio and Jong-Wha Lee: *How does Foreign Direct Investment Affect Economic Growth?* Journal of International Economics, 45: 115-135 (1998)

[6] UNCTAD (2010): *World Investment Report* 2010, United Nations, New York.

[7] For a detailed exposé of the global databases on FDI, see: Sarah K. Lowder and Brian Carisma, Financial resource flows to agriculture: A review of data on government spending, official development assistance and foreign direct investment; FAO-ESA working paper No. 11-19; December 2011 - www.fao.org/economic/esa. This paper presents a detailed analysis of existing databases on FDI with a critique of their strengths and shortcomings.

FIGURE 2
Trends in FDI-African case study countries, 1980-2010

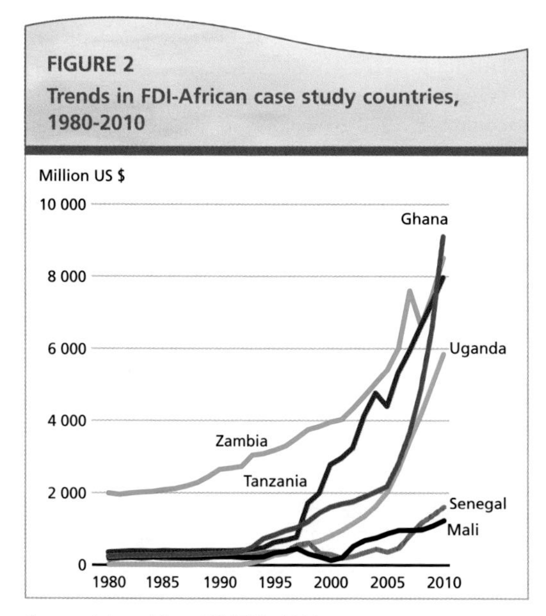

Source: Adapted from UNCTAD, 2009

FIGURE 3
Trends in FDI for Asian case study countries, 1980-2010

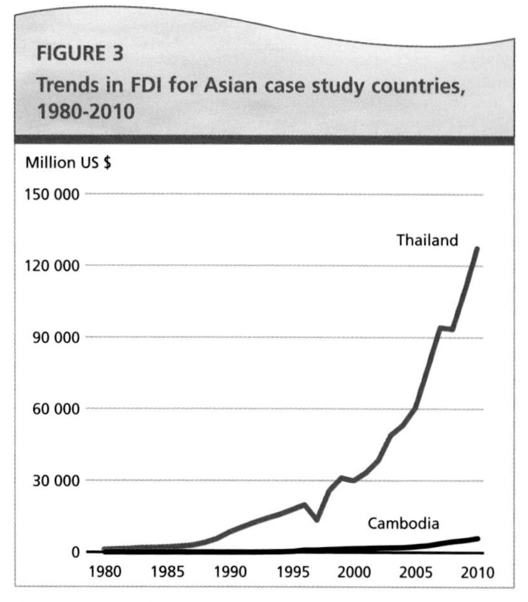

Source: Adapted from UNCTAD, 2009

patterns as the aggregate. The African countries in our case study all show increasing and upward trend in FDI flows during the mid-1990s. Zambia, Ghana, United Republic of Tanzania and Uganda all exhibits steep upward trends while for Senegal and Mali, the trend growth is rather stable. In terms of value, Ghana surpassed Zambia in 2010 as an important destination of FDI inflows primarily due to the recent discovery of petroleum. Figures 3 and 4 also depict trends in FDI for case study countries in Asia and Brazil, which also indicates that growth in FDI started during the 1990s.

From our data, it is clearly evident that FDI has made significant contribution to growth in many developing countries over the 1980-2010 periods. Using our case study countries as example (Table 1 and Figure 5), the longterm contribution of FDI to GDP is as high as 83.8 percent in Zambia. Senegal (6.4 percent) and Uganda (8.9 percent) are the only two out of eight countries in which FDI's contribution to growth has been less than 10 percent.

Over the period 2000-2010, FDI has contributed in excess of 20 percent to GDP in the following countries: Brazil (22 percent), Cambodia (43 percent), Ghana (30 percent), United Republic of Tanzania (32 percent), Thailand (34 percent) and

FIGURE 4
Trends in FDI flows to Brazil, 1980-2010

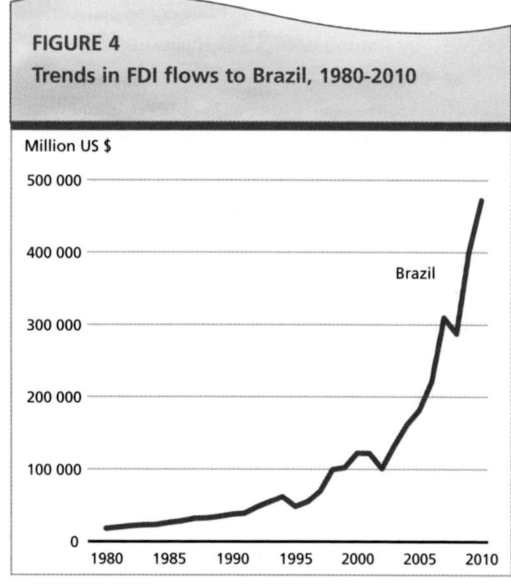

Source: Adapted from UNCTAD, 2009

Uganda (22 percent). In the case of Zambia, FDI has made very significant contribution to GDP even at relative low levels compared to other part of the world. FDI flows to Zambia, Africa's top copper producer, hit a record US$2.4 billion in the first half of 2010 from US$959 million the previous year due to a mining and manufacturing boom with expected creation of 33 140 jobs. Between January

TABLE 1
Contribution of FDI stocks to GDP

Country study	1980-1990	1990-2000	2000-2010	1980-2010
		(%)		
Brazil	11.0	11.5	21.7	14.7
Cambodia	4.3	18.9	43.4	22.2
Ghana	6.6	13.2	30.2	16.7
Mali	14.5	10.3	12.9	12.6
Senegal	5.5	7.1	6.6	6.4
U.R. Tanzania	5.6	11.8	31.9	16.4
Thailand	5.3	14.2	34.2	17.9
Uganda	0.3	4.7	21.8	8.9
Zambia	72.6	96.6	82.3	83.8

Source: Adapted from UNCTAD, 2009

FIGURE 5
Contribution of FDI stock to GDP - case study countries, 1980-2010

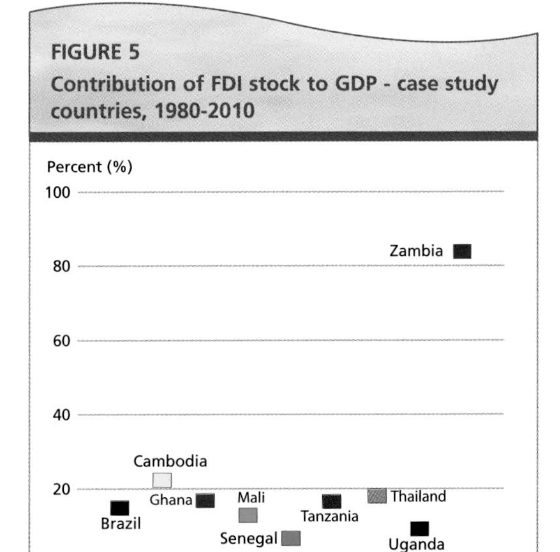

Source: Adapted from UNCTAD, 2009

and June 2010, FDI flows into manufacturing, much of it from China, totalled US$768 million, followed by mining with $593 million and the energy sector with US$565 million.

In the Latin American region, Brazil, the only case study country for that region is often described as one of the hottest destinations in the world inbound FDI. Many multinational companies are seeking to enter with new or expand existing FDI projects due to Brazil's market size, growing middle class and the country's demonstrated ability to yield high rates of returns on investment with many attractive linkages of spillover effects. In fact, according to UNCTAD's Global Investment Trends Monitor, Brazil was the tenth largest recipient of FDI in 2010 with over US$30 billion in new inbound FDI projects up from the 13th slot and US$22 billion in new inbound FDI a decade ago[8].

In the case of Cambodia, where FDI's contribution to GDP was 22.2 percent over the long term 1980-2010, as a result of significant reforms undertaken during the 2000-2010 period, FDI and local investment approvals increased by about 160 percent in 2011, and continued attracting new entrants such as Japanese investors. FDI approvals for Japanese investors accounted for US$6.4 million in fixed assets (three projects) compared with none in 2010. Cambodia's top five investors in were the United Kingdom, China, Viet Nam, Malaysia and Republic of Korea.

A total of 87 projects worth US$5.6 billion were approved by the Council for the Development of Cambodia during 2011. Most of the investments were directed at the key sectors like construction and tourism, real estate, banking and product exports. Garment exports appear to have benefitted from a shift of labour intensive industries from China to lower wage cost countries like Cambodia. Cambodia experienced an 18 percent increase in the number of new investments in garment factories. In addition, milled rice exports have been experiencing a huge expansion recently, recording annual growth of 250 percent and reaching 180 000 tonnes during 2010 and 2011. Despite the floods, rice production is anticipated to increase on the back of increased yields in both wet and dry season production and increased planted-areas. Milled rice exports were also supported by the establishment of new investments in mills that increased milling capacity.

Under normal market conditions, a key ingredient for attracting FDI is the level and development of agricultural capital stock available in a country. This is usually referred to as capital formation and is conventionally defined as the

[8] http://blogs.worldbank.org/psd/brazil-s-new-fdi-frontier-north-and-northeast-regions

FIGURE 6
Average annual growth in capital stock, 1975-2007

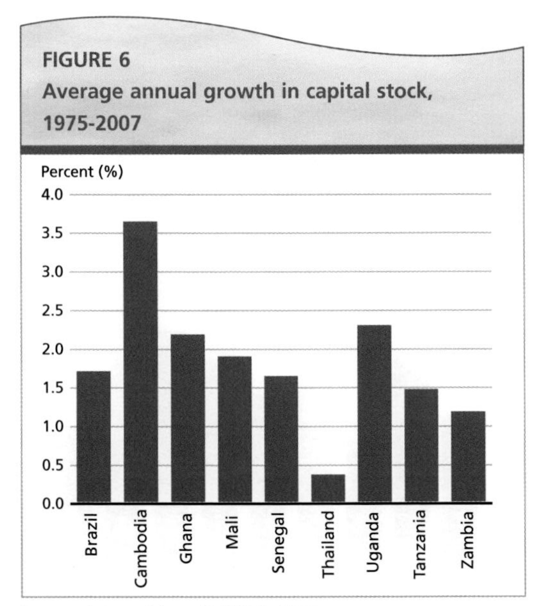

Source: Adapted from UNCTAD, 2009

FIGURE 7
Average annual growth in the value of land asset, 1975-2007

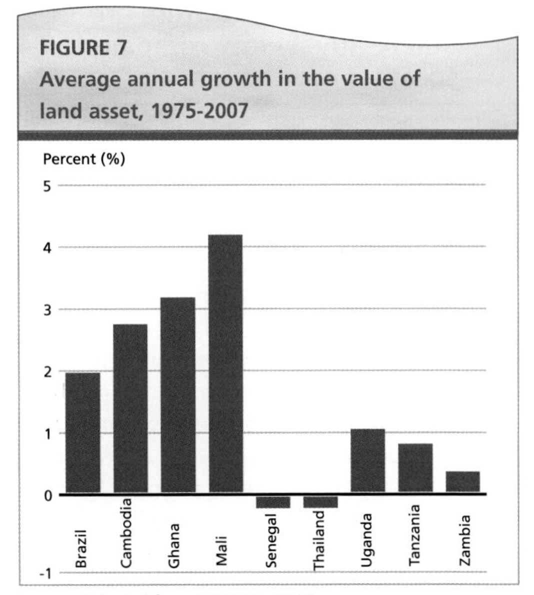

Source: Adapted from UNCTAD, 2009

stock of tangible, durable fixed assets owned or used by resident enterprises for more than one year. This includes plant, machinery, vehicles and equipment, installations and physical infrastructures, the value of land improvements, and buildings. Statistically it measures the value of acquisitions of new or existing fixed assets by the business sector, governments and households less disposals of fixed assets. Estimates of agricultural capital formation or stock are currently available from FAO for 206 countries[9]. Using these estimates, the long-term average annual grow rates of the value capital stock and two of its components – value of land improvements and machinery and equipment are provided for the case study countries from 1975 to 2007.

Figure 6 presents the average annual growth rates in capital accumulation for all nine case countries over the period 1975-2007. Cambodia has experience the largest long-term growth in capital accumulation of 3.6 percent per annum, followed by Uganda 2.2 percent per annum and Ghana 2.1 percent per annum. The remaining six countries all experience growth rates in capital accumulation of less than 2 percent per annum. For Brazil and Thailand which are both highly efficient agricultural producing countries, the low level of average long-term capital accumulation might be revealing of the fact that having had successful developments in agricultural capital stock over time, the pace of capital accumulation is now slowing down.

In terms of land development (Figure 7), with the exception of Senegal and Thailand both of which exhibits negative long-term growth in land development, all the other eight countries exhibit longterm annual growth rates ranging from a low of 0.3 percent to 4.1 percent per annum over the 32 year period from 1975-2007. In the case of Thailand, the rate of land development may have reached a saturation point hence the downward trend being influenced by more stringent environment and land degradation policies. For Senegal, the negative trend on land developments is more a reflection on the scope for more investment in land improvements.

In the case of growth in investment in machinery and equipment (Figure 8), Thailand exhibits the strongest annual long-term growth of around 4 percent followed by Uganda, Cambodia and Mali. In the case of Brazil which is a major user and producer of machinery and equipment, the modest 2 percent long-term annual growth suggests a levelling-off or saturation of investments in the stock of new agricultural machinery and equipment over the long run.

9 http://faostat3.fao.org/home/index.html

3. FDI flows to agriculture are still relatively low compared to other economic sectors

Although they have experienced a large surge recently, FDI flows to Agriculture are still relatively low compared to other economic sectors. Within the broad agricultural sector, FDI is concentrated mainly on the downstream activities (processing, manufacturing, trade and retail), leaving primary agriculture to demise in public sector funding. FDI flows to agriculture tend to increase during periods of both extreme high and low commodity prices.

FIGURE 8
Average annual growth in stock of machinery and equipment, 1975-2007

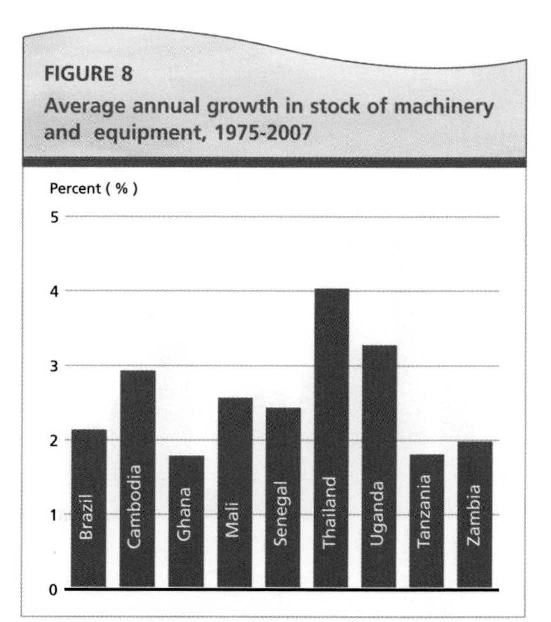

Source: Adapted from UNCTAD, 2009

Available data on global FDI flows to agriculture are generally incomplete due to poor reporting, collection and dissemination efforts coupled with secrecy due to the sensitive nature of most of the investments. In the ensuing analysis, data from both UNCTAD and fDI Market databases are used. The data from UNCTAD categorize FDI to agriculture as those related to crops, livestock, fishing, forestry and hunting. These are further sub-categorized as primary and processed (food, beverages and tobacco). The UNCTAD data run from 1980-2008. In the case of the fDI Market data, FDI to agriculture covers all activities related to food, beverages and tobacco. The system reports only Greenfield investments[10] and the data run from 2003 to 2011[11].

Figure 9 depicts the evolution of trends in the share of agriculture in total FDI inflows. Despite its importance, global FDI flows to agriculture have never exceeded 8 percent since the 1980s. The period between 1996 and 2000 was the worst recorded since the 1980s as the share of FDI to agriculture was at its lowest – at less than 2 percent. Although it has risen since, during 2006-08 it stood at a modest 4.6 percent of total FDI flows globally.

Within the FDI inflows to agriculture, the lion's share has been invested in manufacturing and higher-stage processing sectors including the food retail sector, while inflows to primary agriculture have remained below 15 percent (Figure 10). However, it should be noted that for the two databases used in this analysis, it is the end-stage activity that is reported, i.e. if a company invests in land to grow relevant crops, process and produce biofuel or juice, this would be reported as investment in processing. In this case, it is really difficult to assess the trends in broad terms, except with very detailed micro-level data at the firm or enterprise level.

From both Figures 9 and 10, one can observe that the period 1996-2000 from Figure 9 was the one with the lowest level of FDI inflows to the agriculture sector; but from Figure 10 it represents the period in which primary agriculture experience its highest share (12.2 percent) of FDI inflows

[10] Green Field Investment is a form of FDI where a parent company starts a new venture in a foreign country by constructing new operational facilities from the ground up. The alternative "Brown Field Investments" occurs when a company or government entity purchases or leases existing production facilities to launch a new production activity or expand existing activities.

[11] Although the database managers are doing their best to record all investments, some investments may not have been known and therefore the figures should be treated as estimates.

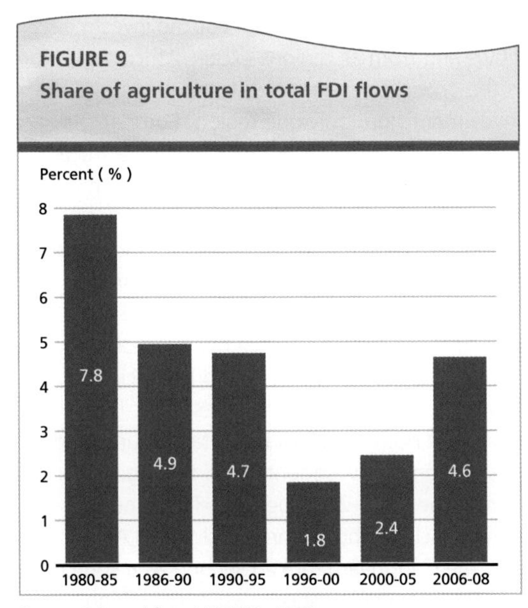

Source: Adapted from UNCTAD, 2009

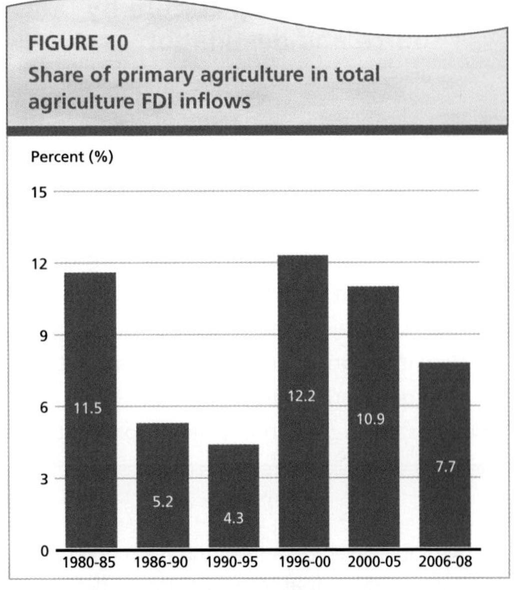

Source: Adapted from UNCTAD, 2009

almost three-fold increase. It should be recalled that this was a period of low and declining global agricultural commodity prices. However, in the recent period of higher commodity prices, we are witnessing a similar trend (Figure 11). Thus, can one conclude that during both periods of extremely high and low commodity prices, FDI flow to agriculture increases.

Figure 12 presents more recent data on FDI flow to agriculture at the global level[12]. The period of the global food crisis 2008-9 witnessed the largest inflow of FDI into agriculture totalling US$25 billion, almost doubling the level five years earlier in 2003. This lends further support to the evidence from using the UNCTAD data. FDI inflow to agriculture seems to have peaked in 2009, however, its level in 2011 is higher than the average for the entire period from 2003-11. Although the recent spike in FDI flows to agriculture has renewed emphasis on private sector investment as the important and missing element to overcome food insecurity and poverty in many developing countries, the trend has reversed after 2009. Similar behaviour has been experienced during earlier global food and economic crises.

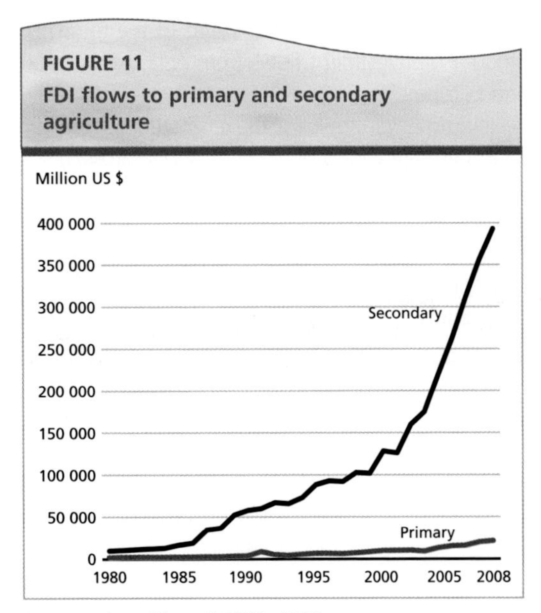

Source: Adapted from UNCTAD, 2009

[12] Data used in this section are from the FDI Markets database – www.fdimarkets.com. Under this database, agricultural investments flows are defined as investment flows into the food, beverages and tobacco sectors.

4. FDI flows to agriculture by source and destination

In terms of FDI flows to agriculture by source and destination – with the exception of Africa, where most of the investment flows originated from outside the continent or region, a major characteristic of investment flow into agriculture is that the destination of larger share of the investments flows are to the same region from where it originated.

At the global level, data from the fDI markets database suggests that total investment flows into agriculture between 2003 and the first half of 2011 amounted to US$143.3 billion. Although, the growth in investment flows to agriculture almost doubled from US$13.6 to US$25.4 billion between 2007 and 2009, it had however completely reversed to its pre-2007 level by the first half of 2007. This is attributed to the huge amount of investment flows from Asia, America and Europe.

Investment inflows for the case study countries vary widely by amount and sources (Table 2). Brazil received most of the investments flows and from all regions except Africa. Mali is the only country reported to have received investment from only one region (Europe). Amongst the African countries, Ghana attracted most investments followed by Uganda, Zambia, United Republic of Tanzania, Mali and Senegal.

Below is a summary of regional agricultural investment flows by source and destination. Europe was the source of 48 percent of the 143 billion recorded to have been invested in agriculture since 2003. It is also the recipient of about 37 percent of the investment flows, making it both the most important source and destination of investment flows into agriculture during the 2003-2011 periods. The Americas (which includes both North and South America and the Caribbean) ranked second as a source but third as a destination with Asia ranking second as a destination but third as a source of investment flows. Although Africa was the source of only 0.7 percent of the total investment flows, it was the destination of about 8 percent surpassing Oceania which received only 2.2 percent of inward flows over the period (Figure 13).

In terms of investment inflows, for Africa (Figure 14), the most important recipient countries are Nigeria (United Kingdom and Netherlands top two sources), South Africa (Switzerland and

TABLE 2
Agro-investment inflow for case study countries (total for period 2003-11)

Recipient countries	Sources of agro-investment inflows
Brazil	America (US$4.2 billion); Asia (US$3.3 billion); Europe (US$2 billion); Oceania (US$65.3 million)
Cambodia	Asia (US$159.7 million); Europe (US$50 million)
Ghana	America (US$203.5 million); Asia (US$31.5 million); Europe (US$1.1 billion)
Mali	Europe (US$47.4)
Senegal	America (US$25 million); Europe (US$10.4 million)
U.R. Tanzania	Africa (US$21.8 million); America (US$6.2 million); Europe (US$136.4 million)
Thailand	America (US$143.8 million); Asia (US$1 billion); Europe (US$460 million); Oceania (US$49.7 million)
Uganda	Africa (US$157.8 million); Asia (US$90 million); Europe (US$53 million)
Zambia	America (US$52 million); Asia (US$155 million); Europe (US$47.4 million)

Source: computed from FDI markets (www.fdimarkets.com)

FIGURE 12
FDI flows to agriculture

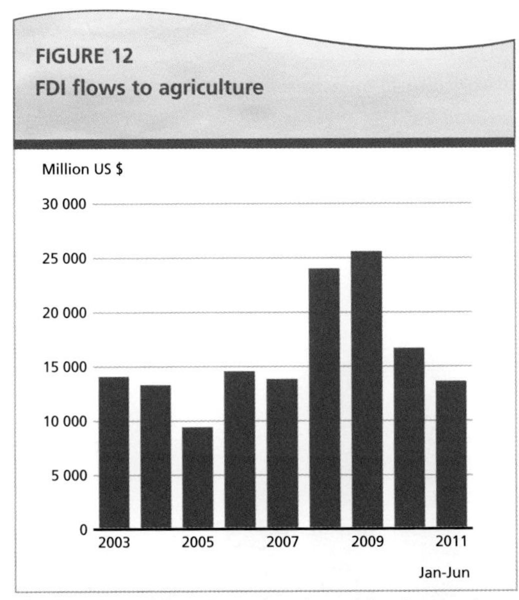

Source: computed from FDI markets (www.fdimarkets.com)

FIGURE 13
Agricultural investments by source and destination (cumulative, 2003-2011)

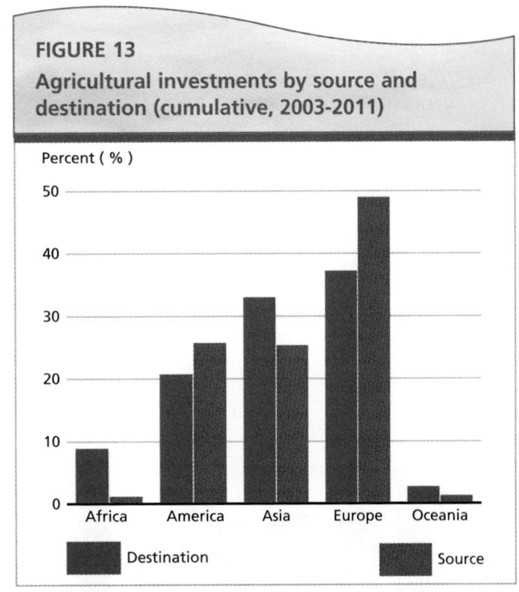

Source: computed from FDI markets (www.fdimarkets.com)

Netherlands), Ghana (United Kingdom and United States), Egypt (Saudi Arabia and Switzerland) and Angola (United States and United Kingdom).

For agricultural investment flows originating from Africa (Figure 15), the main source countries are South Africa with major investments of US$211.4 million in Africa; US$36.9 million in the Americas; US$179 million and US$5 million in Europe. Egyptian investment is less diversified with investments of US$300 million in Sudan and US$14 million in Jordan. Kenya invested US$107 million in Uganda; US$22 million in the United Republic of Tanzania and US$34.4 million in Germany. Investments from Tanzania went to Mozambique (US$30.4 million) and Uganda (US$30 million).

Agricultural investment flows into the Americas (Figure 16) originated principally from with the continent. Brazil and the United States are the two top destination countries, with Argentina, Canada and Mexico relatively less important destinations. Amongst the important investor countries from outside America into America with investments of over US$1 billion are China (investments of US$4.1 billion), Switzerland (US$3.7 billion), United Kingdom (US$2.1 billion), France (US$1.2 billion) and Japan (US$1.1 billion).

The Americas was also a very important source of outward flow of agricultural investments (Figure 17) providing about a quarter of all the investments during 2003-11. The United States was by far the largest investor country at the global level with investments in excess of US$29 billion. Brazil, Canada and Mexico also provided investments in excess of US$1 billion.

For Asia (Figure 18), agricultural investments where principally destined for China (US$14.2 billion); India (US$5.8 billion); Vietnam (US$4.1 billion); Turkey (US$4.0 billion) and Indonesia (US$3.6 billion).

Investment flow from Asia totalled US$35.5 billion during the 2003-11 periods. The principal share came from Japan (US$6.3 billion), China (US$4.7 billion), Saudi Arabia (US$4.5 billion) and Thailand (US$4 billion). China was the recipient of the lion's share as indicated above (Figure 19).

In the case of Europe, investment inflows amounted to US$52.6 billion with Russian Federation, Poland, United Kingdom, Romania and Spain the principal recipient countries (Figure 20).

For outward flows, the investments totalled US$69.4 billion with the United Kingdom attracting the largest share US$14.1 billion, followed by Switzerland, Germany, Netherlands and France (Figure 21).

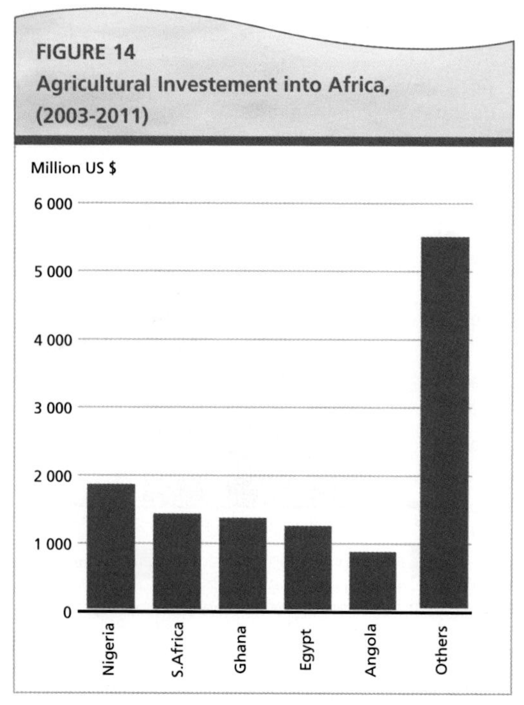

FIGURE 14
Agricultural Investement into Africa, (2003-2011)

Source: computed from FDI markets (www.fdimarkets.com)

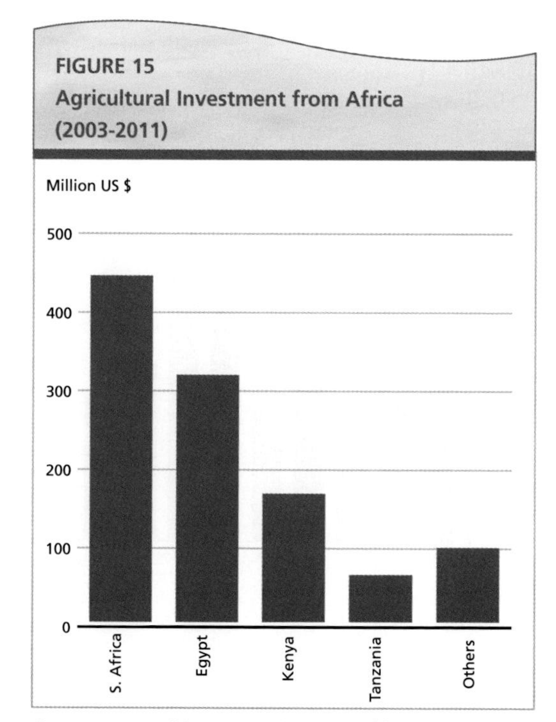

FIGURE 15
Agricultural Investment from Africa (2003-2011)

Source: computed from FDI markets (www.fdimarkets.com)

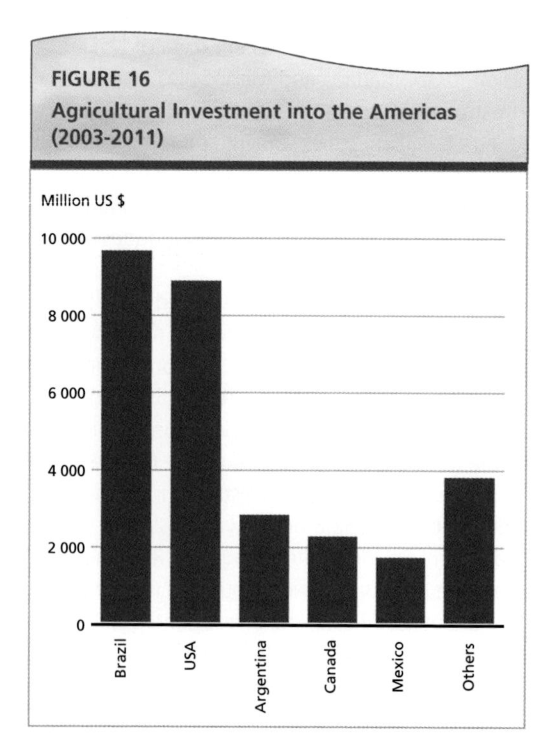

FIGURE 16
Agricultural Investment into the Americas (2003-2011)

Source: computed from FDI markets (www.fdimarkets.com)

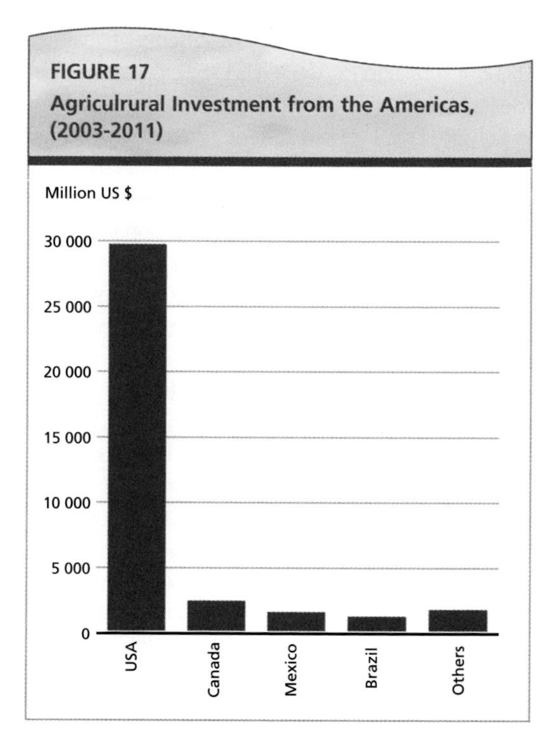

FIGURE 17
Agriculrural Investment from the Americas, (2003-2011)

Source: computed from FDI markets (www.fdimarkets.com)

FIGURE 18
Agricultural Investment into Asia (2003-2011)

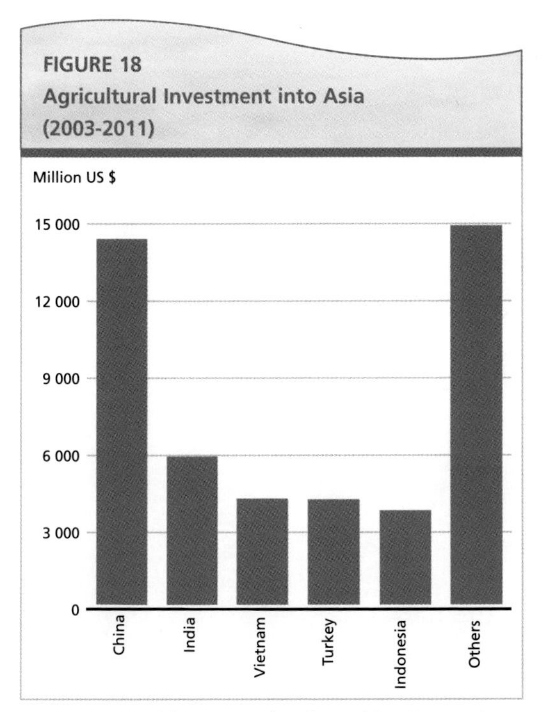

Source: computed from FDI markets (www.fdimarkets.com)

FIGURE 19
Agricultural Investment from Asia (2003-2011)

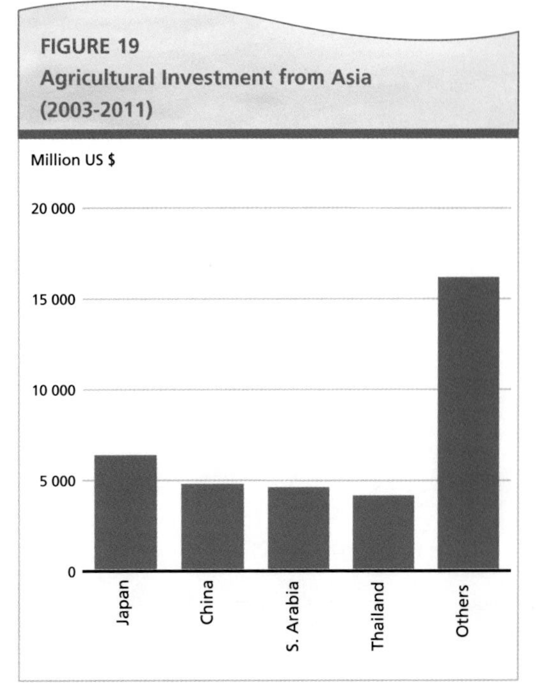

Source: computed from FDI markets (www.fdimarkets.com)

FIGURE 20
Agricultural Investment into Europe (2003-2011)

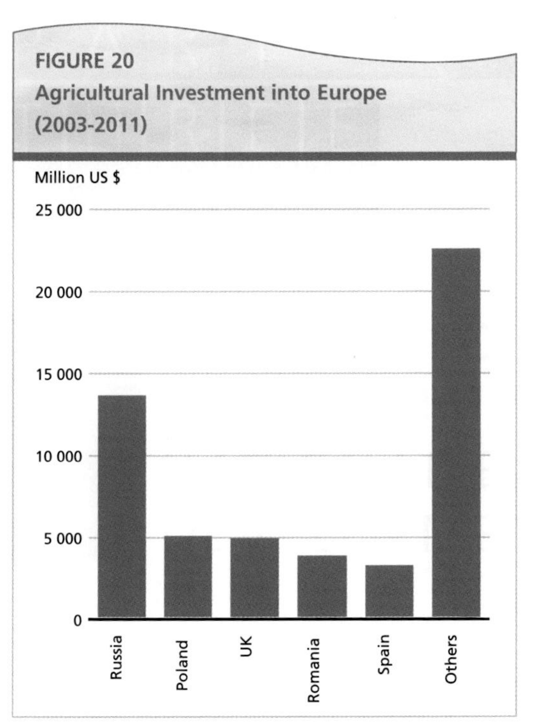

Source: computed from FDI markets (www.fdimarkets.com)

FIGURE 21
Agricultural Investment from Europe (2003-2011)

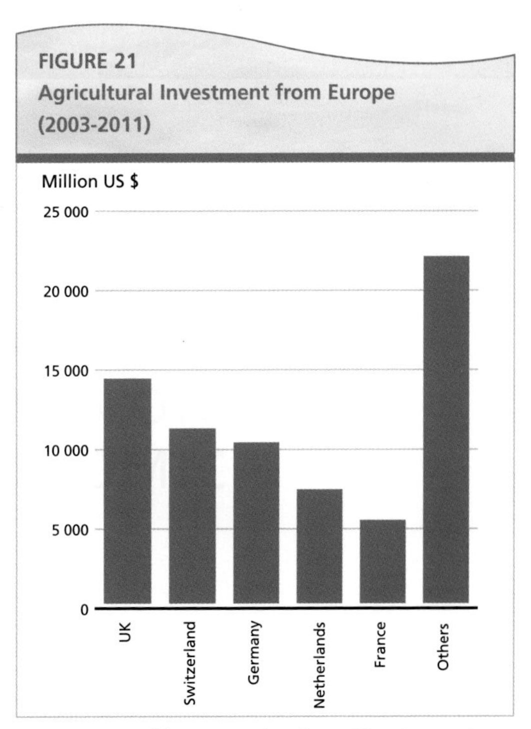

Source: computed from FDI markets (www.fdimarkets.com)

Brazil:
Improving the business climate for FDI[1]

1. Introduction

This Chapter uses Brazil[2] as an example to explore the usefulness of a methodology that both determines and improves the attractiveness of a country to foreign investors. This section describes the overall organization of the chapter. Section two quantifies and describes FDI (Foreign Direct Investment) in Brazil, its contribution to the financing of the agricultural sector, and the importance of Transnational Corporations (TNCs). Section three reviews some critical policies and actions that have contributed to Brazil's agriculture sector performance, and presents an overview of the experience learnt from the development of the *Cerrados*. Section four characterizes the business climate in Brazil through various indicators, and describes a model that helps to identify the main factors that affect it using forestry as an example. Section five presents a procedure used by the InterAmerican Development Bank (IDB) whereby a country gets support for improving its business climate. The last section presents the principal conclusions and recommendations.

2. Foreign direct investment in Brazil

This Section describes and quantifies FDI in Brazil, how it contributes to the financing of agriculture sector related investments, and the role of transnational corporations as a source of FDI. This section uses secondary information from UNCTAD and the Brazilian Central Bank (BCB).

2.1 The comparative importance of FDI in Brazil

Brazil is a relatively large recipient of foreign direct investments. Up to 2008, the country had accumulated a stock of over US$288 billion in FDI in all sectors of the economy, which represented 45 percent of all FDI in South American countries, and nearly a quarter of the total invested in Latin America and the Caribbean (LAC) Region (Table 1).

While foreign direct investment makes a significant contribution to capital formation, net FDI flows in Brazil have changed over the years. Figure 1 shows that at least from 1980 to the early 1990s, outward FDI stocks were larger than

FIGURE 1
Inward, outward and net FDI stocks for Brazil

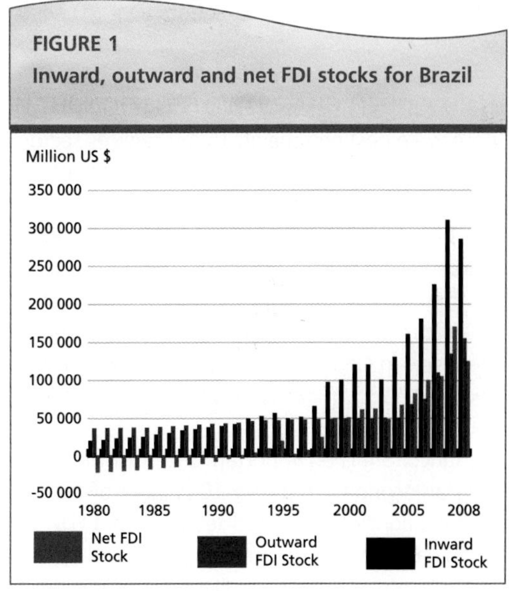

Source: Prepared by the author based on data from UNCTAD, 2009

[1] This chapter is based on an original research report produced for FAO by Jose Rente Nascimento, Senior International Consultant.

[2] Brazil has also become an important direct investor in other countries (Group of Fifteen (G-15)). However, this study concentrates only on the FDI the country receives into its agriculture based sector.

TABLE 1
FDI stock, by regions and economies in Latin America and the Caribbean, 1990, 2000, 2008

Region/economy	FDI inward stock		
	1990	2000	2008
	US$ millions		
Latin America and the Carribean	**110 547**	**502 487**	**1 181 615**
South and Central America	**101 977**	**424 180**	**978 056**
South America	**73 481**	**309 057**	**633 517**
Argentina	7 751*	67 601	76 091
Bolivia (Plurinational State of)	1 026	5 188	5 998
Brazil	37 143	122 250	287 697
Chile	16 107*	45 753	100 989
Colombia	3 500	11 157	67 229
Ecuador	1 626	6 337	11 300
Falkland Islands (Malvinas)	*	58*	..
Guyana	45*	756*	1 422*
Paraguay	418*	1 372	2 398
Peru	1 330	11 062	30 232
Uruguay	671*	2 088	8 788
Venezuela, Bolivarian Republic of	3 865	35 480	41 375

Source: Adapted from UNCTAD, 2009

inward flows. Starting around 1995, inward FDI flows became increasingly larger, albeit with apparent random movements from year to year (see Box 1).

Brazil had accumulated by 2009 a total of US$372 billion in inward FDI stock. Though these were destined mainly to the services sector (Figure 2), agriculture represents an important recipient, especially in recent years. According to UNCTAD (2009), for the period 2005–2007, Brazil received US$421 million, which corresponds to the third largest amount of inward FDI flow into an agriculture sector after China and Malaysia. The Brazilian Central Bank reports an inward FDI stock of US$35 billion up to 2009, as shown in Figure 2.

Agriculture is an important sector of the Brazilian economy, but only 10 percent of the total agriculture-related sector inward FDI stock was destined for primary production. The vast majority of the inward FDI stock of the agriculture-related sector, 90 percent, was made

FIGURE 2
FDI stock until 2009, Brazil

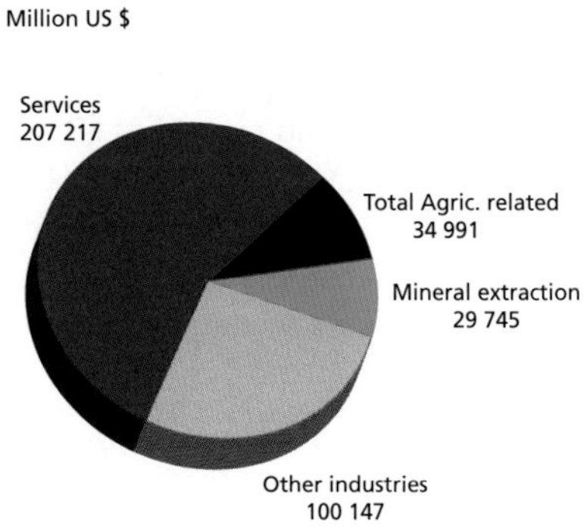

Source: Prepared by the author based on data from Brazilian Central Bank, 2010

BOX 1
A brief history of FDI in Brazil

One of the basic characteristics of the Brazilian economy is a high level of internationalization, with foreign corporations playing a leading role in many sectors. This is not a new phenomenon. FDI inflows and the TNCs' leading role in the most dynamic sectors have been key features of the Brazilian industrialization process from its beginnings. Especially from the early postwar years to the end of the 1970s, TNC affiliates, connected to public and private domestic companies by state planning, were fundamental to developing a diversified industrial structure, convergent with that of high-income countries at least in terms of the sectorial composition of output.

In the 1980s, however, the external debt crisis ended the Brazilian economy's long growth cycle. Brazil started to experience highly volatile GDP growth rates, as well as chronic inflation. FDI inflows stagnated at low levels, with TNC affiliates refraining from large-scale expansion projects.

The resumption of investment during the 1990s meant the return to more aggressive expansion strategies by TNC affiliates. Motivated by changes in economic policy and conditions – liberalization, privatization, and macroeconomic stability, followed by an increase in demand for consumer durables –TNCs began to expand their presence in the Brazilian economy again. From approximately US$1.5 billion annually in the 1980s and early 1990s, FDI inflows increased to an average level of US$24 billion annually (sic) between 1995 and 2000. It is interesting to mention that the inflows continued to grow through the year 2000, despite the Asian crisis of 1997, the Russian crisis of 1998, and even the Brazilian crisis of 1999. Starting in 2001, with a world economic slowdown considerably reducing trade and investment flows, FDI inflows to Brazil declined, reaching a low of US$10.1 billion in 2003. In 2004, the volume of FDI went up again, dipping slightly again in 2005....

Important changes occurred in the sectorial composition of FDI inflows as well. Until 1995, the manufacturing sector accounted for almost 67 percent of all FDI stock in Brazil, whereas in the second half of the decade, the prevalence of the service sector was remarkable, with electricity, gas, water, postal services and telecommunications, financial services, and wholesale and retail trade attracting significant FDI flows. A large part of the investment in these sectors was associated with the privatization process. By 2000, the service sector's share in the FDI stock had increased to 64 percent and that of the manufacturing sector had dropped to 33.7 percent, though manufacturing industries such as food and beverages, automotive, chemicals, metallurgy, and telecommunications equipment continued to receive significant volumes of investment.

Between 2001 and 2006, the service sector continued to account for more than half of total inflows although its share dropped compared to the previous period. The manufacturing sector, in turn, accounted for 38.5 percent of the total inflows during this period. Agriculture and mining also grew in importance, accounting for 7.1 percent of total FDI. (Hiratuka, 2008).

into agriculture-related industries, including tobacco, textiles, food and beverages, leather, wood and pulp and paper industries. Among these, food and beverages received 61 percent of the inward FDI, for a total of US$21.3 billion up to 2009. Forest related industries were second, with US$6.5 billion of inward FDI stock (Figure 3).

Primary agriculture (including livestock) and its related services is the subsector with the greatest amount of the inward FDI stock, followed by silviculture, forest exploitations, and related services. Fisheries, aquiculture and related services received an almost negligible amount of investment (Figure 4).

Box 1 uses examples from the industry to highlight the history of transnational corporations in Brazil's economy since the 1940s. Examples from agriculture are also easy to identify, especially since the 1990s when the presence of TNCs in the sector has grown substantially (Box 2). Transnational corporations such Monsanto (Box 3), and Corn Products, DuPont, Dow chemical, Bunge, to name a few, have been active in the country for decades, some even for more than a century. The importance of TNC presence in the country can be further demonstrated in Table 2 which lists the world's 25 largest TNC suppliers of agricultural inputs, all of which are present in Brazil, except for four (Terra Industries, Inc., Bucher Industries AG, Claas KGaA, Aktieselskabet Schouw & Company A/S and Scotts Miracle-Gro Company).

TNCs in Brazil are present at all stages of the value chain; from suppliers of agriculture and forest inputs, to machine and equipment producers, to agriculture or forest output producer, to processors and industrial firms, to wholesalers, retailers and exporters. For instance, Monsanto (Box 3) provides seeds and herbicides for agriculture production; while Louis Dreyfus Commodities Brazil (Box 4) produces, processes, stores, transports and markets commodities (soybeans, rice, corn, cotton, coffee, sugar and ethanol, citrus fruits, and fertilizers); and ArcherDaniels-Midland Company procures, transports, stores, processes and merchandises agricultural commodities and produces fertilizers and biofuels. International Paper (Box 5) and Stora Enso, two of the largest pulp and paper firms in the world, are good examples of forest TNCs invested in Brazil.

FIGURE 3
Inward FDI stocks in agriculture-related industries until 2009 – Brazil

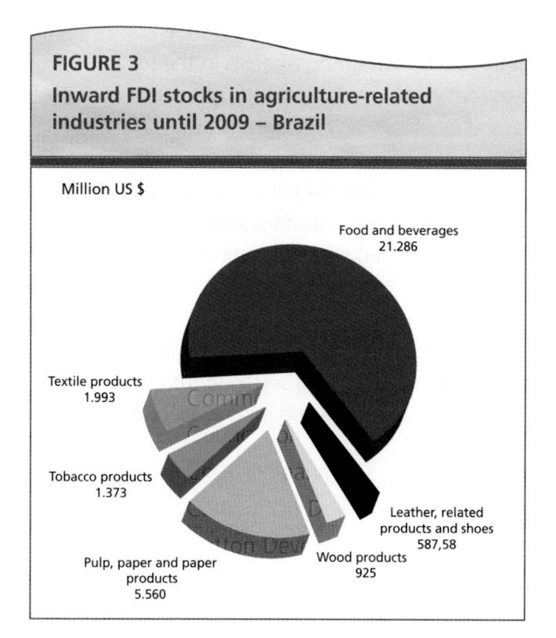

Source: Prepared by the author based on data from Brazilian Central Bank, 2010

FIGURE 4
FDI stocks in agriculture sector (non-industrial) until 2009

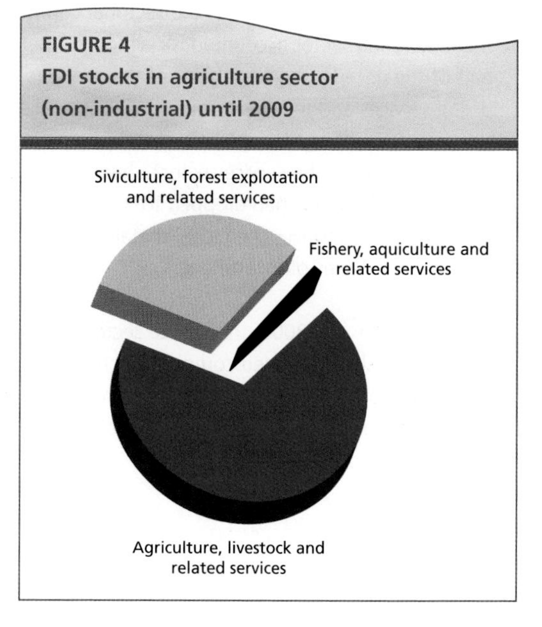

Source: Prepared by the author based on data from Brazilian Central Bank, 2010

BOX 2
TNCs and the Brazilian agrifood sector

Over the last two decades, the Brazilian agrifood system transitioned from a traditional to an increasingly global and industrial model. Fostered by rising incomes, urbanization, economic liberalization, and access to competitive raw materials, multinational food processors and retailers entered or increased their investments in the Brazilian market during the 1990s. Increased … FDI by large, private agribusinesses in Brazil displaced domestic competitors, increased industry concentration, and eliminated many medium and small companies. As a result, the market share of multinational corporations in the domestic food market increased. For instance, Brazilian affiliates of multinational agrifood companies generated 137 000 jobs, almost US$5 billion in exports, and sales of US$17 billion in 2000. Given the total value of food industry shipments in Brazil of US$58 billion, the aggregate market share of foreign companies reached 30 percent in 2000. Among the top ten food processors in the country, eight are multinational firms with foreign headquarters. …Official data show that FDI inflow in the Brazilian agrifood processing industry totaled US$8.2 billion between 2001 and 2004. The top-three food retailers in the country … were then … controlled by two French supermarket chains (Casino and Carrefour) and one US based company(Wal-Mart), with a combined market share of 39 percent.

Concomitant to these structural changes in the post-farm gate stages of the agrifood system, agricultural production also modernized and became increasingly capital intensive and integrated with upstream and downstream supply chain participants. Tightly coordinated agrifood supply chains have been developed by the private sector – in particular, large multinational food processors, fast-food restaurant chains and retailers – to cater to increasingly differentiated domestic and export markets. Farmers in Brazil are increasingly exposed to markets that are much more demanding in terms of food quality and safety, more concentrated and vertically coordinated, and more open to international competition. (Chaddad and Jank, 2006)

3. Lessons from Brazilian agriculture development experience

Tracing the changes in policies and factors affecting the business climate of Brazilian agriculture is a challenge that has been only partially fulfilled in the literature. This section represents an attempt to improve the knowledge base and draw some policy lessons. The first part traces the main policy measures and factors that affected the whole country over the past 40 years. The second part describes the process of agriculture development specifically for the Brazilian Savannah (*Cerrados*), where major investments have transformed it into one of the world's most important food producing regions of the world.

3.1 Business climate history for agriculture investment

Brazil has adopted over the years a wide range of agricultural and macroeconomic policies that had a direct impact on the agricultural sector, including changes to the legal framework, macroeconomic stabilization plans, and the setting up of institutions. The list is long, but the author believes that a few have been pivotal to both increase investment levels and enhance crop production. These are deregulation of the economy, the opening up of domestic markets to world markets, the provision of rural credit, investments in R&D, and minimum guaranteed prices to producers at harvest time.

Looking back over the past century, Brazil's economy was dependent to a large extent on

TABLE 2
The world's 25 largest TNC suppliers of agriculture, ranked by foreign assets, 2007

Rank	Corporation	Home economy	Assets		Sales		Employment
			Foreign	Total	Foreign	Total	Total
			US$ million and number of employees				
1	BASF AG[a]	Germany	44 633	68 897	49 520	85 310	95 175
2	Bayer AG[a]	Germany	24 573	75 634	24 746	47 674	106 200
3	Dow Chemical Company[a]	United States	23 071	48 801	35 242	53 513	45 900
4	Deere & Company	United States	13 160	37 176	7 894	23 999	52 000
5	El Du Pont De Nemours	United States	9 938	34 131	18 101	29 378	60 000
6	Syngenta AG	Switerzland	9 065	12 585	9 281	9 794	21 200
7	Yara International ASA	Norway	8 009	8 541	9 939	10 430	8 173
8	Potash Corp. of Saskatchewan	Canada	6 079	9 766	3 698	5 632	5 003
9	Kubota Corp.	Japan	5 575	12 691	4 146	9 549	23 727
10	Monsanto Company	United States	4 040	12 253	3 718	8 563	18 800
11	Agco Corporation	United States	4 034	4 699	5 654	6 828	13 720
12	The Mosaic Company	United States	3 881	9 164	3 859	5 774	7 100
13	ICL-Israel Chemicals Ltd.	Israel	2 066	4 617	2 092	4 351	
14	Provimi SA	France	1 962	2 237	2 523	2 805	8 608
15	Bucher Industries AG	Switerzland	1 648	1 850	2 058	2 172	7 261
16	Nufarm Limited	Australia	1 191	2 010	925	1 512	
17	CLAAS KGaA	Germany	1 000	2 619	2 884	3 781	8 425
18	Sapec AC	Belgium	826	826	837	837	692
19	Terra Industrires Inc	United States	735	1 888	389	2 360	871
20	Aktieselskabet Schouw & Company A/S	Denmark	695	2 016	1 350	1 598	3 541
21	Genus PLC	United Kingdom	652	851	394	469	2 124
22	Scotts Miracle-Gro Company	United States	591	2 277	470	2 872	6 120
23	Kvemeland ASA	Norway	367	487	649	741	2 717
24	Sakata Seed Corp.	Japan	331	843	140	383	1 711
25	Auriga Industries A/S	Denmark	319	849	624	856	1 615

Source: UNCTAD, 2009

A General chemical/pharmaceutical companies with significant activities in agricultural supplies, especially crop protection, seeds, plant science, animal health and pest management.

Note: Data are missing for various companies. In some companies, foreign or domestic investors or holding companies may hold a minority share of more than 10 percent. In cases where companies are present in more than one agrifood industry, they have been classified according to their main core business.

BOX 3
Monsanto Company

Monsanto Company, together with its subsidiaries, provides agricultural products for farmers in the United States and internationally. It has two segments: Seeds and Genomics, and Agricultural Productivity. The Seeds and Genomics segment produces corn, soybeans, canola and cottonseeds, as well as vegetable and fruit seeds, including tomato, pepper, eggplant, melon, cucumber, pumpkin, squash, beans, broccoli, onions and lettuce. This segment also develops biotechnology traits that assist farmers in controlling insects and weeds, as well as provide genetic material and biotechnology traits to other seed companies. The Agricultural Productivity segment offers glyphosate-based herbicides for agricultural, industrial, ornamental, and turf applications; lawn-and-garden herbicides for residential lawn-and-garden applications; and other herbicides for control of pre-emergent annual grass and small seeded broadleaf weeds in corn and other crops. The company offers its traits products under Roundup Ready, Bollgard, Bollgard II, YieldGard, YieldGard VT, Roundup Ready 2 Yield, and SmartStax; row crop seeds under DEKALB, Asgrow, Deltapine, and Vistive; vegetable seeds under Seminis and De Ruiter; herbicides under Roundup; and corn and cotton under Harness brand names. It also licenses germplasm and trait technologies to seed companies. The company sells its products through distributors, retailers, dealers, agricultural cooperatives, plant raisers, and agents, as well as directly to farmers. Monsanto Company has a joint venture with Cargill, Inc. to commercialize a proprietary grain processing technology under the name Extrax. It also has a collaboration agreement with BASF in plant biotechnology that focuses on high-yielding crops and crops that are tolerant to adverse conditions. The company was founded in 2000 and is based in St. Louis, Missouri.

Source: http://finance.yahoo.com/q/pr?s=MON+Profile

Monsanto arrived in Brazil in 1951 and has its headquarters located in São Paulo, the state where it installed the first factory in São José dos Campos (SP) in 1976. In Brazil, Monsanto produces herbicides and seeds of corn, soybeans, cotton and vegetables, and varieties of cane sugar.

Source: http://www.monsanto.com.br/institucional/monsanto-no-brasil/monsanto-no-brasil.asp

exports of a handful of agricultural products, mostly coffee and sugar. Attempts were made to industrialize the country, for example through import substitution policies introduced in the 1930's, but with limited impact (Abreu and Bevilaqua, 2000). Between 1960 and 1972, various policies adverse to the agricultural sector were applied, such as the overvaluation of the currency, high tariffs for imported industrial products, quantitative restrictions for agriculture exports, discrimination against raw commodities export and preference for industrialized valueadded agricultural products, and policies that sought to make domestic food prices affordable to the growing urban centres.

The Government of Brazil tried to compensate the adverse consequences of these policies with the creation in mid-1960 of a highly subsidized rural credit system. Credit was offered for working capital, investments (machinery, cattle, etc.) and marketing (discounting promissory notes and transport). Some analysts estimate that subsidized credit was directly responsible for a 66 percent increase in agriculture production during the 1970s (Lucena and Souza, NA). This very same decade saw the birth of the Brazilian Agriculture Research Enterprise (EMBRAPA), an R&D institution that has become key for the generation of agricultural technology.

BOX 4
Louis Dreyfus Commodities Brazil

The Louis Dreyfus Commodities Brazil (LDCommodities) is a subsidiary of Louis Dreyfus Commodities, which has more than 160 years the world market for agricultural commodities and has offices strategically distributed in over 50 countries.

In Brazil since the 1940s, the company operates in the production, processing, storage, transportation and marketing of commodities, making its presence felt in the markets for soybeans, rice, corn, cotton, coffee, sugar and ethanol, citrus fruits, and fertilizers.

Listed among the top 10 export companies in Brazil, the LDCommodities is present in the main producing regions of the country, with units in the South, Southeast, Northeast and Midwest. The company is headquartered in Sao Paulo and operates four oil processing plants, three of orange juice, five port terminals, two river port terminals, thirteen sugar mills and ethanol (LDC-SEV) and over 30 grain warehouses, and manage more than 340 000 hectares of land.

With revenues of approximately US$3.4 billion in Brazil (Dec/2009), the LDCommodities generates about 20 000 jobs, reaching 30 000 in harvest times. Besides providing an important contribution to the economy, the company maintains its ongoing effort to support farmers in close relationship with partners and community and commitment to the environment. The LDC-SEV is the second largest company in the world in the processing of sugar cane and production of renewable energy. It was created in October 2009 from the association between the LDC Bioenergy (ethanol and sugar operations of Louis Dreyfus Commodities) and the Brazilian company Santelisa. With 13 branches located in major producing regions of Brazil, the LDC-SEV has a processing capacity of 40 million tons of cane sugar per year and generates about 20 000 direct jobs.

KEY FIGURES:
Offices in Brazil: Regional head office in São Paulo and many others spread around the country; 7 processing plants; 7 ports and river terminals; Around 30 000 hectares of orange plantations; N°1 cotton merchandiser in Brazil.

Processing assets: 4 oilseed crushing plants in Brazil, processing soybeans and cotton into edible oil; meal and lecithin: Ponta Grossa, Paraguaçu Paulista, Jataí and Alto Araguaia; 3 industrial orange processing plants with a combined capacity of more than 60 million boxes per year: Bebedouro, Matão and Engenheíro Coelho.

Logistics assets: Ports and river terminals: Santos (São Paulo state), with three deep draft exporting terminals; Paranaguá (Paraná state), with one deep draft exporting terminal; São Simão (Goiás state) with one river barge terminal; Pederneiras (São Paulo state), with one river barge terminal; Transshipments, conducting logistics operations around seven major export-capable ports along the Brazilian coast; Significant storage capacities for oilseeds (more than 30 warehouses), citrus, cotton and coffee.

Source: http://www.ldcommodities.com

BOX 5
International Paper Company

International Paper Company operates as a paper and packaging company with operations in North America, Europe, Latin America, Russia, Asia and North Africa. Its Industrial Packaging segment manufactures containerboards. Its products include linerboard, medium, whitetop, recycled linerboard, recycled medium and saturating kraft. The company's Printing Papers segment produces uncoated freesheet printing papers, including uncoated papers, market pulp, coated papers and uncoated bristols. Its Consumer Packaging segment offers coated paperboard for various packaging and commercial printing end uses. The company's Distribution segment distributes products and services to various customer markets, supplying printing papers and graphic pre-press, printing presses, and post-press equipment for commercial printers; facility supplies for building services and away-from-home markets; and packaging supplies and equipment for manufacturers, as well as offers warehousing and delivery services. Its Forest Products segment owns and manages approximately 200 000 acres of forestlands and development properties primarily in the United States. The company was founded in 1898 and is based in Memphis, Tennessee.

Source: http://finance.yahoo.com/q/pr?s=IP+Profile accessed on August 21th, 2010

In Brazil International Paper's production system is comprised of two pulp and paper mills in Mogi Guaçu and Luiz Antônio, and a paper mill in Três Lagoas. Together, the three mills produce paper for Brazil and export markets, in addition to products on the Chambril line for conversion and printing. The mill located in Mogi Guaçu, in São Paulo, is the first mill of IP within Brazil and has a production capacity of 440 tonnes of paper per year. Incorporated into the business portfolio of IP in 2007, the Luiz Antônio mill located near Ribeirão Preto, in São Paulo, is capable of producing annually 360 thousand tonnes of paper. In operation since 2009, the Três Lagoas mill in Mato Grosso do Sul state has automated finishing lines, capable of producing up to 140 reams of Chamex paper a minute, non-coated paper production capacity – 200 000 tons a year, and operates some of the most advanced technology on the market. It has had US$300 million invested in it. The newest enterprise of IP in Brazil is the first factory to be built by International Paper out of the United States.

International Paper owns 72 000 hectares of renewable eucalyptus forests used in pulp and paper production. It also has 24 000 hectares of preserved areas, to conserve the original characteristics of the native vegetation. These areas are distributed amongst Mogi Guaçu, Brotas and Luiz Antônio, municipalities in São Paulo State. The necessary care required to guarantee productivity in renewable forests includes research, studies and analysis to improve the eucalyptus species to develop new technologies. The company produces about 16 million cuttings a year which are used in eucalyptus planting. Fire prevention and eco-efficiency in forestry management are also constantly invested in by the company. IP has a Research Centre with laboratories and researchers in different areas, working together and developing more sustainable techniques and processes.

Contract forestry and Partnering: In addition to its own forests, International Paper gets raw material through fostering forests and partnering. In contract forestry, there are about 9 500 hectares in São Paulo and Minas Gerais States. The company supplies cuttings, technical assistance, forestry inventory, soil analysis, a map of the plantation, and recommends fertilizer to local producers. Later, the wood is sold to the company at market prices. So far, 122.7 million cuttings have been donated, grown, on 12 500 hectares of plantation. In its partnering, International paper takes responsibility for expenses in the implantation and maintenance of renewable forests. Later, these amounts are converted into wood for the company.

Source: http://www.internationalpaper.com/BRAZIL/EN/index.html accessed on 20 October 2010

A price boom for agricultural commodities during 1972-1974 triggered the renewal of discriminatory policies against agriculture. Export embargos and price controls were applied, and the unfavourable business climate created by these policies resulted in a decade-long period of substantial reductions of agricultural production and exports (Abreu, 2004) (Lopes, Lopes and Barcelos, 2007). Rural credit, which in principle was to be subsidized, became less attractive. A growing fiscal deficit, foreign debt problems, the 1979 second petroleum crisis, and rampant inflation, eroded the subsidies built into the rural credit system. Credit rates were progressively less favourable to investors and eventually become positive real interest rates in 1984-1985.

An increase in price volatility was perceived contemporary to the erosion of agricultural credit subsidies. The Government of Brazil responded to this uncertainty by reviving price floors at harvest time. Producers could sell their production directly to the government or they could finance short-term storage costs to postpone sale of their outputs to a between-harvest-period when prices were expected to rise and increase revenues. This policy worked reasonably well and increased production, until hyperinflation set in during the 1980s (Lucena and Souza, NA) (Silva Dias and Amaral, 2001).

The late 1980s become a turning point for the Brazilian economy, as policies that had previously led to low agriculture domestic prices and low levels of investment in agriculture started to change. A substantial reduction of support for import substitution, trade liberalization and flexible foreign exchange rates in the mid-1990s improved agriculture prices and the profitability of the sector (Abreu, 2004) (Lopes, Lopes and Barcelos 2007). The Brazilian experience illustrates how allowing international prices to be transmitted to the domestic market (provided trade is fair without dumping or subsidies to foreign producers) injects dynamism to the agricultural sector.

In addition to liberalization, the Government of Brazil promoted the development of a strong agricultural and rural credit programme that targeted small and medium farmers, the National Family Farming Programme (PRONAF). It implemented, as part of the First and Second National Development Plans (PND), an infrastructure investment programme that built a large network of roads to allow transportation of agricultural production from distant frontier areas in the savannahs. This programme also installed power lines, communications facilities, input distributors and producers of machinery including tractors. A comprehensive agricultural and rural extension service, the National System for Rural Extension and Technical Assistance that was initially created in 1954 for the state of Minas Gerais, was expanded during the 1970s to all states. The service was implemented by the Brazilian Enterprise of Technical Assistance and Rural Extension (EMBRATER). In addition, a large network of storage facilities ruled by the Brazilian Storage Company (CIBRAZEM) was also established to buy, store and distribute agricultural production in the major producing areas of the country. Last but not least, the Government of Brazil created the successful EMBRAPA, the Brazilian Agricultural Research Corporation, whose research on agricultural technology would start showing key results a few years later.

EMBRAPA is a public company linked to the Ministry of Agriculture, Livestock and Food Supply, with legal characteristics similar to a private company. The enterprise coordinates the National Agricultural Research System created in 1992, which includes most public and private entities involved in agricultural research in the country. Today EMBRAPA is present in almost all Brazilian states, networking through 38 research centres, 3 service centres and 13 central divisions. In 2008 it had 8 275 employees, including 2 113 researchers, 25 percent with masters' degrees and 74 percent with doctoral degrees. At the end of the 2010, the workforce at EMBRAPA was 9 248 employees, and it received the highest operating income in history, more than US$1.15 billion. EMBRAPA estimates that in 2010 the rate of return on R&D was 39 percent (EMBRAPA 2008 and 2010).

EMBRAPA has generated and recommended more than 9 000 technologies for Brazilian agriculture, reduced production costs and helped Brazil to increase the offer of food while, at the same time, conserving natural resources and the environment and diminishing external dependence on technologies, basic products and genetic materials. It has been a key contributor to the transformation of the Brazil's *Cerrados* (savannahs) area into one of the most agricultural productive regions in the world.

3.2 Agricultural development of the Brazilian Savannah

Several studies have been undertaken in recent years to describe the process of agriculturebased development of the Brazilian Savannahs, a region that from a historical point of view was unimportant to agricultural production. Two of the studies were selected by the author and are included in this section. The first study developed a model (Chart 1) that identified factors that affected domestic production costs, as well as the impacts of domestic, export, and import logistic costs on the competitiveness of agriculture products. The book clearly established the importance of the adaptation and adoption of highly productive technologies. It also stressed the importance of training to improve labour productivity. The authors demonstrate how the costs of logistics impact on the competitiveness of agricultural products, and the large shares that they accrue at all levels from the domestic, regional, to global stages of the market chain. The authors conclude that key areas of intervention include the adoption of technology, training, and a reduction of the cost of logistics.

The second study by Tollini (ND) concentrates on explaining the factors that resulted in the impressive growth of agriculture production in the Savannah region of Brazil. His explanation, summarized in Chart 2, helps to identify key issues and intervention strategies that were

CHART 1
Factors affecting agriculture competitiveness in developing countries

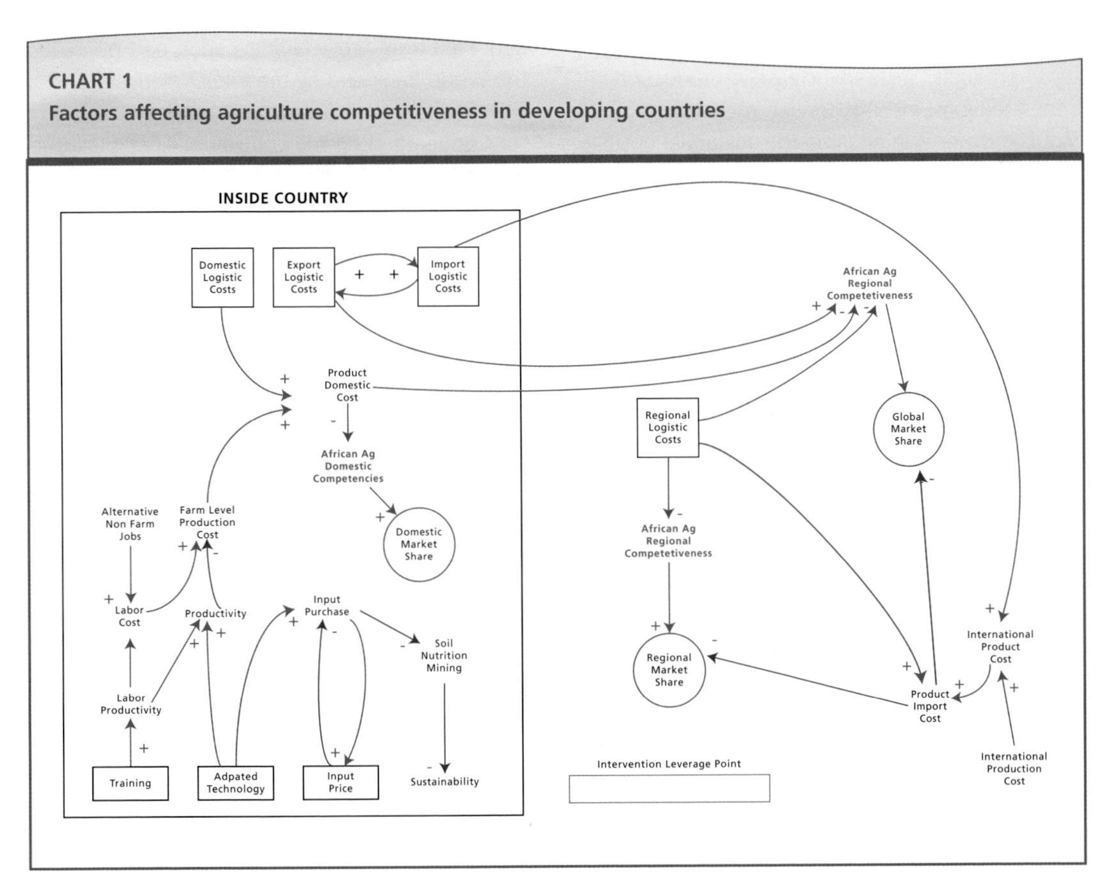

Source: J.R. Nascimento, 2009

instrumental to the transformation of the region. His analysis classified them in two groups: those that affected the supply of agricultural products, and those that affected their demand. He described the real and potential impact of the growing demand for these products on price formation and inflation pressures.

His study concludes that in addition to a growing demand for exports, pulled by the opening of the economy to world markets and flexible exchange rates, agriculture prices in Brazil were being pressured upwards by a growing domestic demand resulting from population and income growth, especially in urban areas. Income growth in urban areas contributed to the increase in demand as poorer members of society became more able to buy more and higher quality food items. Sustained demand pressures generated incentives for farmers to invest in agricultural production growth, while the government understood that the control of inflationary pressures from agriculture products could be addressed by greater growth in the supply of those products. Such an increase in production generated jobs, income, foreign exchange, reduced poverty, in addition to substantial positive externalities.

The heart of the strategy was to improve profits of agricultural investments, so that the growth in supply of agricultural products could be sustained. Government interventions were mainly designed to reduce costs and risks, so that producers and investors need not rely on high prices to make their businesses profitable. Although not explicitly discussed by Tollini, it is clear that the authorities understood the critical role of the private sector.

Tollini highlights the following interventions[3]:

- Investments to improve infrastructure in the areas of transportation, energy, and communications;
- Measures to improve land markets;
- The mobilization of public and private banks in the financing of agriculture production;
- Increased and sustained investments in research and development to overcome the limitations of the *Cerrados* soils and increase productivity;
- Creation of business opportunities for service providers to help in the several operations directly or indirectly associated with agriculture production;
- Mobilization of southern agriculture producers (gauchos[4]) to bring to the *Cerrado* their skills, knowledge, entrepreneurship, and capital; and
- Measures to support the training and education of rural labour and professionals.

Tollini recalls that "a point to note is that Brazil received support of bilateral and multilateral agencies in its effort to promote institutional development. For instance, EMBRAPA has benefited from projects financed in part by the World Bank and by the Inter-American Development Bank. The Inter-American Institute for Cooperation on Agriculture, IICA, also assisted EMBRAPA during its first years with the allocation of some professionals to help with the installation and initial research planning and programming. EMBRAPA was recognized as a good administrator of resources received through these projects, and has been able to benefit from several sequential projects, each adding new objectives as the research programme develops."

It should be noted that numerous business climate measures were introduced at stages and in sequencings that were contingent on context. While the results of these interventions were not always immediate or successful, the policy consistency applied for more than four decades eventually bore fruit. The process was initially slow but gained momentum, and today the Brazilian Savannah is one of the most influential food producing regions of the world.

[3] The supply box in Chart 2 - Factors that contributed to the agriculture-based development of the Brazilian Savannah includes variables and factors normally associated with the rural or agriculture branches of government responsibility at the time. Several other extra sectorial policy instruments were also used by the government in a mostly coordinated effort.

[4] Gauchos are the decedents of early Europeans who migrated to Southern Brazil at the end of 1800s and early 1900s. They have been key for the development of the Brazilian *Cerrados*.

CHART 2
Factors that contributed to the agriculure-based development of the Brazilian Savannah

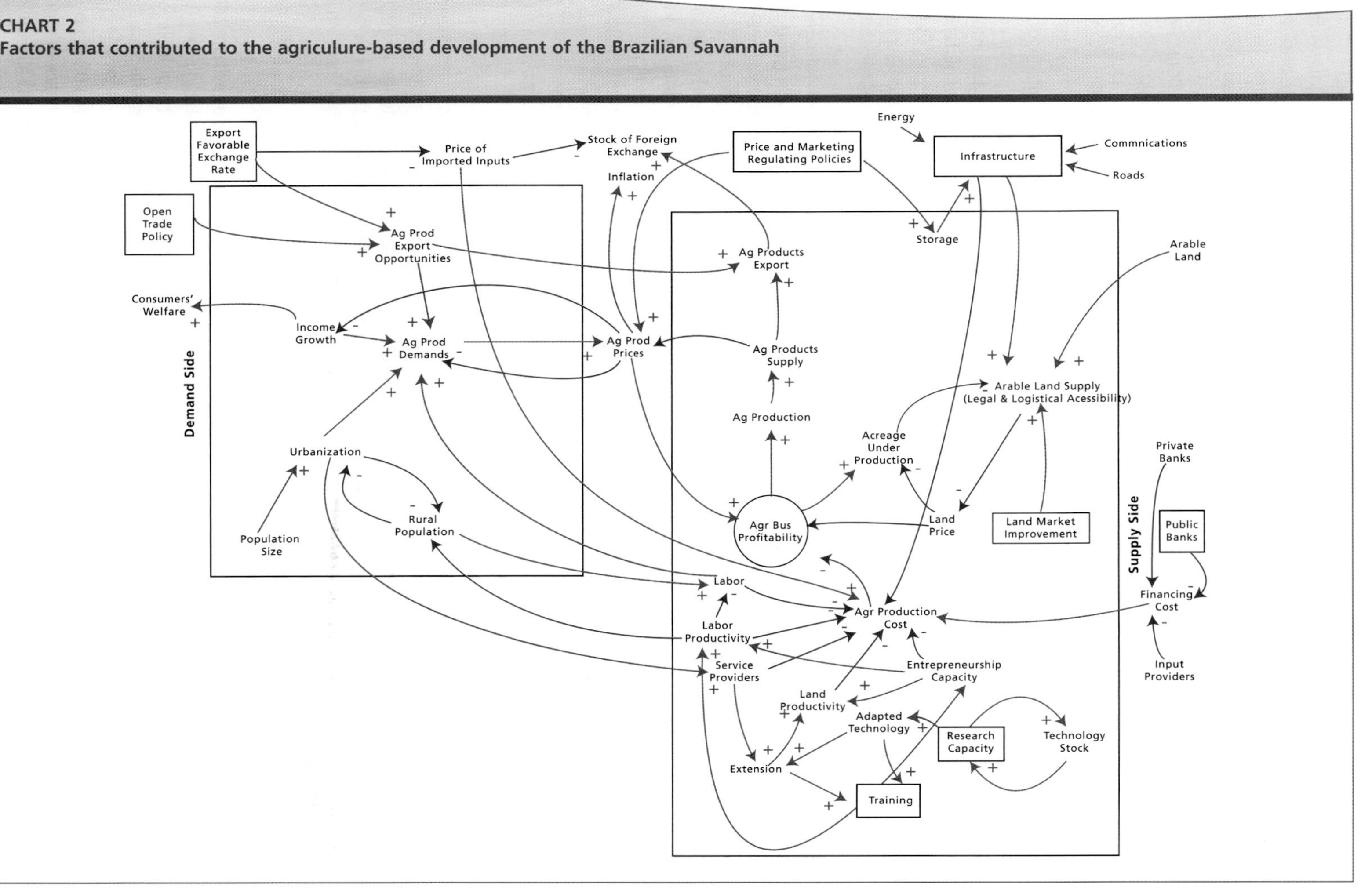

Source: J.R. Nascimento, 2009

4. The business climate for FDI in Brazilian agriculture

Entrepreneurs operate within an environment that determines to a large extent the conditions for profit. Individual firms cannot usually control external factors such as the rules of the game (laws, regulations, tax burden, and their enforcement), input and output markets, or others that directly affect their costs, revenues, and profits. Commercial success depends on the business climate that a given country can offer to investors (OECD, 2003).

This section is divided into two parts. The first relates the performance of the Brazilian business climate as perceived by various organizations. The second part introduces a model that tries to identify the factors and relationships affecting the success of agricultural businesses in Brazil. The model presents a framework that may help governments to development strategies for improving the business climate of their respective countries and considers forestry as an example.

4.1 The enabling environment for FDI in Brazil

Table 3 shows various indices that, taken together, illustrate a general perception of how the enabling environment for FDI in Brazil performs and ranks relative to other countries. It should be emphasized are not entirely independent from each other because they sometimes use similar variables. Nevertheless, these show that Brazil has relatively low scores, and often ranks half way through their tables. The table is useful not only in suggesting to investors the challenges they may need to face in Brazil, but also in highlighting areas where measures could be applied to further the business climate of the country.

4.2 A business climate model for investment in agriculture

The model presented here has been developed to better understand the conditions that prevail for investors seeking to invest in agriculture and forest-based sustainable businesses. It assumes that the more attractive a country is for agriculture and forest-based sustainable business investments, the more profitable investments are likely to be. The profitability of these businesses depends on the costs investors have to face and the expected benefits from their operations. As background, readers are invited to explore Box 6 which includes a checklist prepared by OECD for attracting FDI in an economy in general.

The model (Chart 3) proposes that the costs investors face, and the expected benefits from their operations, are affected by three groups of factors: supra-sectorial, inter-sectorial, and intra-sectorial. The supra- and the inter-sectorial factors are also called extra-sectorial conditions, as they not part of the agriculture or forestbased sector.

Supra-sectorial factors

Supra-sectorial factors influence the performance of firms in all the sectors of the economy, and include macroeconomic conditions and political risks. The supra-sectorial group consists of: (i) Gross Domestic Product growth; (ii) exchange rate stability; (iii) interest rates; (iv) tax burdens; (v) free trade; and (vi) political risks.

Two hypotheses exist that relate these factors with each other and demonstrate how they affect the profitability of agriculture or forest-based businesses. Thus, the model states that profitability is expected to increase with faster GDP growth, with an exchange rate that is more stable; and/or as the economy opens up (positive arrows). Equally so, profitability of agriculture or forest-based businesses is expected to increase as interest rates get smaller, the tax burden is less expensive; and/or the political risks diminish (negative arrows).

Inter-sectorial factors

Inter-sectorial factors are those managed by other sectors of the economy but which have substantial impacts on the cost and benefit structures of agriculture or forest-based businesses. The model identifies eight: (i) economic infrastructure; (ii) social infrastructure; (iii) credit accessibility; (iv) licences and permits; (v) environmental restrictions; (vi) capital treatment; (vii) labour; and (viii) rule of law (see Table 4).

TABLE 3
Performance of Brazil for selected indices

Index name	Brazil's score and rank	Brief description
Ease of Doing Business Ranking	Rank: 127 out of 183	The *Ease of Doing Business Ranking* is reported yearly by The World Bank, a financial assistant to developing countries. The Doing Business Ranking provides measures of business regulations and their enforcement across countries by measuring specific regulatory obstacles to doing business, such as protection of investors, protection of property rights, employment issues, and contract enforcement capabilities. The highest ranked country has the most favourable environment for conducting business in the world. Data collected in 2010. Source: The World Bank. http://www.doingbusiness.org/data/exploreeconomies/brazil
Global Competitiveness Report	Score: 4.23 out of 7 Rank: 56 out of 133	The *Global Competitiveness Report* is compiled yearly by the World Economic Forum, an independent international organization based in Geneva, Switzerland. The rankings provide a description of the economic competitiveness based on twelve pillars of competitiveness for countries at all stages of development. Some of the factors included come from publicly available data, but the majority comes from a survey the World Economic Forum sends to over 11 000 business executives worldwide. The highest ranked countries are the most competitive. Data collected in 2009. Source: http://www.weforum.org/pdf/GCR09/GCR20092010fullreport.pdf
Human Development Index	Score: 0.699 out of 1. Rank: 73 out of 182.	The *Human Development Index (HDI)* which looks beyond GDP to a broader definition of well-being. The HDI provides a composite measure of three dimensions of human development: living a long and healthy life (measured by life expectancy), being educated (measured by adult literacy and enrolment at the primary, secondary and tertiary level) and having a decent standard of living (measured by purchasing power parity, PPP, income). The index is not in any sense a comprehensive measure of human development. It does not, for example, include important indicators such as gender or income inequality and more difficult to measure indicators like respect for human rights and political freedoms. What it does provide is a broadened prism for viewing human progress and the complex relationship between income and wellbeing. Data: 2010. Source: UNDP. http://hdrstats.undp.org/en/countries/profiles/BRA.html
Index of Economic Freedom	Score: 55.6 out of 100. Rank: 113 out of 179.	*The Index of Economic Freedom* is reported annually by the Heritage Foundation, a research and educational institute. The Index of Economic Freedom analyses a wide range of issues including trade barriers, corruption, government expenditures, property rights, and tax rates to generate an overall ranking of economic freedom. The highest ranked country is the country with the least number of restrictions and constraints on businesses. Data collected in 2010. Source: http://www.heritage.org/Index/Ranking.aspx
Economic Freedom of the World	Score: 6.0 out of 10.0. Rank: 111 out of 141	The index published in *Economic Freedom of the World* measures the degree to which the policies and institutions of countries are supportive of economic freedom. The cornerstones of economic freedom are personal choice, voluntary exchange, freedom to compete, and security of privately owned property. Forty-two variables are used to construct a summary index and to measure the degree of economic freedom in five broad areas: (i) size of government; (ii) legal structure and security of property rights; (iii) access to sound money; (iv) freedom to trade internationally; and (v) regulation of credit, labour and business. Data collected in 2007 Source: Fraser Institute. http://www.fraserinstitute.org/research-news/research/display.aspx?id=13006
Corruptions Perception Index (CPI)	Score; 3.7 out of 10. Rank: 69 out of 178 countries studied.	The *Corruptions Perception Index (CPI)* is reported annually by Transparency International, an international civil society organization. The CPI ranks countries in terms of the degree to which corruption exists in the misuse of public power for private benefit among public officials and politicians. CPI is a composite index determined by expert assessments and opinion surveys. The highest ranked country is the country with the least amount of perceived corruption. Index units, 10=least corrupt, 0=most corrupt. Data collected in 2010. Source: http://www.transparency.org/policy_research/surveys_indices/cpi/2010/results

TABLE 4
Brief description of the inter-sectorial factors

Factors	Brief description
1. Economic infrastructure	Includes availability of economic infrastructure services at competitive prices and quality such as those provided by roads, communications, energy, ports, railroads, airports.
2. Social infrastructure	Includes availability of social infrastructure services at competitive prices and quality related to human development such as education; health; water, sewage and waste disposal.
3. Credit accessibility	Includes the sophistication of financial and capital markets, availability of credit at competitive terms as well as other capital markets instruments.
4. Licences and permits	Includes bureaucratic procedures and legal requirements to open, operate, and even close firms and that take much time, efforts and other resources to comply with.
5. Environmental restrictions	Unfounded or useless environmental restrictions that increase firms' costs without generating environmental benefits.
6. Capital treatment	Includes barriers and restrictions to the movement of capital into, out of, or within the country.
7. Labour	Includes the costs generated by labour legislation, the level of general productivity and the availability of skilled workers at competitive prices.
8. Rule of law	The existence of favourable legislation, enforcement, and justice services. Includes clear definition and protection of property legislation; respect to the letter of contracts, and timely justice at reasonable cost.

Source: Adapted from Nascimento and Tomaselli, 2007

As in Supra-Sectorial Factors, a positive arrow indicate that factors that positively affect profits, including economic infrastructure, social infrastructure, credit accessibility, favourable capital treatment; competitively priced and productive labour; and rule of law effectiveness increases (decreases). Negative arrows indicate factors that improve profitability with lower incidence.

Intra-sectorial factors

Intra-sectorial factors are those managed by public or private actors within the agriculture or forest-based sector of the economy. These factors are, by definition, under the control of stakeholders in the sector. The model identifies five: (i) agriculture or forest products domestic market; (ii) agriculture and forest productivity; (iii) availability of agriculture and forest vocation lands; (iv) favourable supports; and (v) adverse actions (see Table 5).

Except for Adverse Actions, all intra-sectorial factors shift profitability in the same direction. For example, the bigger the market for agriculture and forest products, including those used as input for export products or directly sold overseas, the more potential exists for profitable agriculture and forest businesses (trade integration or free trade agreements may be effective policies). Productivity growth is also a critical factor. Productivity depends *inter alia* on the availability and adoption of appropriate technology; production inputs such as seeds, fertilizers, machinery; skilled labour and professionals; and supporting services. Research, technical assistance, adaptation of technologies, and other innovations are vital to increase productivity. The availability of agricultural and forest vocation lands (FVL)[5] are also factors that affect the attractiveness of a country. The greater the land area a country has that can potentially be used for agriculture or forest production, the greater the contribution of this factor to the intra-sectorial conditions that favour successful agriculture or forest businesses. While the existence of FVL is a positive sign, these lands

[5] Forest Vocation Lands are those that, due to their physical site features such as soil, topography, and the rainfall they receive, should be kept under forest cover or other sustainable land use if soil or water related negative externalities are to be avoided. FVL classification does not depend on the type of cover the land actually has, nor does it depend on the requirements it may have for agriculture, crop or forest production. Therefore, lands with no forest cover or use can still be classified as FVL if their physical features so indicate; while lands covered with forest may not be FVL. (J.R. Nascimento, 2005).

BOX 6
Checklist for Foreign Direct Investment Incentive Policies

Policies for attracting FDI should provide investors with an environment in which they can conduct their business profitably and without incurring unnecessary risk. Experience shows that some of the most important factors considered by investors as they decide on investment location are:

- A predictable and non-discriminatory regulatory environment and an absence of undue administrative impediments to business more generally.
- A stable macroeconomic environment, including access to engaging in international trade.
- Sufficient and accessible resources, including the presence of relevant infrastructure and human capital.

The conditions sought by foreign enterprises are largely equivalent to those that constitute a healthy business environment more generally. However, internationally mobile investors may be more rapidly responsive to changes in business conditions. The most effective action by host country authorities to meet investors' expectations is:

- Safeguarding public sector transparency, including an impartial system of courts and law enforcement.
- Ensuring that rules and their implementation rest on the principle of nondiscrimination between foreign and domestic enterprises and are in accordance with international law.
- Providing the right of free transfers related to an investment and protecting against arbitrary expropriation.
- Putting in place adequate frameworks for a healthy competitive environment in the domestic business sector.
- Removing obstacles to international trade.
- Redress those aspects of the tax system that constitute barriers to FDI.
- Ensuring that public spending is adequate and relevant.

Tax incentives, financial subsidies and regulatory exemptions directed at attracting foreign investors are no substitute for pursuing the appropriate general policy measures (and focusing on the broader objective of encouraging investment regardless of source). In some circumstances, incentives may serve either as a supplement to an already attractive enabling environment for investment or as a compensation for proven market imperfections that cannot be otherwise addressed. However, authorities engaging in incentive-based strategies face the important task of assessing these measures' relevance, appropriateness and economic benefits against their budgetary and other costs, including long-term impacts on domestic allocative efficiency. (OECD, 2003)

have to be accessible to investors through secure and flexible mechanisms that allow for long-term investments.

Forestry as an example

Based on the above definitions, and using forestry as an example, the following identifies factors that influence the businesses environment of forestry, and how they impact investment profitability. The methodology is called Forest Investment Attractiveness Index (IAIF, from the Spanish acronym) and computes an index that measures the business climate for forest-based investments. The IAIF's purpose is to flag the

CHART 3
Factors influencing the attractiveness of FDI in agriculture and forest business

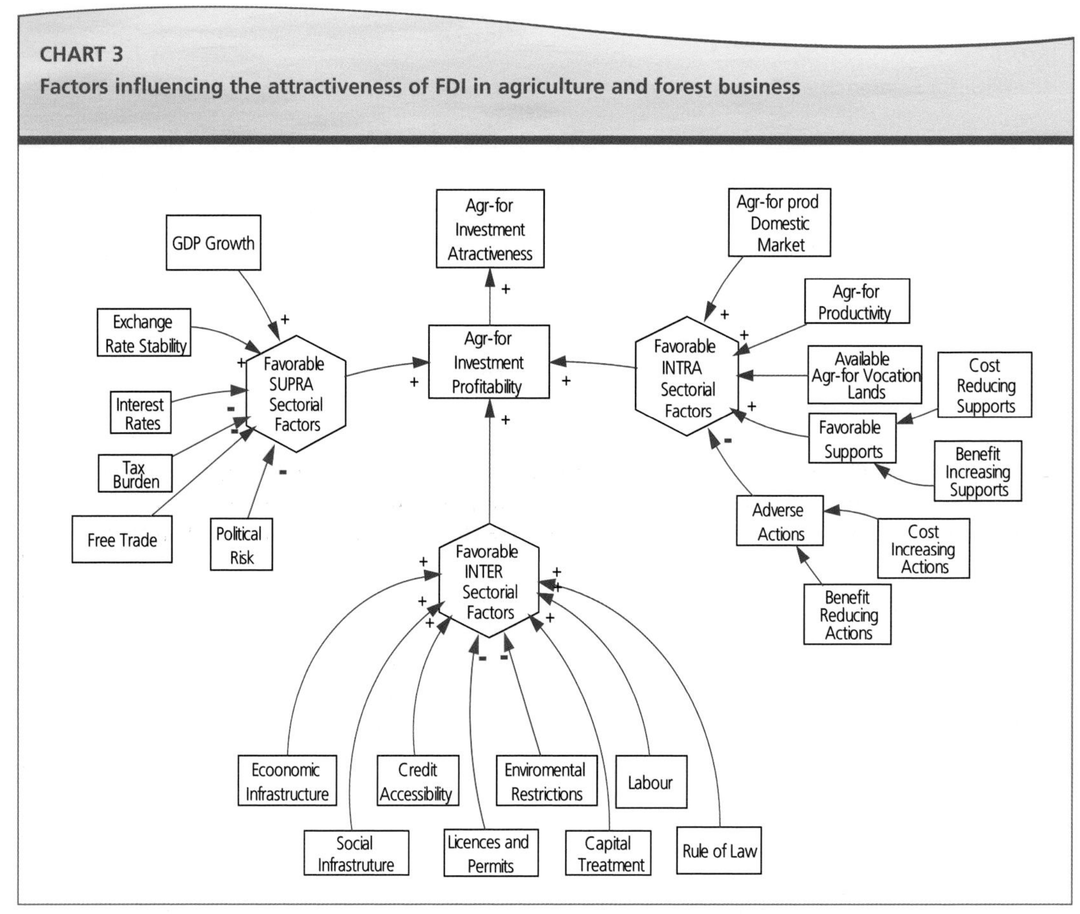

Modified from Nascimento and Tomaselli, 2007

TABLE 5
Brief description of the intra-sectorial factors

Factors	Brief description
1. Agriculture or forest products domestic market	Includes the size of the domestic consumption of inputs and outputs of the agriculture and forest based sector. It also includes the domestic consumption associated with the export of outputs from the sector.
2. Agriculture and forest productivity	Includes the land productivity of agriculture or forest based businesses. It is directly associated with the technologies used for production in the country.
3. Availability of agriculture and forest vocation lands	Includes the size of lands in the country that are arable, or are forest vocation lands. Agriculture production is often, but not always, more competitive in arable lands than forest production, while the opposite is true for forest vocation lands. (J.R. Nascimento, 2005).
4. Favourable supports	Includes policies and measures taken the public or private sectors that reduce costs or increase benefits for investors.
5. Adverse actions	Includes policies and measures taken the public or private sectors that increase costs or decreases benefits for investors.

Source: Adapted from Nascimento and Tomaselli, 2007

factors that affect, lead to success, and attract private direct investment, be it domestic or foreign. The IAIF allows: (i) to compare the performance of countries in the same year and the trend over time, (ii) to assist investors to pre-identify the countries where sustainable forest business will most likely be successful, and (iii) to clarify which Supra, Inter and Intra factors affect the business climate. The IAIF methodology considers 80 variables that make up a total of 20 indicators. It has been applied to countries seeking support from the Inter-American Development Bank using data from 2004 to 2006. Table 6 shows the detailed IAIF results for Brazil for indicators and sub-indices in 2006.

Brazil, according to this index, is the most attractive country for investment in forest-based businesses in Latin America and the Caribbean. However, it scores only 60 out of a total of 100 points possible, implying that the country has room for improvement. By comparing the performance of each index with the theoretical possible score as shown in the last column of the Table 6, analysts can easily identify the indicators with the greatest potential for improvement. For instance, the IAIF indicates that inter-sectorial factors such as Labour, Licences and Permits, Property Rights, and Capital and Foreign Investment Flow can more than double their performance, while Intra Sectorial factors such

TABLE 6
Brazil's performance according to the Forest Investment Attractiveness Index (2006)

Indicators / Subindex / IAIF	Rating in 2006	Max. rating possible	Potential growth in %
GDP Growth Rate	75	100	34
Passive Real Interest Rate	97	100	3
Exchange Rate Stability	100	100	0
Trade Openness	58	100	72
Political Risk	67	100	50
Tax Share of GDP	53	100	90
Supra-Sectorial Subindex	75	100	34
Economic infrastructure	62	100	61
Social Infrastructure	79	100	26
Licences and Permits	50	100	100
Labour	39	100	156
Capital Market	55	100	82
Property Rights	50	100	100
Capital and Foreign Investment Flow	50	100	100
Agricultural Policies	57	100	76
Planting and Harvesting Restrictions	52	100	91
Inter-Sectorial Subindex	55	100	82
Forest Resources	40	95	138
Favourable Support	37	100	168
Domestic Market	95	100	5
FVL	80	100	25
Adverse Actions	42	100	137
Intra-Sectorial Subindex	59	99	68
IAIF	60	99	65

Source: Annex 9

as Favourable Support, Forest Resources and Adverse Actions can be almost three times better.

It is beyond the scope of this study to calculate the most recent score Brazil can obtain in the corresponding indicators for agriculture-related investment attractiveness. Such calculation should be undertaken periodically for the design, monitoring and evaluation of interventions. In addition, a simultaneous calculation of these indices for various countries can allow for comparison, help investors to identify those countries that are better suited for establishing their businesses, and ultimately foster an environment of healthy competition among countries.

5. Business climate improvement process to attract FDI into the agriculture sector

Through IAIF, Brazil has learned how it benchmarks relative to other countries, and understands how various factors foster or deter from business ventures. The analysis showed that Brazil, with its abundant natural resources, has room to further its enabling environment. With this information at hand, the challenge for the Government of Brazil is how to improve the investment climate, and thus increase the inflow of FDI. The object of this section is to present a methodology for this purpose that was prepared by the Inter-American Development Bank, called the Forestry Investment Business Climate Improvement Process (PROMECIF). PROMECIF uses the results of IAIF, both its indicators and sub-indices at all stages of the process, either as elements of analysis, intervention design, simulations, or as indicators for monitoring and ex-post evaluation. Although IAIF and PROMECIF are designed for the specific purpose of forest-based investments, they may also be used in other sectors of the economy.

The methodology seeks to help countries improve their business climate through the implementation of a process that is both systematic and cyclical. First, the country confirms its intent on taking steps to make necessary adjustments, carries out a diagnosis of the situation, defines the strategy, and then it designs, implements, monitors and evaluates an Action Plan.

5.1 Overall process

PROMECIF is a cyclical process that seeks to identify, develop, implement, monitor and evaluate actions that pertain to factors that affect the attractiveness of a country for foreign investors. The process is divided into three interdependent phases (Chart 4). Since the purpose of this section is to explain how this process can be useful to understand Brazil's situation, the following deals mostly with Phase II.

Phase I – Country identification and change commitment

Phase I consists of three stages: (i) promotion, (ii) identification, and (iii) setting up of the Coordinating Committee. In the promotion stage IAIF results are presented to stakeholders. The results show the country's performance in absolute terms or relative to other countries or subregions, and signals the critical factors that affect the investment climate for sustainable forestry businesses. It is at this stage, and motivated by those involved in the private sector, that the government may be persuaded to apply the PROMECIF methodology, which is formalized by the signing of a commitment (identification phase). This stage is completed with the constitution of a Coordinating Committee (CC) that organizes the implementation of phases II and III of PROMECIF. The CC should allow for stakeholder participation, and should be located, whenever possible, within the scope of national institutions promoting competitiveness.

Phase II – Diagnostic and strategy definition

The outcome of phase II is the definition of a strategy to improve the business climate for forest-based business investments, and the process includes a Diagnostic stage and an Action Plan.

CHART 4
PROMECIF phases cycle

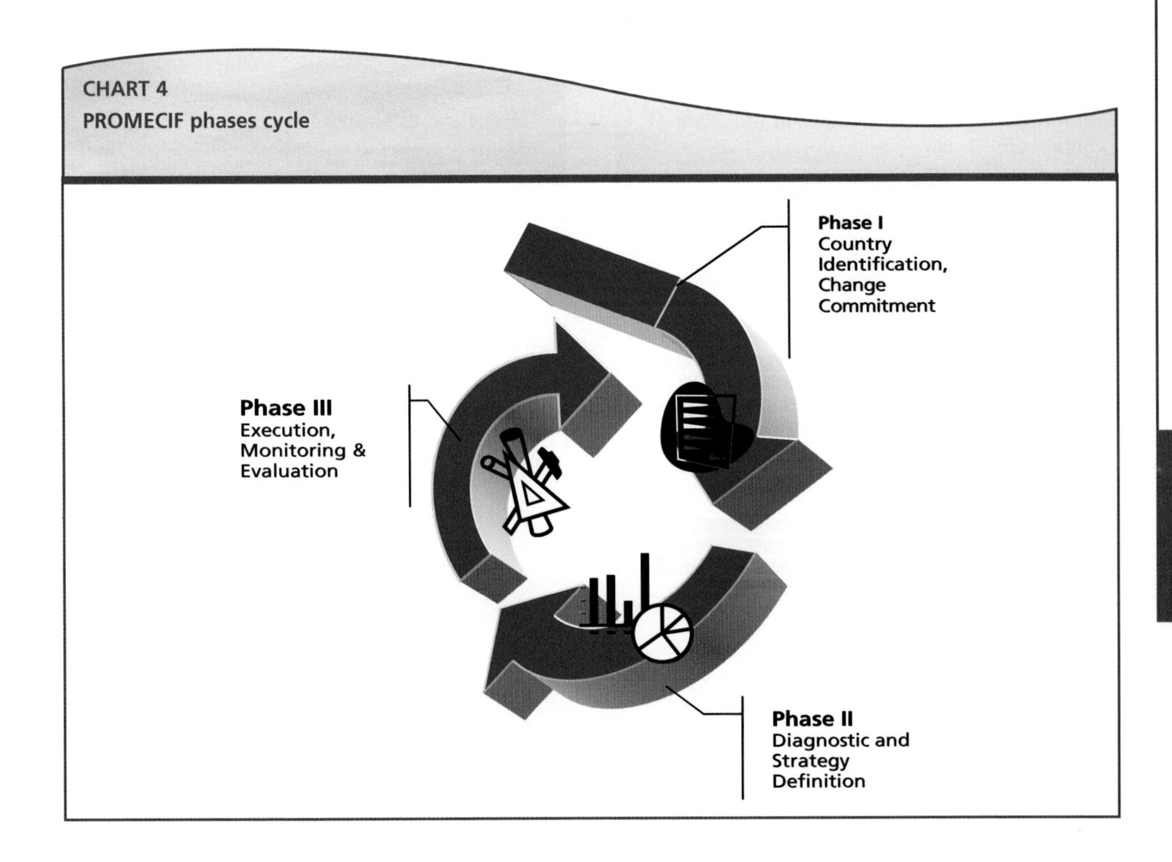

FIGURE 5
Comparison between IAIF of Brazil and selected countries

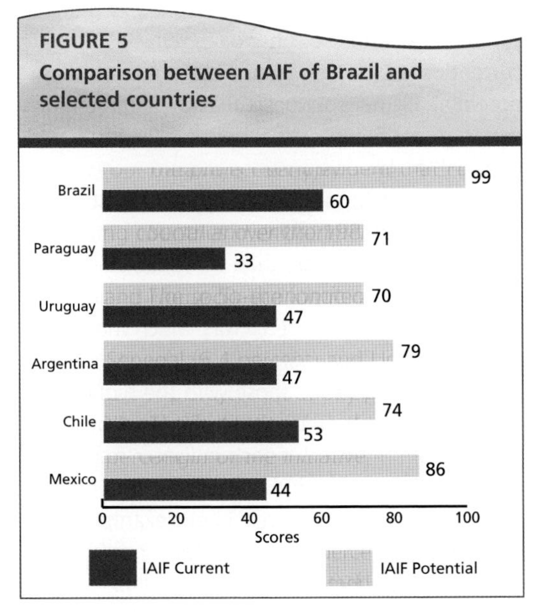

Source: IAIF 2006

FIGURE 6
Subindexes contributions to Brazil's IAIF 2008 score

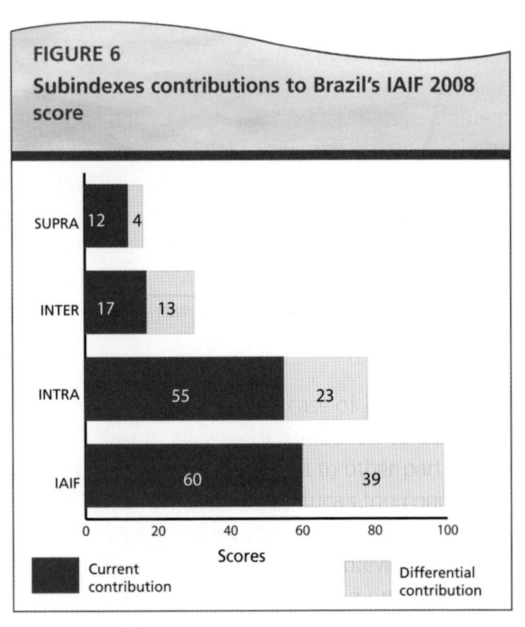

Source: IAIF 2006

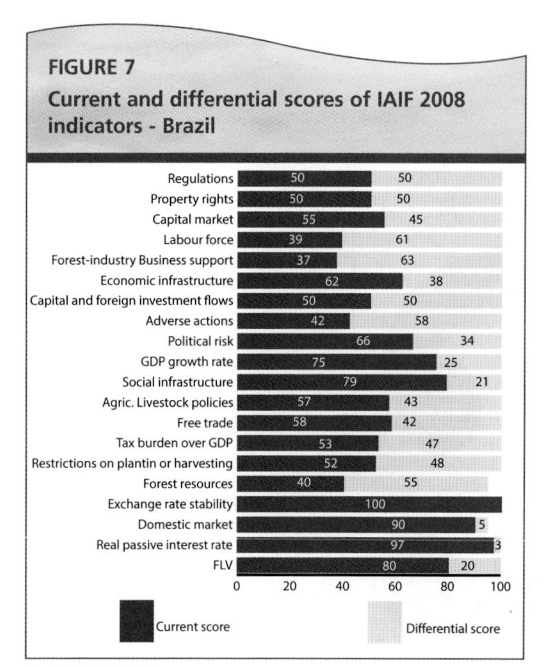

FIGURE 7

Current and differential scores of IAIF 2008 indicators - Brazil

Source: IAIF 2006

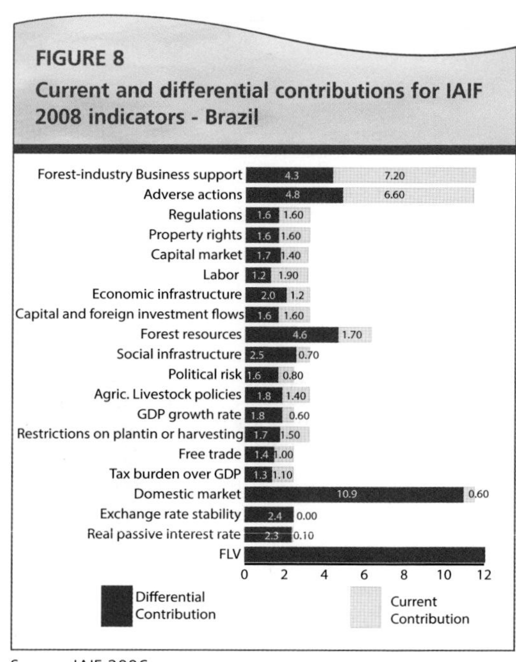

FIGURE 8

Current and differential contributions for IAIF 2008 indicators - Brazil

Source: IAIF 2006

BRAZIL

The Diagnostic

The diagnostic aims to characterize current trends, and the situation of the sector today and in the future as envisaged by stakeholders. It uses IAIF and its indicators and models to analyze the causes and effects that generate, and are generated by, each factor. Systems dynamics methodology is suggested for the identification and inspection of factor interactions. As already mentioned, IAIF shows indicators for the country and how the country ranks relative to others. Due to its simplicity, clarity, accuracy, measurability and validity, IAIF lends itself to countless forms of analyses, and to an understanding of the situation from different angles. However, IAIF is less useful for defining the processes to identify the desired future situation.

The Analysis

The Analysis starts with the results of IAIF for the country in question. IAIF can identify convergences between countries or highlight contrasts. The country may be compared to others with similar IAIF performance, with the top best ranked, with countries that have similar GDPs or which are geographically closed. Figure 5 shows as an example how Brazil compares to the five best ranked countries in IAIF (2006).

In addition to overall scores, an analysis of Supra, Inter and Intra sub-indices will help to identify the factors that deserve greater attention and have higher potential. Figure 6 shows a simplified example of the type of analysis that can be used with the sub-indices, where their current contribution is compared to its maximum potential. Figure 6 suggests that the interventions with higher potential are those at inter- and intra-sectorial levels.

Every indicator and factor used to construct the IAIF should be analysed, i.e. each of the 20 indicators and the more than 80 variables that make up these indicators. Priority for action should be given to those factors which make the greatest contribution to increasing IAIF performance, or rather those with the highest potential to generate impact. A measure of the growth potential for each indicator may be obtained by calculating the spread between the current and potential scores. In our example, Figure 6 shows that the IAIF analysis should concentrate on the indicators of the Inter- and Intra-sub-indices. Figure 7 helps to further identify the priority indicators within these sub-indices and how they compare to each other.

The indicators in Figure 7 need to be weighted according to how important the sub-index in which they are located affects the final IAIF score. The weighted results of Figure 8 indicate that the factors are, in order of decreasing priority: forest industry business support, adverse actions, labour, forest resources, trade licences and permits, and property rights. It is beyond the scope of this study to undertake a complete application of the PROMECIF, therefore further step of the process are only a brief described. [6]

Complementary Analysis

The factors identified must be studied in detail to improve our understanding of how the investment climate is determined. It should explain the processes that led to the current situation, and thus suggest actions that inhibit or promote investments. Complementary studies identify and fill gaps in data, information and analysis, and help the CC to identify problems and opportunities, the future situation, and the strategies and specific actions that would be required to achieve the desired situation. The diagnostic stage is concluded when all the elements required for defining a strategy and action plan are in place.

Defining Strategies

Based on the diagnostics and identification of problems and opportunities, alternative intervention strategies can be designed to improve the business climate for forestry investments. It is important that members of the CC, acting within their respective competencies, adopt the recommended interventions. They can act directly or articulate their actions with other authorities.

Action Plan

The Action Plan is a set of strategic interventions or actions that make supra-, inter- or intra-sectorial factors more favourable to forest businesses. The methodological tool recommended for the preparation and implementation of the Plan of Action is the Logical Framework.

Phase III - Implementation, Monitoring and Evaluation

Implementation of PROMECIF starts as soon as the Plan of Action is validated by the CC. The process begins by identifying the most appropriate source of funding for each strategic action. Once funding is obtained, a detailed analysis and design of the project that takes into consideration the requirements of the funding source(s) is carried out. Once approved, the plan is implemented, monitored and evaluated by the implementing agency, the CC and other independent entities. Once the execution is completed, an evaluation provides lessons for future projects. Post-evaluation may help to decide whether further actions are needed to achieve the desired situation, which in turn kick-starts a new cycle of PROMECIF.

5.2 Improving business climate critical factors

This section provides a brief discussion of some of the critical issues and factors that affect the business climate for FDI in agriculture and the forest sector (Chaddad and Jank, 2006) (OECD, 2009). The discussion is structured using the same 'supra', 'inter', and 'intra' classification of factors used in the above.

i. Improvement in Supra-Sectorial Factors

Supra-sectorial factors are those which affect the whole economy and therefore unlikely to be changed to accommodate the needs of any particular sector. Nevertheless, an evaluation of these is necessary both to illustrate their impacts in a given sector and for policy debate. Supra-sectorial factors that may be considered include exchange rates, interest rates and taxation.

[6] The PROMECIF has been fully applied in Panama, Paraguay and Ecuador with the financial and technical support of the Inter-American Development Bank.

BOX 7
Restrictions to land ownership by foreigners in Brazil

Brazil's National Security Council does not allow foreigners to own land located within 150 km of the national borders. In addition, the Brazilian Chamber of Deputies approved in October 2009 legislation that would further restrict foreign ownership of land along Brazil's borders, and within the Amazon. This legislation is not yet binding because it requires the approval of the Brazilian Senate and that of the President. In August 2010, the Nation's General Attorney issued a directive, approved by the President, which limits the size of properties that foreigners are allowed to own.

- Exchange Rates: Brazil adopted in 1998 a flexible exchange rate that allows for a greater competitiveness in the international markets and, as a consequence, increased exports and the flow of FDI for agribusinesses.
- Interest Rates: The implementation of Plan Real in 1994 put a downward pressure on both inflation and interest rates. Though they continue to be high in real terms due to spreads associated with risk premiums, their persistent decline of the last 10 years has stimulated investments agriculture.
- Tax burden: In the last eight years the internal public debt has almost doubled, reaching around US$1 trillion, and taxes have escalated. The tax burden is considered a major factor affecting the competitiveness of the Brazilian businesses, and has deterred investors.

ii. Improvement in Inter-Sectorial Factors

The inter-sectorial factors that have most affected investments in agriculture and forestry and highlighted include economic infrastructure, social infrastructure and environmental restrictions.

- Economic infrastructure: Transport and energy are very relevant and het Brazil has in general neglected its transportation network (highways, railroads, ports, airports, waterways) and the energy sector (bioenergy, hydroelectricity, oil) requires substantial investments for prices to remain competitive into the future.
- Social infrastructure: Brazil's performance of the Human Development Index is relatively low, scoring only 0.699 points and ranking

BOX 8
Forest vocational land policies

One solution to minimizing conflicts between agriculture and environmental protection is the adoption of a Forest Vocation Land (FVL) policy. It identifies lands that are more at risk of erosion and runoff, and requires from landowners that want to use them, the adoption of specific measures to preserve land and water. Lands that are not subject to risk of degradation, the so called non-forest vocation lands, may be put to any use, including forestry. The Forest Vocation Land policy is intuitive, simple and inexpensive to establish and enforce.

73rd in a total of 182 countries evaluated. Low scoring represents low productivity, and therefore reduced competitiveness.

- Environmental restrictions: The conflict between environmental protection and the creation of an enabling environment for investors raises complex issues that are beyond the scope of this Chapter.

iii. **Improvement in Intra-Sectorial Factors**

As described above, intra-sectorial factors pertain to agriculture and forestry, affect the costs, benefits and profitability at various stages of the value chain, and are under the mandate of the agricultural authorities who have the power to address them. The factors discussed here include land property rights and reconciling agricultural and forest uses with environmental protection.

- Property rights: Land property rights that are protected by the State and respected by the Rule of Law are paramount to agricultural investors, where businesses normally take a long time to mature. In relation to foreign investors, governments are sometimes obliged by geopolitical reasons to apply certain restrictions as described, for the case of Brazil, in Box 7.
- Reconciling agriculture with environmental protection: Decisions on whether a particular piece of land should be allocated for agriculture or forestry are always problematic. At a highly competitive commercial level, agriculture and forestry are often mutually exclusive alternatives. In many cases, lands covered with native forests are converted into agricultural land uses, resulting in deforestation. Traditionally, deforestation comprises as a first step the slash-and-burn process, which in itself is a major source of greenhouse gases. Misused land often generates erosion, and runoff which deteriorates the quality of the environment, reduces natural fertility of the soils, and pollutes waters. All these situations exemplify the need for clear rules of the game so that the decision about land in the country can be made taking private and social considerations into account, such as the Forest Vocational Land policy (Box 8).

6. Conclusions and recommendations

Brazil is a relatively large recipient of foreign direct investments, accumulating 45 percent of all FDI in South America and nearly a quarter of the whole of Latin America and the Caribbean. The share of the total inward FDI flows that is allocated to agriculture in Brazil varies every year, but has an average of 20 percent in the period 1998–2007. Up to 2009, agricultural related industries (agricultural processing) received 90 percent of the total, with food and beverages capturing US$21.3 billion or 61 percent. Within primary production agriculture, the largest recipients are crops and livestock, followed by forestry.

Transnational corporations have had an important role in the economic history of Brazilian agriculture. Monsanto, DuPont, Dow Chemicals and Bunge, have been active in the country for decades. Today, only 4 of the world's largest 25 agricultural TNCs have no operations in Brazil. TNCs in Brazil operate at all the stages of the value chain, from the supply of inputs including the production of machinery and equipment, to primary production, processing, wholesale, retail and export levels.

This chapter highlighted some of the key policies and actions that have contributed to investment and production growth in Brazilian agriculture over the years, notably in bringing agriculture-based development to the savannah region. These include:

i. The National Programme for Family Agriculture (PRONAF), a large agricultural and rural credit programme that gave access to credit to a large number of small and medium farmers;

ii. The Brazilian Agricultural Research Company (EMBRAPA), which with a new paradigm of research development has yielded technological packages adequate to the country's major agricultural ecosystems;

iii. A dynamic partnership policy that included international investment support for the development of the inter-land, especially the savannahs of the central-west region of Brazil that became the most important agricultural producing area of the country;

iv. The National System for Rural Extension and Technical Assistance, implemented through the Brazilian Enterprise of Technical Assistance and Rural Extension (EMBRATER), a comprehensive agricultural and rural extension service created in 1954 for the state of Minas Gerais, which was expanded in the late 1970s to all states.

v. The First and Second National Development Plans (PND), an infrastructure investment programme that was implemented during the 1970s. The programme built a large network of roads that allow the transport of agricultural production from otherwise remote areas of the savannahs, power lines, communications facilities and a network of factories that produce and distribute agricultural inputs, machinery, and tractors;

vi. The Brazilian Storage Company (CIBRAZEM), a large network of storage facilities that buy, store and distribute agricultural produce.

Several of these public initiatives have been dismantled over the years as the private sector takes over their roles, notably for the production of inputs and machinery. However, others are still strong such as EMBRAPA and PRONAF.

This chapter also showed how the business climate of a country can be measured using as an example the Forest Investment Attractiveness Index (IAIF). IAIF also explains how private direct investment in forestry is affected by various factors, both inside and outside agriculture. IAIF is applicable for the specific purpose of investments in forestry. It allows static comparisons between countries or how attractiveness changes in any particular country over the years. It assists investors in the identification of countries where a sustainable forest business is most likely to be successful, and flags Supra, Inter and Intra sectorial factors that affect the business climate.

The chapter presented a process used by countries seeking to improve their business climate that is called PROMICEF. Lastly, it presented a brief analysis and recommendations for improving key factors that require attention from stakeholders seeking to improve the business climate for investments. These include at Supra-sectorial level: exchange rates, interest rates and tax burdens; at Inter-sectorial level economic and social infrastructure and environmental restrictions; and at the Intra-sectorial level, the availability of forest vocation lands and solving conflicts related to environmental concerns.

References

ABRAF. 2010 ABRAF *Statistical Yearbook – Base Year 2009.* Brasilia, Brazil: Brazilian Association of Forest Plantation Producers.

Abreu, M.P. 2004. *The Political Economy of High Protection in Brazil before 1987*. Working Paper SITI 08A, Buenos Aires: INTAL - ITD.

Abreu, M.P. 2004. *Trade Liberalization and the Political Economy of Protection in Brazil since 1987*. Working Paper - SITI 08B, Buenos Aires: IDB-INTAL.

Abreu, M.P. & Bevilaqua, A.S. 2000. Brazil as an Export Economy, 1880-1930. In: *An Economic History of Twentieth-century Latin America*, by Enrique Cardenas, Jose Antonio Ocampo and Rosemary Thorp, 32-54. New York: Palgrave.

Albuquerque, F. 2009. *Supermercados suspendem compra de carne de frigoríficos acusados de desmatamento na Amazônia*. http://www.agenciabrasil.gov.br/noticias/2009/06/15/materia.2009-06-15.3487571221/view (accessed 15 June 2009).

Almeida Togeiro, L. & dos Santos Rocha, S. 2008. Brazil: Are Foreign Firms Cleaner than Domestic Firms? In *Foreign Investment and*

BRAZIL

Sutainable Development: Lessons from the Americas, by Kevin P. Gallagher, Roberto Porzecanski, Andrés López, and Lyuba Zarsky Editors, 36-39. Washington, DC: Heinrich Böll Foundation North America.

Batista, J.C. 2008. *The Transport Costs of Brazil's Exports: A Case Study of Agricultural Machinery and Soybeans*. Washington, DC: InterAmerican Development Bank.

Borregaard, N., Dufey, A. & Winchester, L. 2008. Chile and Brazil: Does FDI Promote or Undermine Sustainable Forest Industries?" In *Foreign Investment and Sutainable Development: Lessons from the Americas*, by Kevin P. Gallagher, Roberto Porzecanski, Andrés López, and Lyuba Zarsky Editors, 28-31. Washington, DC: Heinrich Böll Foundation North America.

Braga Nonnenberg, M.J. & Cardoso de Mendonca, M.J. 2004. *Determinantes dos investimentos diretos externos em paises em desenvolvimento*. Rio de Janeiro: IPEA.

Brazilian Central Bank. 2010. *Investimento estrangeiro direto*. http://bcb.gov.br/?INVEDIR (accessed August 2010).

Brazilian Forest Service. 2009. *Florestas do Brasil em Resumo. Dados de 2005 - 2009*. Brasilia, DF: Brazilian Forest Service.

Chaddad, F. & Jank, M. 2006. The Evolution of Agricultural Policies and Agribusiness Development in Brazil. *Choices*, 2006: 85-90.

Cristini, M. 2010. *Foreign Direct Investment (FDI) in MERCOSUR Agribusiness Sector* . NA: NA.

DCED. 2008. *Supporting Business Environment Reforms: Practical Guidance for Development Agencies*. NP: Donor Committee for Enterprise Development.

Dethier, J.-J. & Effenberger, A. 2011. *Agriculture and Development: A Brief Review of the Literature*. Policy Research Working Paper 5553, Washington, DC: The World Bank.

EMBRAPA. 2008. *About Us*. 20 October 2008. http://www.embrapa.br/english/embrapa/about-us (accessed 24 August 2011).

—. 2010. *Balanco Social 2010*. 2010. http://bs.sede.embrapa.br/2010/ (accessed 29 September 2011).

—. 2008. *Ciência, gestão e inovação: dimensões da agricultura tropical*. Brasilia, DF: EMBRAPA-Assessoria de Comunicação Social.

FAO. 2009. *The State of agricultural commodity markets 2009: high food prices and the food crisis -- experiences and lessons learned*. Rome: FAO.

—. 2000. *World Soil Resources Report 90: land resource potential and constraints at regional and country levels. Rome*. Rome: FAO.

Fergie, J.A. & Satz, M. 2007. *Harvesting Latin America's Agribusiness Opportunities. The McKinsey Quartely*. Sao Paulo & Buenos Aires: Mckinsey & Company.

Getulio Vargas Foundation and INCAE Business School. 2008. *Forest Investment Atractiveness Index - 2006 Report*. Inter-American Development Bank, Rio de Janeiro, Brazil and San Jose, Costa Rica: Getulio Vargas Foundation and INCAE.

Group of Fifteen. 2010. *A Survey Of Foreign Direct Investment In G-15 Countries* . Working Papers Series, Volume 7, NA: G-15.

Gwartney, J.D. & Lawson, R. 2009. *Economic Freedom of the World: 2009 Annual Report*. Vancouver, BC: Economic Freedom Network.

Hiratuka, C. 2008. *Foreign Direct Investment and Transnational Corporations in Brazil: Recent Trends and Impacts on Economic Development. Discussion Paper Number 10*. NA: Celio Hiratuka, 2008. Foreign Direct Investment and Transnational Corporations in Brazil: Recent Trends aThe Working Group on Development and Environment in the Americas.

Jaumotte, F. 2004. *Foreign Direct Investment and Regional Trade Agreements: The Market Size Effect Revisited. IMF Working Paper WP/04/206*. Washington DC: International Monetary Fund.

Josling, T. 2011. Agriculture. In *Preferential Trade Agreement Policies for Development: A Handbook*, by Jean-Pierre Chauffour and Jean-Christophe Maur, 143-159. Washington, DC: World Bank Publishing.

Lopes Vidigal, I., Rezende Lopes, M. & Campos Barcelos, F. 2007. Da Substituição da Importação a Agricultura Moderna." *Conjuntura Economica*, November 2007: 56-66.

Lucena Batista, R.& Jesus Souza, N. 2000. Politicas agrícolas e desempenho da agricultura brasileira, 1950/2000." NA, NA: 14.

Meadows, D. 1999. *Leverage Points: places to intervene in a system*. Asheville, USA: Sustainability Institute.

Nascimento, J.R. 2005. *Forest Vocation Lands and Forest Policy: When Simpler is Better*. RUR-05-03 . Washington, DC: Interamerican Development Bank, Sustainable Development Department.

Nascimento, J.R. 2009. *Framework for the Sustainable Development of African Savannah: The Case of Angola (Final Report)*. Brasilia, Brazil: FAO, TCAS.

Nascimento, J.R. & Mota-Villanueva, J.L. 2004. *Instrumentos para el desarrollo de dueños de pequeñas tierras forestales*. RE2-04-005. Washington, D.C: Inter-American Development Bank.

Nascimento, J.R. & Tomaselli, J. 2007. *Como Medir y Mejorar el Clima para Inversiones en Negocios Forestales Sostenibles*. RE2-05-004 Serie de Estudios Economicos y Sectoriales. Washington, DC: Inter-American Development Bank.

Nery, N. 2009. *Moratória da Soja na Amazônia é renovada e quer ampliar ações*. 28/7/2009. http://www.estadao.com.br/noticias/geral,moratoria-da-soja-na-amazonia-e-renovada-e-quer-ampliar-acoes,409777,0.htm.

OECD. 2009. *Agricultural Policies In Emerging Economies 2009: Monitoring And Evaluation*. Paris: OECD.

OECD. 2005. *Agriculture Policy Reform in Brazil*. Policy Brief, Paris: OECD, 2005.

—. *Checklist For Foreign Direct Investment Incentive Policies*. Paris: OECD, 2003.

—. *Environment and the OECD Guidelines for Multinational Enterprises Corporate Tools and Approaches*. Paris: OECD.

—. 2005. *Environmental Requirements and Market Access*. Paris: OECD.

—. 1999. Foreign Direct Investment and the Environment. Paris: OECD.

OECD. 2008. *Rising agricultural prices: causes, consequences, and responses*. Policy Brief, Paris: OECD Observer.

Ramos, S.Y. 2009. *Panorama da politica agricola brasileira: a politica de garantia de precos minimos*. Documentos EMBRAPA Cerrados, Planaltina, DF: EMBRAPA Cerrados.

Rocha, S. & Togeiro, L. 2007. *Does Foreign Direct Investment Work For Sustainable Development? A case study of the Brazilian pulp and paper industry*. Discussion Paper Number 8, NA: Working Group on Development and Environment in the Americas.

Silva Dias, G.L. & Moitinho Amaral, C. 2001. *Mudancas estruturais na agricultura brasileira: 1980-1998*. Serie desarrollo productivo 99, Santiago de Chile: CEPAL.

Spehar, C.R. *Opportunities and Challenges for the Development of African Savannahs Using the Brazilian Case as Reference. Consultancy Report for FAO. Mimeo*. Rome: FAO, SD.

STCP. 2005. *Índice de Atracción A la Inversión Forestal (IAIF) – Informe final (Rev.01) para el Banco Interamericano de Desarrollo*. Washington, DC: Inter-American Development Bank.

World Bank. 2005. *World Development Report 2005: A Better Investment Climate for Everyone*. Washington, DC: World Bank Publishing.

—. 2004. *World Development Report: A Better Investment Climate for Everyone*. Washington, DC: World Bank Publishing.

Tollini, H. *Integrating Brazilian Savannahs to the Production Process: Lessons Learned*. NP: FAO, ND.

UNCTAD. 2009. *World Investment Report 2009. Transnational Corporations, Agriculture Production and Development*. New York and Geneva: United Nations.

US State Department. 2010. 2010 Investment Climate Statement - Brazil. *US State Department. Diplomacy in Action*. 3 2010. http://www.state.gov/e/eeb/rls/othr/ics/2010/138040.htm (accessed 8/3/2010).

—. 2010. Background Note: Brazil. *US State Department. Diplomacy in Action*. 2 5, 2010. http://www.state.gov/r/pa/ei/bgn/35640.htm (accessed 3 August 2010).

World Bank. 2009. *Awakening Africa's Sleeping Giant: Prospects for Commercial Agriculture in the Guinea Savannah Zone and Beyond*. Washington, DC: FAO and The World Bank.

—. 2009. *Brazil at a Glance*. Washington, DC: The World Bank.

—. 2009. *Doing Business 2010*. Brazil. Washington, DC: The World Bank and the International Finance Corporation.

—. 2008. *World Development Report 2008: Agriculture for Development*. Washington, DC: World Bank.

World Economic Forum. 2009. *The Brazil Competitiveness Report* 2009. Geneva: World Economic Forum and Fundação Dom Cabral.

—. 2009. *The Global Competitiveness Report 2009-2010*. Geneva: World Economic Forum.

—. 2009. *World Economic Forum. (2009). The Brazil Competitiveness Report 2009. Geneva: World Economic Forum and Fundação Dom Cabral.* Geneva: World Economic Forum and Fundação Dom Cabral.

TANZANIA:

Analysis of private investments in the agricultural sector of the United Republic of Tanzania[1]

1. Introduction

In Africa, international concerns have been raised by recent foreign, large-scale land acquisitions over the impacts on small farmers and food security. There are fears that local concerns are not emphasized in investment contracts and international investment agreements, and that domestic laws are inadequate to redress this imbalance. However, given the limited information on the nature and impact of these investments, this chapter attempts to highlight the key issues. The study examines the extent, nature and impact of international (private) investments in the agricultural sector of the United Republic of Tanzania. It achieves this by analysing the policies, legislation and institutions and other related issues affecting international investment generally. Agriculture and land are then examined in more specific detail. It traces the evolution of investment and divestiture policies, and highlights the primary practices and policies – including business models – influencing the investment climate in the country. The investment status of certain agricultural commodity sectors is then identified and the areas of the value chains that are more attractive to investors are examined. Finally, the study proposes options for policy-makers and investors to ensure that the nation's food security and the rights of resource-poor farmers are not compromised by these large-scale land investments.

Agriculture is the backbone of the Tanzanian economy; it contributes significantly to the production of food and raw materials for industries, employment generation and foreign exchange earnings. In 2009, agriculture contributed about 27 percent to the GDP; second only to the services sector (Figure 1). Given the economic significance of this sector, investment (both public and private), is seen as a way of spurring economic growth. The role played by foreign direct investment (FDI) in stimulating production, bringing in new technology and capital for investment, contributing to the balance of payments and opening up employment is generally recognized.

Since the mid-1980s, the Tanzanian economy has undergone a gradual and fundamental transformation that has redefined the role of government and the private sector. Under the current prevailing environment, most of the production, processing and marketing functions have been assigned to the private sector, while the government has retained regulatory and public support functions. These macro changes have,

FIGURE 1
Shares of GDP by type of economic activities, 2009 current prices

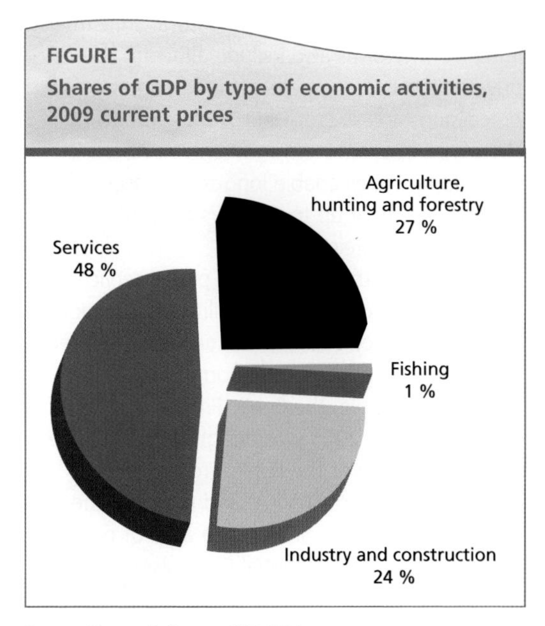

Source: Economic Survey, URT 2010

[1] This chapter was prepared by Suffyan Koroma, Economist, Trade and Markets Division, FAO and Bede Lyimo, Tanzanian national Consultant.

and will continue to have, a profound impact on the agricultural sector in which already agricultural input and output prices have been decontrolled, subsidies have been removed, and the monopolistic tendencies of cooperative and marketing boards have been significantly reduced. The government and stakeholders in agriculture are working to achieve by the year 2025 an agricultural sector that is modern, commercial, highly productive and profitable, and which utilizes natural resources in an overall sustainable manner and acts as an effective basis for inter-sectoral linkages.

2. Overview of the regulatory and incentive framework

Tanzania's investment climate has improved considerably following strategies geared towards greater private sector participation in the economy, and an improved regulatory and legal framework – in particular the Tanzania Investment Act 1997, which sets out clear criteria for all potential investors and encourages private sector financing, together with the establishment of the Tanzania Investment Centre as a *one-stop* facilitation institution. Parastatal reforms were designed to diminish the dominance and monopolistic characteristics of state-owned enterprises as part of wider structural adjustment initiatives. Reforms also include allowing the private sector to compete in marketing and processing of cash crops in the increasingly liberalized economic environment. Revisions in the land law rules enable long-term leasehold property rights of up to 99 years for both domestic and foreign investors.

Changes in the provision of public services, coupled with greater predictability, consistency and transparency of the investment environment have attracted positive attention in recent years. For example, the Public Procurement Act implemented in 2005 was designed to enhance the transparency of the Public Procurement Regulatory Authority (PPRA) and promote the participation of local firms in the area of public procurement.[2] A presidential Commission, the PRSC, was formed to oversee the transfer of property rights from state to private sector. After the transfer, the respective ministries were required to follow up implementation of the contracts entered into between the government and private buyers.

Financial reforms have enhanced the investment climate, enabling 26 licensed banks (both foreign and domestic), to be fully operational in the country. In addition, non-bank financial institutions (e.g. telephone money transfer services, etc.) are licensed to conduct business in the country.[3] However, non-residents of the United Republic of Tanzania cannot generally borrow directly from local banks but foreign investors may acquire credit in the country for inputting capital locally or importing capital goods to be used inside the United Republic of Tanzania. While normal banking regulations are followed, few overseas investors borrow from local banks because of the high interest rates which range from 14 to 24 percent for ordinary borrowers (although larger firms can negotiate lower rates).[4] A credit reference bureau is expected to become operational during the second half of 2012, which will facilitate expansion of credit to MSMEs (Micro, Small and Medium Enterprises).

The introduction of private banks has enabled the freeing-up of interest rates. The Foreign Exchange Act of 1992 removed foreign exchange restrictions and was implemented by the Bureau de Change Regulations of the same year. The Act has greatly alleviated shortages of foreign exchange. Under the Capital Markets and Securities Act 1994 (amended in 1997), capital that supports product or factor markets can be freely exchanged. Through this instrument, the Dar-es-Salaam Stock Exchange was opened to foreign investors with a maximum limit for

2 Tanzania Investment Climate report, available at http://www.state.gov/e/eeb/ifd/2006/62039.htm

3 One very popular scheme in Eastern Africa is the M-PESA – M for mobile; PESA, Swahili for money, is the product name of mobile (SMS) based money transfer system.

4 Investment Climate Tanzania, available at http://www.state.gov/e/eeb/ifd/2006/62039.htm

foreign participation set at 60 percent.[5] This law is currently under review to bring it in line with international standards, and to comply with the International Organization of Securities Commissions (IOSCO) Multilateral Memorandum of Understanding.[6]

As the government cannot fund capital investment, or provide new equity to revive enterprises in many cases, these reforms are expected to strengthen the development of capital markets so as to enable investors/companies to raise funds and increase the public accountability of businesses.[7]

The regulatory framework allows for unconditional transferability, via authorized banks and in freely convertible currency, of net profits; the repayment of foreign loans; charges in respect of foreign technology, etc.[8] Regulation of the financial sector is the responsibility of the Bank of Tanzania whose authority was enhanced by the Banking and Financial Institutions Act of 1991.

Tanzania's competition policy seeks to mitigate restrictive business practices which ultimately result in high prices, poor quality and limitations on the availability of certain products. The policy promotes free trade and access to markets by prohibiting anti-competitive behaviour and the abuse of dominant market position.[9] The objective of the competition policy has been identified as being "to address the problem of the concentration of economic power arising from market imperfections, monopolistic behaviour in economic activities and consequent restrictive business practices".[10] The Fair Competition Commission (FCC) was established by the Fair Trade Practices Act of 1994 (amended in 2000), to monitor compliance with competitive equality standards.

Tanzania's BEST programme (Business Environment Strengthening for Tanzania) was designed to reduce the difficulties associated with operating a business in the country; improve government services, and reformulate the regulatory framework. Other programmes to enhance agricultural productivity include the Agricultural Sector Development Programme (ASDP) and the Integrated Road Projects (IRP) to open up transport networks including rural roads in key agricultural areas.[11] An Export Credit Guarantee Scheme (ECGS) has also been set up by the government in partnership with the Bank of Tanzania which is responsible for administering the scheme.[12] Investors in EPZs (export processing zones) have benefited from this mechanism as well as buyers and exporters of crops.

3. Investment in the United Republic of Tanzania

In 2009, the government continued to make reforms aimed at reducing the costs of doing business through the Tanzania National Business Council and the programmes under Business Environment Strengthening for Tanzania (BEST) and Business and Property Formalization (BPF). According to the World Bank Report *Doing Business*, the United Republic of Tanzania made progress on indicators related to business contracts and employment. In general, the country's ranking rose slightly from 127 out of 181 countries reducing the costs of doing business in 2008, to 126 out of 183 in 2009.

In 2009, a total of 572 projects valued at Tshs.2 970 730.10 million were registered with an employment potential of 56 615 people, compared to 871 projects worth about Tshs.8 billion, with

5 Investment Climate Tanzania, available at http://www.state.gov/e/eeb/ifd/2006/62039.htm

6 http://www.cmsa-tz.org/lagislation/legisla_pipeline.htm

7 http://www.psrctz.com

8 Wetzel H, FY 2004 Country Commercial Guide for Tanzania, International Market Research Reports, US Foreign and Commercial Service and US Department of State.

9 National Trade Policy Background Papers: Trade Policy for a Competitive Economy and Export-led Growth, Ministry of Industry and Trade, Dar-es-Salaam 2003, at p. 81.

10 National Trade Policy Background Papers: Trade Policy for a Competitive Economy and Export-led Growth, Ministry of Industry and Trade, Dar-es-Salaam 2003, at p. 81.

11 Investment Opportunities – Tanzania: Investors Guide', Agricultural Sector Profile, Tanzania Investment Centre, Dar-es-Salaam.

12 SADC Trade Industry and Investment Review 2006: Country Profiles - Tanzania, available at http://www.sadcreview.com/country_profiles/frprofiles.htm

TABLE 1
Distribution of private investment projects in the United Republic of Tazania, 2009

Sector	Number of projects	Capital invested (million Tshs.)	Employment potential
Manufacturing	183	654 472	14 143
Tourism	151	519 259	7 302
Construction	81	922 467	3 360
Transport	61	303 849	5 659
Agriculture	27	45 626	15 114
Human Resources	25	174 226	6 597
Services	16	25 026	814
Financial Institutions	8	65 662	665
Economic Infrastructure	6	80 535	1 495
Telecommunications	5	39 000	88
Broadcasting	4	77 376	1 098
Energy & Natural Resources	5	21 112	88

Source: Economic Survey, URT 2010

employment potential of 109 521 people in 2008. Out of the total projects registered in 2009, 407 were new, while 165 were listed for rehabilitation and expansion. A total of 284 projects were owned by local investors; 149 were owned by foreign investors and 139 were joint venture projects. The distribution of the projects was as shown in Table 1:

3.1 Foreign Direct Investment

Foreign direct investment serves as an important complement to domestic investment as a source of external capital. Domestic savings in developing countries such as the United Republic of Tanzania are small. It is widely accepted that the successful impact of FDI flows into the country hinges on the level of progress in education, technology, infrastructure, and financial markets.[13] This means that comprehensive policies are needed, such as export promotion schemes, or those which promote local technological competence (such as training), to better harness technology transfers brought about by FDI. National treatment is accorded to all FDI in the United Republic of Tanzania.[14]

As part of its efforts to improve the investment climate, the government continued to enforce the application of Environmental Impact Assessments (EIA) before executing any large-scale investment projects. Thus, in 2009, more than 90 percent of the projects were given operational certificates after meeting the EIA requirements and standards.

A marked consequence of the improved investment climate has been the increased flow of foreign direct investments (FDI) into the country since 1995, as evidenced in Figure 2. The United Republic of Tanzania's FDI inflows were at their highest level of US$744 million in 2007, making it one of Africa's leading FDI target countries. However, in 2009, FDI declined by 14.5 percent to US$650 million, due to the impact of the global financial and economic crises. Figure 2 maps out the evolution of FDI between 1995 and 2009. Despite the impressive sums of FDI inflow

[13] Msuya E (2007).'*The Impact of Foreign Direct Investment on Agricultural Productivity and Poverty Reduction in Tanzania*', Kyoto University Munich Personal RePEc Archive (MPRA) Paper No. 3671, available at http:// mpra.ub.uni-muenchen.de/ 3671/

[14] WTO Trade Policy Review - Doc WT/TPR/S/171/TZA, available at http://www.wto.org/english/tratop_e/tpr_e/s171-02_e.doc

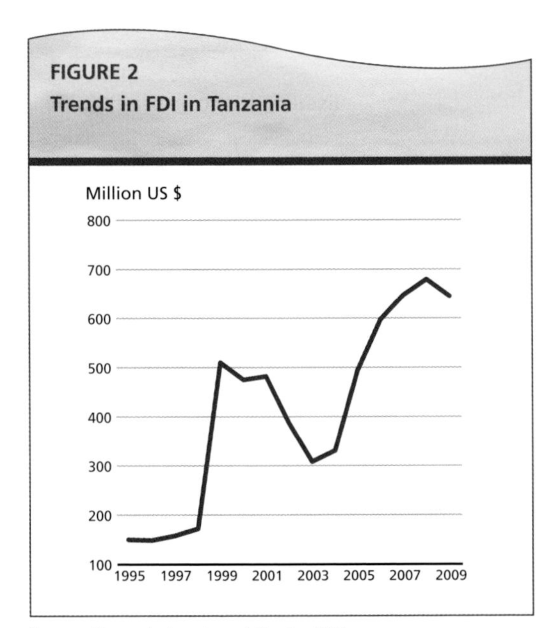

Source: Tanzania Investment Centre (TIC)

into the country, Tanzania Investment Centre (TIC) reports indicate that agriculture has not attracted a large share of this foreign investment. However, this can also be explained by the fact that projects that involve agro-processing like beverages, etc. (which have attracted considerable investment in the United Republic of Tanzania) are not included under agriculture.

Tanzanian agriculture is dominated by smallholders with low levels of productivity, but also limited education, skills and experience, and insufficient access to credit and input. Their low performance, small-scale and weak institutional arrangements therefore do not make them a viable option for joint ventures with foreign investors.

Furthermore, only a small fraction of those that are sufficiently organized and structured to support foreign investments (for example the sugar, barley and sisal subsectors), or larger commercial enterprises, are able to attract a greater percentage of FDI.[15] Empirical evidence suggests that smallholder producers with links to larger estates and foreign firms benefit from increased productivity and efficiency.[16]

Large-scale, foreign-owned farming operations in the United Republic of Tanzania include Brooke Bond (tea) from the United Kingdom, Ilovo (sugar) from South Africa and Africa Plantations (coffee) from Zimbabwe.[17] South Africa, Kenya, United States, United Kingdom, Germany, India, Thailand, Canada and Italy comprise 90 percent of the foreign investments in the country.[18] Although South Africa is perceived as the leading foreign investor in the country in terms of quantity of investment,[19] the United Kingdom is the largest investor with respect to employment, projects and investment value (Table 2).

Although both public and private sector investment are rising, as evidenced by the trends in capital formation (Table 3), public sector capital formation has declined from 39 percent in 2002 to 26 percent in 2009.[20] Total private sector fixed capital formation experienced an eight-fold increase during the same period. Comparatively, the United Republic of Tanzania is one of sub-Saharan Africa's primary FDI target countries; however, statistics demonstrate that FDI has been primarily directed at mining and quarrying activities, with much smaller levels of FDI going towards the agricultural sector. Investments in

[15] Msuya E (2007).'The Impact of Foreign Direct Investment on Agricultural Productivity and Poverty Reduction in Tanzania', Kyoto University Munich Personal RePEc Archive (MPRA) Paper No. 3671, available at http:// mpra.ub.uni-muenchen.de/ 3671/

[16] Msuya E (2007).'The Impact of Foreign Direct Investment on Agricultural Productivity and Poverty Reduction in Tanzania', Kyoto University Munich Personal RePEc Archive (MPRA) Paper No. 3671, available at http:// mpra.ub.uni-muenchen.de/ 3671/

[17] 'Investment Opportunities – Tanzania: Investors Guide', Agricultural Sector Profile, Tanzania Investment Centre, Dar-es-Salaam.

[18] Wetzel H, FY 2004 Country Commercial Guide for Tanzania, International Market Research Reports, US Foreign and Commercial Service and US Department of State.

[19] Mkono N. and Wilms BJ, 'Gateway to Foreign Investments in Tanzania', Mkono and Co Advocates, Dar-es-Salaam, available at http://www.iflr.com/?Page=10&PUBID=33&ISS=20856&SID=595031&TYPE=20

[20] Capital formation is the transfer of savings from households and the government to the business sector resulting in increased output and economic expansion.

TABLE 2
Top 10 foreign investors *

Foreign Investor/country	Investment value (US$ million)	Employment created	No. of projects
United Kingdom	1 115	232 030	595
Kenya	958.21	37 511	249
India	825.88	-	-
South Africa	466.58	14 243	111
Netherlands	426.58	-	-
China	383.23	-	-
United States, Germany, the UAE and Botswana	696.3	-	-

* ImaniLwinga, 'UK, Kenya Leading Investors in Tanzania', The Sunday Observer, 3 June 2007, available at: http://www.ippmedia.com/ipp/observer/2007/06/03/91766.html

agriculture are categorized into 'new' and 'old' (potential for expansion and rehabilitation), which includes privatized entities. At present however, while the investment level has fallen to between 16 and 18 percent of the agricultural GDP, the government contribution has been much higher (Table 3).[21] Foreign investment has traditionally been the major source of funding. Capital investment, FDI and joint venture projects have increased, although it should be noted that the sectors that attract the most FDI interest are mining, manufacturing and tourism.[22]

[21] WTO Trade Policy Review - Doc WT/TPR/S/171/TZA, available at http://www.wto.org/english/tratop_e/tpr_e/s171-02_e.doc

[22] Wetzel H, FY 2004 Country Commercial Guide for Tanzania, International Market Research Reports, US Foreign and Commercial Service and US Department of State.

TABLE 3
Capital formation by public and private sectors at current prices

Sector	2002	2003	2004	2005	2006	2007	2008	2009
					Tshs. million			
Public Sector:								
Central Gov't	568 022	753 610	953 157	1 039 910	1 134 578	1 352 763	1 628 172	1 921 243
Parastatals	59 405	72 745	119 245	162 413	141 822	141 570	148 299	157 197
Institutions++	72 900	89624	120 042	138 362	141 822	144 659	152 971	163 067
Total Public Sector	700 327	915 979	1 192 444	1 340 685	1 418 222	1 638 992	1 929 442 214	2 241 506 986
Private Sector	1 050 490	1 372 817	1 903 078	2 618 878	3 465 267	4 480 021	5 344 872	5 779 463
Total Fixed Capital	1 750 817	2 288 796	3 095 522	3 959 563	4 883 489	6 119 013	7 274 314	8 020 970
Increase in Stocks	44 596	43 387	57 845	64 405	74 292	90 728	106 943	152 252
Total Capital Formation	1 795 413	2 332 183	3 153 367	4 023 968	4 957 781	6 209 741	7 381 257	8 173 221

++ Includes non-profit making organizations
Source: National Bureau of Statistics

In the agricultural sector, however, there have been an increasing number of investors targeting mostly the biofuels sector as shown below in Table 4. The range of commodities has been widened to include dynamic products for export. The main non-traditional commodities which have attracted investments in recent years include sugar, seaweed, maize, poultry, mushrooms, vegetables and cut flowers, beef ranching, fruits, sesame, honey and moringa trees. However, the bulk of recent investments have been geared towards biofuels – jatropha, oil palm and sugarcane, etc.

4. Divestiture policy

A greater degree of privatization was advocated to revive the ailing parastatals under government control, increase government revenue, increase employment and to broaden ownership and participation in management of these enterprises. Privatization enables the sale of non-performing assets, and has been seen to increase production (for example, the sugar industry production levels rose from 96 227 metric tonnes during 1988/89 to 229 617 metric tonnes during 2004/05. In the United Republic of Tanzania, opportunities for new investment were unable to be seized as a result of the government's focus on maintaining production and solvency.[23] Increased privatization measures, through the Public Corporations Act 1992 (as amended in 1999) promoted private sector participation in the economy and encouraged local ownership in the newly privatized, state-owned enterprises by reserving a certain amount of shares for sale to Tanzanians.[24] The Parastatal Sector Reform Commission (PSRC) coordinates the government's restructuring and privatization efforts of state-owned enterprises and government shares in privately owned companies.

The parastatal divestiture seeks to fuel investment in agricultural firms through the enhancement of production of these enterprises; the objectives of the Tanzanian divestiture policy as regards agricultural parastatals are to "increase efficiency, productivity, and quality of goods, and services through capital injection, new and improved management and technology."[25]Unfortunately, several problems present themselves as regards Tanzanian parastatals. Acquisitions have led to a number of lay-offs; the cutback of benefits to existing workers has meant that the latter have slowed down the divestiture process until their remuneration is clarified.[26] Inadequate legal

[23] http://www.psrctz.com

[24] Mkono N and Wilms BJ, 'Gateway to Foreign Investments in Tanzania', Mkono and Co Advocates, Dar-es-Salaam, available at http://www.iflr.com/?Page=10&PUBID=33&ISS=20856&SID=595031&TYPE=20

[25] Agricultural and Livestock Policy (1997), Ministry of Agriculture, Dar-es-Salaam.

[26] Agricultural and Livestock Policy (1997), Ministry of Agriculture, Dar-es-Salaam.

TANZANIA

TABLE 4
Selected agro-based Investments in the United Republic of Tanzania by source country, sector and estimated jobs

Date	Source country	Investment (US$)	Estimated jobs	Sector
June-11	United Kingdom	45 000 000	150	Beverages
September-09	Republic of Korea	6 200 000	156	Food & Tobacco
May-09	United Kingdom	20 400 000	261	Beverages
April-09	United Kingdom	20 000 000	211	Beverages
December-08	United States	12 000 000	126	Beverages
August-08	Kenya	21 800 000	287	Food & Tobacco
June-08	United Kingdom	50 000 000	150	Beverages
November-04	Belgium	1 000 000	5	Food & Tobacco

Source: FDI Markets

safeguards to guarantee continuity of divested interests further reduce investor confidence, as does the lack of policy guidelines that protect local expertise in the newly privatized parastatals. Indeed, there was not always sufficient encouragement of local ownership; in any event, the majority of the local population has neither the capital, nor the ability to access credit to allow them to participate in the divestiture process.[27] To overcome these problems towards a sustainable and accelerated divestiture process, the following policies were identified:[28]

- MAFSC will advocate for divestiture of production and commercial oriented parastatals.
- Since MAFSC parastatals are land based, attaching value to the land asset, promotion of employment and wider participation of people including those surrounding the enterprises being divested, will be considered.
- The government will continue to invest in strategic areas which have failed to attract investors.

In 1994, agriculture-related state enterprises were put under the PSRC. There are no restrictions on foreign participation in the newly privatized enterprises; the tender evaluation criteria are published in the tender invitation and related documents.[29] Furthermore, in line with reducing its role to one of enforcing regulations, the Tanzanian Government withdrew its majority stakes in the parastatals and instead focused on the promotion of a competitive economic environment, controlling restrictive trade practices and setting up appropriate regulatory frameworks.[30]

The particular approach taken for the state-owned enterprise depends on the characteristics of the entity; for example, should the company suffer severe losses which cannot be recovered, it is likely to be liquidated. If the enterprise is still commercially viable, restructuring of the entity is likely to be carried out by the new owner.[31] The choice of divestiture method is selected according to the objectives of the particular privatization, following an assessment of factors including *inter alia* financial viability, the nature of the industry and technology involved, whether a certain degree of public ownership is economically desirable, and its past performance.

These considerations will affect the degree of investment and the types of investors attracted. Investors can therefore enter the market through the following divestiture methods: trade sales and joint ventures (the most common privatization method), public share offers, public auction, private placement, buy-outs by management and/or employees, privatization funds as purchase vehicles for wider share ownership. In some cases, the sale of shares is not possible and requires ownership to remain the same, for example through the use of lease and management contracts. These methods are not conducive to investment and do not offer investors much incentive to make creative, long-term restructuring of the parastatals; these options have not proved to be as successful for the United Republic of Tanzania as anticipated, in situations where managers do not have a large ownership stake in the enterprise.[32] The PSRC itself recognizes the time taken for the divestiture process to be completed – however, the divestiture process itself should not serve as an impediment to investors.

Investment opportunities are visible at the implementation phases of the divestiture, and the ways in which the government can make the sale or its negotiating position stronger depends on the method of divestiture. The government can strategically attract investors by following the approaches identified in the Tanzania Investment Centre website:

[27] Agricultural and Livestock Policy (1997), Ministry of Agriculture, Dar-es-Salaam.

[28] Agricultural and Livestock Policy (1997), Ministry of Agriculture, Dar-es-Salaam.

[29] 'An Investment Guide to Tanzania: Opportunities and Conditions', United Nations Conference on Trade and Development (UNCTAD) and International Chamber of Commerce (2005).

[30] http://www.psrctz.com

[31] http://www.psrctz.com

[32] http://www.psrctz.com

- preparation of sales memoranda, prospectuses or other suitable documents for the targeted investors;
- marketing the offer, including targeted advertising and industry and trade searches to identify buyers;
- pre-qualifying buyers, inviting bids;
- assessing bids or proposals against predetermined selection criteria.

In furtherance of these goals of broad ownership, shares are sold to the general public (local and foreign) and through management and employee buy-outs. The latter is encouraged, at a discount or on deferred terms, as a matter of policy.[33] Other mechanisms by which the government seeks to achieve wider share-

[33] http://www.psrctz.com

BOX 1
Incentives under the Parastatal Sector Reforms

To promote broader ownership
arranging for deferred payments by new indigenous owners out of profits;
employee share ownership schemes with a discounted price;
deferred payment schemes for such shares with loans from banks backed by government guarantee or pledge of securities;
retention by government, through Privatization Trust, of blocks of shares for wider sale at future dates. Alternatively the core private buyers could be required to divest part of their shareholdings at a later date. The approach towards pricing such shares needs to be agreed in advance;
tax incentives to share purchasers, on a case-by-case basis, comparable to those received by incoming investors in new businesses.

To promote domestic investment
the putting together of consortia combining a core investor, a technical partner as appropriate, and indigenous investors wherever feasible;
pre-qualification of bidders to ensure, inter alia, that ownership will not be too concentrated; and transparency in indicating criteria on which bids will be evaluated, including preferences for widening the entrepreneurial and ownership base, but always emphasizing the need for sustaining competitiveness.

To protect the interests of investors, consumers and employees
legislation to curb restrictive trade practices and regulate the use of monopoly power;
equal access to investment incentives whether in new enterprises or in divested businesses; and
equal employment opportunities and security of employment legislation.

Special incentives for new businesses
assisting displaced employees to use any retrenchment grant for business start-up;
training, technical support and advisory services;
assistance in obtaining loans, equity investment; and
relocation support.

Source: PSRC website

ownership involve specific strategies such as discounts or reductions of certain fees and taxes, lowered purchase prices for domestic investors, including deferred payments. In order for domestic entrepreneurs to have access to sufficient credit, the Entrepreneurship Development Fund, together with merchant banks and financial institutions are working together to facilitate funding and provide advisory services to emerging businesses.

The divestiture method adopted as the useful for stimulating investments (particularly from foreign sources), is the joint venture/trade sale to another company either in its entirety or parts of the enterprise. This new investment is expected to inject financial and technical resources to stimulate production, and improve marketing and management. Investment in agricultural firms could also be in the form of the sale of shares through the stock market, auction or private placement.[34]

Foreign direct investment into the United Republic of Tanzania is a comparatively recent phenomenon. As regards traditional commodities, a fairly common method of FDI entry has been through mergers and acquisitions, where multinational or foreign enterprises acquire complete ownership or majority shares in local establishments.[35] Historically, parastatals enjoyed monopolies in these traditional commodities, but the reformed economic climate – which facilitated acquisitions – enabled the unproductive and poorly managed parastatals to be privatized to multinational corporations. Acquisitions enable the possibility of capitalizing on existing local networks and suppliers as well as existing local and regional markets. Foreign direct investment has also come into the United Republic of Tanzania through what are termed 'green-field' investments in non-traditional sectors such as fishing, or cotton ginning; these types of investments are where foreign parent companies enter developing countries and construct new operational facilities. These investments are noted for their creation of new jobs locally.

[34] http://www.psrctz.com

[35] Ngowi, 'Foreign Direct Investment Entry Modes in Tanzania', Tanzanet Journal Volume 3(1) 2002, at pp 1-2.

5. The TRIMs agreement

The role of the WTO Agreement on Trade-related Investment Measures - TRIMs Agreement, which aims to negate the trade restricting and distorting effects of investment measures that applies to the goods trade only, in shaping the investment policy of the country should be considered. As a Member State of the WTO, the United Republic of Tanzania is prohibited from applying investment measures that are contrary to the provisions of GATT that seek to eliminate quantitative restrictions or that violate the principle of national treatment.[36]Local content requirements, trade balancing, exchange requirements, use of local raw materials or technology transfer requirements are examples of stipulations used to regulate foreign investments. The TRIMs Agreement fleshes out, in its Annex, the types of investment measures which run counter to the principle of national treatment.

The category of TRIMs that are inconsistent with the obligation of national treatment provided for in paragraph 4 of Article III of GATT 1994 include those which are mandatory or enforceable under domestic law or under administrative rulings, or compliance with which is necessary to obtain an advantage, and which require:

- the purchase or use by an enterprise of products of domestic origin or from any domestic source, whether specified in terms of particular products, volume or value of products, or of a proportion of volume or value of its local production; or
- that an enterprise's purchases or use of imported products be limited to an amount related to the volume or value of local products that it exports.

Similarly, TRIMs that are inconsistent with the obligation of general elimination of quantitative restrictions provided for in paragraph 1 of Article XI of GATT 1994 include those which are mandatory or enforceable under domestic law or

[36] Article 2.

under administrative rulings, or compliance with which is necessary to obtain an advantage, and which restrict:

- the importation by an enterprise of products used in or related to its local production generally, or to an amount related to the volume or value of local production that it exports;
- the importation by an enterprise of products used in or related to its local production by restricting its access to foreign exchange to an amount related to the foreign exchange inflows attributable to the enterprise; or
- the exportation or sale for export by an enterprise of products, whether specified in terms of particular products, in terms of volume or value of products, or in terms of a proportion of volume or value of its local production.

Many developing countries are disadvantaged by the prohibition of local content requirements in investment schemes, which require that a business must buy or use a minimum amount of locally originating materials. This measure is useful to developing countries wishing to use this mechanism to "encourage domestic economic activities benefiting from raw materials, discouraging wastage of foreign exchange, ensuring linkages of FDI with economic activities and encouraging economic empowerment."[37] While there is no mandatory requirement in the United Republic of Tanzania to use local raw materials (which would run counter to the TRIMs Agreement), investors are nevertheless encouraged to use local materials whenever possible.[38] Similarly, there is no legal requirement for investors to undertake technology transfers, although this, together with training local personnel, is encouraged.

Least-developed countries such as the United Republic of Tanzania are granted time concessions of seven years after the entry into force of the Agreement which expired in 2002. Developing country members are allowed to temporarily deviate from the terms of the agreement as regards balance of payments matters in accordance with the relevant provisions of GATT 1994 (Article XVIII), the Understanding on the Balance-of-Payments Provisions of GATT 1994, and the Declaration on Trade Measures Taken for Balance-of-Payments Purposes.[39]

The intention of the government is to implement measures to enhance socio-economic development in the context of TRIMs pertaining to equity requirements, local content requirements, technology transfer and export performance; however, it has not as yet introduced such measures.[40]

6. The incentives framework

Investment opportunities available in the United Republic of Tanzania are divided into two categories: the Lead Sector where businesses can import capital goods associated with the investment at 0 percent duty, and the Priority Sector where businesses can import related capital goods at a 5 percent rate. Relevant to this study, the former includes agriculture, livestock and export processing zones, and the latter includes natural resources such as fishing. Both these sectors qualify for VAT deferment until the business begins its operations; and further, a tax holiday for the first five years is granted together with a capital allowance of 100 percent.[41]

To qualify for a Certificate of Incentives issued by the Tanzania Investment Centre, minimum investments should be valued at least at US$100 000 for projects owned by Tanzanian

[37] National Trade Policy Background Papers: Trade Policy for a Competitive Economy and Export-led Growth, Ministry of Industry and Trade, Dar-es-Salaam 2003, at p. 122.

[38] 'An Investment Guide to Tanzania: Opportunities and Conditions' United Nations Conference on Trade and Development (UNCTAD) and International Chamber of Commerce (2005).

[39] Article 4.

[40] National Trade Policy Background Papers: Trade Policy for a Competitive Economy and Export-led Growth, Ministry of Industry and Trade, Dar-es-Salaam 2003, at p. 81.

[41] WTO Trade Policy Review - Doc WT/TPR/S/171/TZA, available at http://www.wto.org/english/tratop_e/tpr_e/s171-02_e.doc

BOX 2
Required application procedures

The intended project should aim at foreign exchange generation and savings, import substitution, the creation of employment opportunities, linkage benefits, technology transfer, expansion of goods production, etc. The feasibility study should contain: a clear statement of investment costs [foreign and local expected capital expenditure], how the proposed investment will be financed, specific sources of finance for the project, terms and conditions of the loan, sources of technology, project financial and economic analysis, market study, project capacity, production processes, environmental impact assessment, employment generation, proposed implementation schedule.

- Three completed copies of TIC application forms (issued with a fee of US$100);
- In cases of expansion/ rehabilitation, a copy of audited account for the past three years;
- A copy of the company's Memorandum and Articles of Association;
- A certified copy of the Certificate of Company Incorporation;
- A brief investor profile;
- Three copies of the project's Business Plan/Feasibility Study;
- Evidence of sufficient financial capital to implement the project;
- Evidence of land ownership for the location of the project;
- Project implementation schedule;
- Covering letter.

Source: TIC Website

citizens and US$300 000 for those owned by foreigners or for joint ventures.[42] While foreigners are required to apply to the TIC for permits, locals are not subject to this stipulation in order to invest. A processing fee of US$750 is required to accompany investment certificate applications.

In Zanzibar, which operates under a different law (the Investment Act 1986), the minimum level of investment varies according to sector. For agriculture the minimum foreign direct investment necessary to benefit from incentives is US$500 000 for foreigners and equivalent of US$50 000 for citizens. In the fisheries sector, the minimum for foreigners is US$1 million, while it stands at equivalent of US$100 000 for citizens. These discrepancies within the framework and structure are in the process of being harmonized following the greater integration of tax issues under the EAC (East African Community) framework.[43]

6.1 Tax exemptions

The rationale behind the package of tax relief incentives set up by the government is to allow investors to recover their initial expenditure while their businesses take time to get off the ground before having to pay taxes.[44]The Investment Act sets up a structure for tax incentives as well as non-fiscal benefits. Under section 17 of the statute, TIC may grant TIC Certificate of Incentives which confers benefits such as automatic work permits for five foreign nationals.[45] While there

[42] http://www.tic.co.tz

[43] WTO Trade Policy Review - Doc WT/TPR/S/171/TZA, available at http://www.wto.org/english/tratop_e/tpr_e/s171-02_e.doc

[44] http://www.tanzania-gov.it/modules.php?name=News&file=article&sid=44

[45] It should be noted that although additional permits can be sought, approval is often difficult. There is an abundance of unskilled and inexpensive labour in Tanzania; but due to lack of training, the local workforce often does not occupy managerial or administrative positions.

is no maximum set for the number of foreign nationals working on a particular project, they are more likely to be granted working permits where it can be shown that the required expertise cannot be found locally.[46] The Certificate also confers benefits such as ease of obtaining residence and work permits, and industrial and trading licenses. Land rent on commercial agricultural farms, livestock ranches and forests is set at a nominal fee of 200 Tshs. per acre annually. Furthermore, it also grants the right to transfer out of the United Republic of Tanzania the entire amount of profit, capital and foreign exchange earned; royalty fees and similar charges; and payment of emolument and other benefits to foreign personnel.[47] It should be noted that capital transfers still require approval by the Bank of Tanzania.[48]

The Certificate of Incentives provides investors with tax exemptions, particularly import duties and certain VAT exemptions on project, capital and deemed capital goods;[49] capital expenditure allowances; a special rate of corporation tax set at 30 percent, a withholding tax rate on dividends set at 10 percent and zero tax on loan interest in the priority sectors.[50] The Investment Act makes provisions for additional benefits and incentives in order to promote 'strategic or major investment' projects of over US$20 million, determined at the discretion of the minister for those that are considered to be strategic to the economy.

A problematic clause has been identified in the Tax Revenues Appeals Act 2000 section 12(3) which declares that where a person objects to a tax assessment, the amount which is not in dispute or one third of the assessed tax (whichever is greater) must be paid. This has been noted to give rise to a sense of unpredictability, and restrains cash flows, together with claims that unsubstantiated tax assessments are made to meet revenue targets and do not reflect the income of the businesses that are assessed.[51] In this regard, there have been recommendations for a clear and simple tax appeals process, which indicates clear timeframes for each stage, to avoid abuse of the system through stalled payments, and with payments made only for undisputed assessments.[52]

In Zanzibar, incentives under the 1986 Investment Act charge zero duty for capital goods during the beginning stages of operation of the business, although a service charge of 5 percent is still levied. For the first five years, tax holidays are granted at the discretion of the responsible minister.

The United Republic of Tanzania has signed a number of bilateral treaties which promote FDI by preventing double taxation with Canada, Denmark, Finland, India, Italy, Norway, Sweden, United Kingdom and Zambia, and with pending treaties subject to ratification with Kenya, South Africa, Republic of Korea, Uganda and Zimbabwe. While an EAC double taxation treaty was signed in 1997, the absence of ratification on the part of Uganda poses an impediment to intra-regional transactions, raising the tax rate by 50 percent.[53]

6.2 Export Processing Zones Programme

Frequently, EPZs are used to attract FDI in countries where infrastructure is a challenge; industrial parks are then developed separately

46 'An Investment Guide to Tanzania: Opportunities and Conditions', United Nations Conference on Trade and Development (UNCTAD) and International Chamber of Commerce (2005).

47 http://www.tic.co.tz

48 SADC Trade, Industry, and Investment Review 2006. Country Profiles - Tanzania, available at http://www.sadcreview.com/country_profiles/frprofiles.htm

49 http://www.tic.co.tz

50 MkonoN and Wilms BJ, 'Gateway to Foreign Investments in Tanzania', Mkono and Co Advocates, Dar-es-Salaam http://www.iflr.com/?Page=10&PUBID=33&ISS=20856&SID=595031&TYPE=20

51 Blue Book on Best Practice in Investment Promotion and Facilitation – Tanzania, United Nations Conference on Trade and Development (UNCTAD) and Japanese Bank for International Cooperation (JBIC) 2005.

52 Ibid.

53 Blue Book on Best Practice in Investment Promotion and Facilitation – Tanzania, United Nations Conference on Trade and Development (UNCTAD) and Japanese Bank for International Cooperation (JBIC) 2005.

BOX 3
Tax measures for agriculture

Income Tax Act 2004

- 100 percent first year capital allowance for plant and machinery used in agriculture, including irrigation tools and equipment. The measure is aimed at attracting investment in agricultural technology.
- 100 percent deduction for capital expenditure on land clearance, excavation of irrigation canals, cultivation of perennial crops and planting trees on agricultural land to prevent soil erosion. Formally these are capital expenditures and would be subject to long time deductions.
- Costs incurred in the course of environmental conservation for farming land, animal husbandry, fish farming or restoration of the land to normalcy after use are allowable deductions in assessing taxable income.
- Agricultural businesses are not subject to the equal quarterly instalment payment requirement for income tax purposes but are required to pay taxes at the end of the third and fourth quarter after harvest.
- Agricultural research and development expenditures are also deductible as expenses for income tax purposes.

Value Added Tax Act, 1977

- Unprocessed agriculture and livestock, including unprocessed meat, unprocessed fish and all unprocessed agricultural produce is VAT-exempt.
- Industries producing inputs for agriculture and fishing such as pesticides and fertilizers are zero-rated to enable producers to reclaim input VAT-incurred in the course of production. This measure is aimed at generating enabling environment for investment in the production of agricultural inputs. Imported inputs remain exempt from VAT.
- Processed tea (black tea) and packaged tea are exempt from VAT, to provide a competitive edge to local tea producers.
- Small agricultural producers whose produce is exported may receive a VAT rebate through their cooperative union or associations.

Customs and Excise Tariff Act, 1976

- Agricultural inputs and implements are subject to zero import duty.
- There is no excise duty on wine and brandy manufactured from locally produced grapes. This measure is aimed at expanding the market for wine and hence wine production.

Stamp Duty Ordinance

- Reduction of the stamp duty rate for conveyance of agricultural land to a nominal amount of Tssh.500, in order to reduce costs in conveying land ownership.
- Stamp duty on receipts has been abolished for all receipts including on sale of agricultural produce.

Vocational Education and Training Act (VETA)

- Exemption granted from Skills Development Levy for employment in agricultural farms.

Local Government Finances Act, 1982

- Agricultural produce cess limited to 5 percent of the farm gate price and within the district of production.
- Voluntary contributions collected on agricultura1 produce by local authorities accepted only if introduced by the village community for specific projects implemented by the village or villages.

Source: Ministry of Finance http://www.mof.go.tz/mofdocs/news/taxationreg.htm

to encourage domestic production. In the United Republic of Tanzania, rather than pursue two separate initiatives, the country opted to develop multi-facility economic zones (MFEZs) which would combine domestic production and export-oriented industries in one facility. Through cost-sharing with the private sector, and implementation of the regulatory environment envisaged by the BEST programme combined with efficient administration, MFEZs could provide the best possible business environment within a limited geographical area. Furthermore, if the initial MFEZ proved successful, new strategically placed MFEZs could be established, and MFEZ status and facilities could be extended to other areas of large-scale economic activity.

In 2006, the EPZA developed two types of zones. The first was the standard EPZ, which required companies to export 80 percent of production and the second - the special economic zone (SEZ). In SEZs, companies have no export requirements; they can sell to the local market, and do not have to be in manufacturing. In 2008, the EPZA developed a five-year plan to merge EPZs and SEZs and create economic development zones (EDZs) to incorporate the incentives of both EPZs and SEZs.

Another plan being formulated is one whereby "township" economic zones will be created, mirroring China's approach to industrial organization.

The incentives offered in EPZs are not dependent upon "zone" incentives but rather on the amount of exports. Companies that export most of their output receive more incentives than those servicing the domestic market. In general, the incentives are the same as those given by the TIC, but the infrastructure component could be expected to make the difference in attracting investment. One important difference between the EPZA and the TIC can be found in the area of regulations. The 2006 EPZ Act specifies incentives available to the EPZA to attract investors, whereas the 1997 Investment Act does not. This may mean that TIC is having a more difficult time in assuring investors of incentives, compared to the EPZA.

Existing EPZs have, for the most part, been developed through local, private investors, and a few joint ventures. The developers are responsible for infrastructure within the zone, with the government having responsibility for providing the necessary connections to infrastructure outside of the zone. Like the Dar-es-Salaam Port and the Mtwara Corridor Development Project, EDZ development is suffering delay because of the lack of a PPP (Public Private Partnership) policy and operating guidelines.

Currently, there are three EPZ sites and one SEZ ready for lease. There are 18 companies operating under EPZ status in industrial parks, and 15 single factory units with EPZ status. Export Processing Zone enterprises are nearly evenly divided between local and foreign companies. The foreign companies are primarily from China, Denmark, India and Japan. The majority of companies are in engineering, followed by textiles, agroprocessing and mineral processing. In addition, there are 14 sites designated for EDZ development. Priority is being given to the zones at the ports of Mtwara and Tanga, at the coastal town of Bagamoyo (50 km north of Dar-es-Salaam), and at the northern, inland town of Arusha. Bagamoyo, with a completed feasibility study and master plan, is farthest along in terms of development, and is the top priority of the EPZA. It is envisioned that this EDZ will encompass 9 000 hectares, which is large when compared to the standard 2 000 hectares set aside for other EDZs. The Bagamoyo EDZ will be one of the first "township" style EDZs, and will include the construction of a new port and airport.

The benefits pertaining to EPZs are offered to export industries but are not dependent on location within a specific geographic zone. Companies benefiting from this scheme are required to export 70 percent of the goods they produce and a minimum of US$100 000 in order to qualify.[54] Interestingly, exporters previously established cannot qualify, leaving the EPZ package available only to new export companies.[55] The EPZ parks can be useful for export processing where there is a dearth of adequate infrastructure in the rest of the country; however, the operations

[54] 2006 Investment Climate Tanzania, available at http://www.state.gov/e/eeb/ifd/2006/62039.htm

[55] 2006 Investment Climate Tanzania, available at http://www.state.gov/e/eeb/ifd/2006/62039.htm

of the Zanzibar EPZ Programme of 2002 have been constrained by the lack of adequate infrastructure within the zone itself.[56]

Best practices show that a stronger public sector input into the functioning of EPZ would have a beneficial impact on the success of these ventures; this includes increasing the public and private stakeholder input and participation, reforming legislation and implementing government agencies to assist with the development of these zones. It should be noted that on a fundamental level, the Tanzanian EPZ structure is in line with international practice through its features, such as regulation by an autonomous public corporation (the National Export Processing Zones Authority - EPZA), and a framework enabling public sector development and management of zones.[57] Further, the regulatory framework lucidly sets out the general regulations which describe the rules for setting up an EPZ enterprise and describes the management and monitoring of exports within such a programme.[58] However, certain improvements can be made. Specifically, the United Republic of Tanzania should focus on improving the infrastructure problems such as power and water provision, providing an on-site customs office and management offices.[59] These factors together with poor security services, limited transport access and high rent charges result in the low occupancy of the EPZ.[60]

[56] Diagnostic Trade Integration Study - Tanzania, Volume 1, Integrated Framework for Trade-Related Technical Assistance to Least Developed Countries, 2005.

[57] Diagnostic Trade Integration Study - Tanzania, Volume1, Integrated Framework for Trade-Related Technical Assistance to Least Developed Countries, 2005.

[58] Diagnostic Trade Integration Study - Tanzania, Volume 1, Integrated Framework for Trade-Related Technical Assistance to Least Developed Countries, 2005.

[59] Diagnostic Trade Integration Study - Tanzania, Volume 1, Integrated Framework for Trade-Related Technical Assistance to Least Developed Countries, 2005.

[60] Diagnostic Trade Integration Study - Tanzania, Volume 1, Integrated Framework for Trade-Related Technical Assistance to Least Developed Countries, 2005.

7. Land policies and related issues in the United Republic of Tanzania

The Land Policy of 1995 and the legislation emanating from that policy, i.e. The Land Act No. 4 of 1999, provides the legal basis for the management of land ownership and user rights and settlement of disputes and related matters for all land other than village land. The Village Land Act of 1999 provides for management of land, settlement of disputes and related matters specifically for village land. The two laws, if effectively implemented, provide a robust framework for safeguarding communal and individual rights to land. Land user rights are entrenched in the fundamental principles of the National Land Policy comprising of, among others, the following:

- All land is public land and is vested in the President as trustee on behalf of all citizens;
- Citizens' rights to land are user rights that are recognized in longstanding occupation or use of land as clarified and secured by the law;
- Equitable distribution and access to land by all citizens;
- Regulation of the amount of land that any one person or corporate body may occupy or use;
- Recognition of the fact that an interest in land has a value and that value is taken into consideration in any transaction affecting that interest;
- Payment of full, fair and prompt compensation to any person whose right of occupancy or recognized long-standing occupation of customary use of land is revoked or interfered with to their detriment by the State ….. based on among other things: the market value of real property and cost of acquiring and getting the subject land and capital expenditure incurred for the development of the subject land;
- Provision of efficient, effective, economic and transparent system of land administration; and

- Facilitation of the operation of a market in land and regulation of the operations of that market to ensure that rural and urban smallholders and pastoralists are not disadvantaged.

The Land Act (No 4 of 1999) generally referred to as the Land Act, provides for three types of land holdings: general land; reserved land and village land. The Land Act empowers the President to transfer any area of land from general land to reserve or village land. The Village Land Act (No 5 of 1999), subsequently referred to as the Village Land Act, defines village land and provides for its management. It also provides for the transfer of village land to general land. There are four categories of land user rights in the United Republic of Tanzania: general land, reserved land, village land and hazardous land.

- **General land** defined as all public land which is not reserved land or village land, whereby public land is all the land of the United Republic of Tanzania based on the premise that all land is held by the President.
- **Reserved land** designated under a series of nine separate chapters including the Forests Ordinance (Cap 389), the National Parks Ordinance (Cap 412), and the Land Acquisition Act, 1967 among others.
- **Village Land** defined as including but not limited to:
    - land within villages registered under the Local Government (District Authorities) (Act No 7 of 1982);
    - land designated as village land under the Land Tenure (Village Settlements) (Act No. 27 of 1965); and
    - land the boundaries of which have been designated as village land under any law or administrative procedures at any time before the Village Land Act (No. 5 of 1999) became operational.
- **Hazardous land** defined as land the development of which is likely to pose a danger to life or lead to the degradation of the environment, contiguous land such as mangrove swamps, land within sixty metres of a river bank or shoreline, or specified land.
- **Derivative Rights** are used to provide for land holdings by citizens or group of citizens or their corporate bodies under rights of occupancy or a derivative right. Non-citizens may only obtain a right of occupancy or derivative right for the purpose of investment as prescribed under the Tanzanian Investment Act, 1997. Land to be designed for investment purposes has to be identified, published in the national gazette and allocated to the TIC which proceeds to create derivative rights to investors. A derivative right, referred to as a residential licence, confers upon licensees the right to occupy land in non-hazardous land, including urban and peri-urban area for a period of time for which the residential licence has been granted.

Effective implementation of the Land Act and Village Land Act is premised on adoption of policies and enactment of secondary legislation to provide guidance for corresponding operations in specific functional areas including: land use planning; surveying and mapping services; land valuation and estate agency services; land acquisition and compensation; land registration; land mortgages and sectional properties.

With the exception of the Land Use Planning Act of 2007 and the Land Acquisition and Compensation Act, also of 2007, most of the remaining secondary policies and legislation were drawn up prior to the adoption of the Land Policy of 1995 and the Land Act/Village Land Act of 1999. Further, the United Republic of Tanzania has never had specific legislation on the estate agency function. Specifically, the existing legislation for surveying and mapping, land valuation and land registration require major reforms for alignment with the objectives of the Land Act and the Village Land Act. Initiatives are already underway to update these statutes.

7.1 The Institutional Framework for Implementation

The institutional framework for implementation comprises of two central ministries: the Ministry of Lands Housing and Human Settlements

Development (MLHHSD), responsible for policy formulation and oversight of land administration functions with a network of six zonal offices. Policy implementation is mandated to the Prime Minister's Office, Regional Administration and Local Government, which oversees the operations of Local Government Authorities (LGAs). The LGAs, on their part, coordinate and oversee the operations of village governments and councils who have the legal mandate for land administration and management of village land, where the bulk of land resources are located. In 2010, the number of LGAs was increased from 134 to 168 councils overseeing land management through approximately 14 000 villages. Institutional capacity is a factor of technical capacity embedded in human resources, systems and procedures and equipment and infrastructure for land administration ranging from surveying and mapping facilities to modern ICT-based registries located at the district level.

Available information points to an employment gap of 75 percent of requisite technical staff for land administration in the two ministries. The United Republic of Tanzania has a large pool of potential professional land administrators graduating from Ardhi (Land) University, dedicated to land administration services, with a first year student population of 2 866 in the 2009/10 academic year compared to 2 221 in the 2005/06 academic year [61]. The presence of this large pool has not translated into higher land administration capacity due to limitations in recruitment and limited public-private partnerships in this area. These shortcomings reinforce the implementation weaknesses stemming from rent-seeking tendencies reinforced by lack of transparency in an environment where the central land registry still operates largely as a paper-based system.

8. Recent trends in large-scale land investment in the United Republic of Tanzania

The available information shows that most land acquisition for agricultural investments in Tanzania is largely still at the request stage, for which approval may not have been granted as yet. According to one source, in 2009, a total of 4 million hectares were requested by foreign investors. The largest requests emanated from SEKAB which had reportedly requested a large area in Bagamoyo (that could reach up to 400 000 ha) and 500 000 hectares in Rufiji, for sugar cane production. Sulle and Nelson (2009) argue that a British energy company, the CAMS Group had also acquired 45 000 hectares for sweet sorghum production while another British company, Sun Biofuels, acquired over 8 000 hectares in Kisarawe.[62] Although these figures are large, there is evidence that only a small share of the requested area was eventually acquired.

However, there is an increasing trend of acquisition of land by small and medium farmers in Tanzania, which is apparent in the data on farms that were surveyed and registered during the 2004 to 2010 period. Data published through the Minister for Lands, Housing and Human Settlement Development budget speech for 2009/2010 shows that a total of 623 farms out of a target of 800 farms were registered during the period July 2008 to June 2009. Further, the Ministry was targeting to register a total of 1 000 farms between July 2009 and June 2010. Sixty two percent of the 623 farms registered in 2008/2009 – equivalent to 386 –were located in three regions, i.e. the Coast region (174), the Tanga region (125) and the Morogoro region (87) that happen to be the favourite destinations of large TNC agricultural investors because of the prime arable agricultural land, good climate, reliable rainfall patterns and easy access to surface water resources for irrigation purposes

[61] Ministry of Finance and Economy, United Republic of Tanzania; Economic Survey for 2006, 2007, 2008 and 2009.

[62] Sulle, E. and F. Nelson F. (2009), *Biofuels, land access and rural livelihoods in Tanzania*, in Theting & Brekke, Land Investments or Land Grab? A critical view from the United Republic of Tanzania and Mozambique.

TABLE 5
Status of recent investments and business models

Company	Location District	Crop	Land Requested (ha)	Previous Land Status	Status	Business Model	Notes
Sekab BT (Sweden)	Bagamoya	sugarcane	24 200	govt ranch/TIC land	granted by TIC derivative right in progress land acquisition progress	90% estate; 10% outgrowers or block farming	Only a small portion fo the land is under cultivation deal is questionable due to the withdrawal of financial support from Scandanavian pension fund on grounds that the deal risk displacing local communities
	Rufjii	sugarcane	250-500 '(000)	village land			
FELISA (Belgium)	Kigoma 1 Kigoma 2	Oil palm	10 000	TIC land bank village land	land acquisition in progress under negotiation	Hybrid estate & outgrowers	
Sun Biofeuls	Kisarawe	Jatropha	50 000	village land 12 villages	transferred from village to general land	estate and possibly outgrower	
Diligent (Netherlands)	Arusha	Jatropha	none	n/a	n/a	contract farming	
BioMassive AB (Sweden)	Lindi	Jatropha		village land	66 year leashold agreement with Lindi district council		Company has avoided payment of land citing that contract stipulates that it can only pay for acreage cultivated not total acquired. The company is still trying to raise funds for the project.
BioShape (Netherlands)	Kilwa	Jatropha		village land	50 year lease signed between company and villagers		contract area is one-third of surface area of the district
KRC (South Korea)	Rufjii	Agro-processing	325 117		Joint venture b/w KRC and Tanzania govt thru Rufjii Basin Development Authority		half of the land will be developed and givenb to the local farmers and the rest will be used by KRC to process cocking oil wine and starch include plans for a food processing centre and technology transfer (irrigation)
AgriSol (USA)	Mpanda & Kigoma			refugee camps Katumba - 80 317 ha Mishamo - 219 800 ha Lgufu - 25 000ha	MOU signed b/w Mpanda district government and AgrSol		75% US and 25% Tanzanian - held by former senior cabinet minister. Deal does not conform to Tanzanian law on shareholding can be equal to more but not less

Source: : Compiled by authors from various sources

where necessary. No data was given regarding the average acreage of farms involved. Table 5 presents the status of selected recent investment, their nature and the proposed type of business models.

The United Republic of Tanzania is facing a rising incidence of conflicts over land and water rights between medium commercial farmers and smallholder subsistence farmers and between farmers and traditional pastoralists as well as between pastoralists and tourism sector investors. The migration of pastoralists to new pasture lands in regions that are still characterized by regular long rainy periods also highlight the issue of changing patterns in informal land use that is already a source of conflict and clashes. Conflict over water rights amongst smallholder farmers, between smallholders and commercial farmers and between smallholders and pastoralists has become increasingly common. The extent of the problem is apparent in difficulties on the part of the government regarding the allocation of water rights between competing national objectives, in particular irrigation farming, *vis-à-vis* power generation.

At the national level, the authorities have discerned the sensitivities associated with land ownership and user rights and the need for more careful responses to requests for land for agricultural investment. In January 2011, the Government of the United Republic of Tanzania issued directives on handling of requests for allocation of land for investment in biofuel production. These guidelines address the issues of protecting the land rights of local communities while taking advantage of opportunities for new linkages with the global market. They provide a comprehensive package for acceptable agricultural investment in biofuel production. Among other things, the package limits large-scale land acquisition to a maximum of 20 000 hectares, and includes mandatory provision for outgrower schemes, local processing and reservation of 25 percent of allocated land for production of food crops in response to the food security threat.

The unfolding experience from the ongoing preparations for implementation of SAGCOT (Southern Agricultural Growth Corridor of Tanzania) also provides a useful basis for leaders to understand the issues involved in current international interest in agricultural investments and the consequences of land acquisition generally. The statements emanating from political circles and the responses from the international community reveal a gap in understanding that is being bridged in favour of the adoption of existing best practices, focusing on inclusion of the interests of hitherto voiceless rural communities.

The perspective emerging from the official "Investors Guide for SAGCOT" limits the area of land involved to 350 000 hectares over a period of 20 years and involving an investment of US$ 2.5 billion. Indeed international literature on farm sizes shows that large farms worldwide barely exceed 50 000 acres per farm.[63]

> *" ... Mwanza – In May 2010, cotton stakeholders in Tanzania resolved to implement contract farming throughout the country's western cotton growing area (WCGA), starting this season... the farming model to be employed in Tanzania entails formation of farmer-business groups (FBGs) comprising between 50 and 90 smallholders ..... the number of FBGs that have joined contract farming between 2008 and 2011 has increased by 353 percent from 47 groups, with 2 241 farmers in 2008 to the current 587 groups with 37 951 farmers".* (The Citizen on Sunday, Special Report, 16 January 2011).

Results from piloting of the Mwanza cotton project show that yields per acre have gone up from 341 kg per acre to 487 kgs per acre in pilot areas. Consequently, project stakeholders including the Tanzania Cotton Marketing Board with funding from GATSBY, with TECHNOSERVE providing technical services, have agreed to scale up production to include 30 ginneries serving as processors/marketers. The ginneries will interact with smallholders through Farmers Business Groups (FBGs), comprising between

63 http://thegulfblog.com/2010/04/23/largest-dairy-farm-in-world

50 and 90 farmers under contract farming arrangements. The ginneries will provide access to upstream production inputs including pesticides and fertilizers to be recovered from sales. The scheme's structure links farmers to specific ginneries to avoid side-selling by farmers. The role of ginneries is underwritten by a Cotton Development Trust Fund (CDTF) supported by the Tanzania Gatsby Trust and the Tanzania Cotton Board.

The earlier acquisition or proposals for acquisition of land in the United Republic of Tanzania were problem prone, akin to most others initiated in other African countries. However, for very practical reasons, many of those proposals were not carried through and have eventually fallen apart as awareness rises inside and outside the country, and there are concerted movements against these initiatives. Table 5 above presents a selection of cases of significant land requests in Tanzania over the past five years.

9. Issues and implications for large-scale land investments in the United Republic of Tanzania

The Tanzania Investment Centre (TIC) plays a hands-on role in facilitating land access, and formal approval for the investment is needed from the TIC (financial viability), the Ministry of Agriculture (agricultural viability), the Ministry of Lands and Housing Development (land registration) and the Ministry of Environment (environmental impact assessment). Coordination and communication among government agencies tasked with different aspects of the investment process is poor, hampered in part by government departments' preference to report positive outcomes only, without sharing problems and setbacks.

The United Republic of Tanzania has to undertake new and/or strengthen ongoing reforms of its investment climate. Table 6 presents a clear picture of the ease of doing business in the United Republic of Tanzania and points the way in areas where more attention is needed.

On the investor side, private investors have the advantage of being able to act as a single legal entity with a cohesive set of values. Investors can only lease and use 'general land', not 'village land'. Land can be transferred from 'village' to 'general' status with the permission of the local community. Prospective investors start at the national level, with the Tanzania Investment Centre, the one-stop-shop that facilitates investment in the United Republic of Tanzania, where they are required to demonstrate the financial viability of the proposed project in order to get a Certificate of Incentives. From here they go to the district level, as advised and facilitated by the TIC. In the simple case they take up previously identified and surveyed land, registered with the TIC "land bank", but if all or part of the proposed land area is still 'village land', negotiations with local communities are necessary. The investor must have the request for land transfer approved in turn by the village council (senior village representatives), the village assembly (comprising all adult residents of a village) and the district council land committee. In principle the land transfer must also be vetted by the Ministry of Lands, Housing and Human Settlements Development.

Many companies have shown interest in acquiring lands that are underdeveloped 'general' lands. For instance, a Swedish company requested 400 000 hectares for sugarcane production in the Wami River basin in Bagamoyo District. Evidence suggests that, if the deal went ahead, about 1 000 small-scale rice farmers on these lands would need to move, and would not be eligible for compensation as the land is 'general' not 'village' land. The process of negotiation over village land tends to be slow, in large part because of the lack of precedent and guidance. In one case, for instance, the investor FELISA completed the process, securing approval for 350 hectares from two village assemblies, but later received a message from one of the villages withdrawing the offer as the land had apparently already been allocated to another individual. Intervention by local authorities resolved the issue in FELISA's favour, and arrangements have been made for community infrastructure investment and an oil palm outgrowing scheme, which have

TABLE 6
Ease of doing business in the United Republic of Tanzania

	DB 2012 Rank	DB 2011 Rank	Change in Rank
Starting a Business	123	122	⬇ -1
Dealing with Construction Permits	176	177	⬆ 1
Getting Electricity	78	80	⬆ 2
Registering Property	158	155	⬇ -3
Getting Credit	98	96	⬇ -2
Protecting Investors	97	93	⬇ -4
Paying Taxes	129	123	⬇ -6
Trading Across Borders	92	115	⬆ 23
Enforcing Contracts	36	33	⬇ -3
Resolving Insolvency	122	120	⬇ -2

Source: http://www.doingbusiness.org/

convinced villagers of the value of the investment. However, there are no formal documents to bind either party to these agreements.

There is a legal requirement that villagers be compensated fairly by the government when village land is transferred to general land. In practice however, investors themselves tend to pay compensation directly to the villagers. There are substantial differences in opinion and confusion over the amount of compensation and the entitled beneficiaries. Given the lack of an active land market in the United Republic of Tanzania, market-based per hectare rates have little meaning. Some companies compensate for the value of the resources on the land, such as trees and grazing, rather than the land per se. Access to water resources is of particular concern to both villagers and investors, as well as other competing interests (downstream users, conservation, etc.), and is a source of conflict in some instances – conflict that is difficult to resolve in the absence of clear regulations or guidelines from the government on sustainable levels of water abstraction.

10. Existing business models for large-scale land investment

Most documented large-scale land investment in Table 5 above is based on a single simple model of concentrated production within a single plantation unit, operated for maximum efficiency. But an emerging trend among governments is that investors contribute to local development not only through job provision, environmental protection and social investments, but also through direct involvement of local farmers and small-scale businesses in the supply chain as in the case of KRS from the Table. Apart from considerations linked to the longstanding farm size efficiency debate, the choice of production models may have major implications for the distribution of project benefits. Maximizing local benefits may require developing collaborative business models, from properly negotiated contract farming with small-scale producers through to joint ventures (shared equity) with legally recognized community organizations.

The Government of the United Republic of Tanzania is taking first steps to promote the involvement of local investors and smallholders. The government is developing standards for biofuels investments that include provisions for the involvement of local small-scale producers in some variant model as outgrowers, contract or mixed schemes. Most outgrower schemes and other inclusive approaches to production are, however, voluntary rather than a response to government regulation. Investors seek to create more robust business models and to pre-empt local conflict and international criticism through building up local participation from the start. Examples of mixed business models in the United Republic of Tanzania include that of the bioethanol company SEKAB, which proposed a gradual transition from a single ownership plantation to franchised block-farming for sugarcane for 500 000 hectares in Rufiji, United Republic of Tanzania. The biodiesel company Diligent is sourcing jatropha oil entirely from a network of small-scale farmers, under loose contractual terms. But the vast majority of documented projects continue to be run as large plantations based on concessions or leases. As large areas of land are commonly offered on very favourable terms, an incentive is created for establishing company-managed plantations rather than promoting contract farming approaches. Even "local content" provisions requiring prioritization of the local workforce in recruitment, common in extractive industry contracts, appear rare for agriculture investments. There is enormous scope here for governments to develop systems of incentives to promote more inclusive business models among large-scale investors.

Market outlets for agricultural produce are another key issue. The production of crops for export to the investor's home country is a key driver in many recent land acquisitions, particularly those led by foreign governments concerned about their food security. Several host countries are at present highly dependent on food imports, and in some cases recipients of food aid. The United Republic of Tanzania still imposes export bans on key food items; how this will play out with these investments is an area to watch. While these investments have been widely criticized in national and international media, a counterargument is that agricultural investment will bring yield increases that will benefit food security in the host country as well as the investor country. Reconciling food security in both home and host countries requires careful policy responses. This issue deserves to be dealt with in contracts, yet most of the current investment contracts in the United Republic of Tanzania are silent on the matter.

The extent to which national legal frameworks protect local land claims varies among countries, but is often limited. Local people may enjoy use rights over state land but land titles, whether individual or collective, are extremely rare in rural areas. Overall, the current wave of FDI flows and land acquisitions is taking place in contexts where many people have only insecure land rights – which makes them vulnerable to dispossession. However, in the case of Tanzania's Land Act 1999 and Village Land Act 1999, steps have been taken to strengthen the protection of local land rights, including customary rights, through initiatives for village land registration, regardless of the fact that all land is either vested with the state in trust for the nation or state owned.

But even where legal protection may be conditioned to "productive use" – for instance in the United Republic of Tanzania, lacking a clear definition of what constitutes "productive use" and given the ensuing broad administrative discretion, these requirements may open the door to abuse, and undermine the security of local land rights. This is particularly so for those groups whose resource use is often not considered as "productive enough", as is often the case in pastoral communities or even in cases of village forests that serve as a source of firewood and traditional medicine for agricultural communities.

11. Impact of FDI on agriculture in the United Republic of Tanzania

Recent data or studies on the impact of FDI on agriculture and food security in the United Republic of Tanzania are difficult to assess, especially as the trend in large-scale land acquisition is too recent for its full effects to be observed. However, the existing evidence suggests that impact of FDI on the United Republic of Tanzania's economy is very noticeable in the industries in which FDI is concentrated. According to the Tanzania Board of Trade, in the case of mining, FDI has served as an engine of growth and has helped to increase gold exports, which amounted to US$703.7 million in 2006, contributing about 42 percent of the total export value for the country. Gold exports remain dominant in total non-traditional exports, followed by manufactured goods and fish and fish products. The increase of FDI inflow has also contributed to the modernization of the various industries. Foreign investors have restructured privatized enterprises, thus boosting their competitiveness and contributing to the transfer of technology and skills.

However, although the impact is strongest in the industries in which FDI is concentrated, it has mixed implications for the entire economy. The scale of this impact is still small and a number of desired results are not occurring (such as linkages to the local economy thus impacting poverty reduction, or strengthening local science and technology capacities). In most cases, FDI currently has little impact on the employment situation, as it is directed towards capital-intensive sectors. Likewise, there is considerable public concern that impact on government revenue generation has remained minimal and measures have been initiated to address this concern, through negotiations with mining companies for higher royalties and public share ownership in publicly traded companies. One of the outcomes of these initiatives is the cross-listing of African Barick Gold (ABG) at the Dar-es-Salaam Stock Exchange (ABG is formally listed at the London Stock Exchange). Thus, after initial successes with FDI, the challenge for the United Republic of Tanzania is now to push FDI towards new frontiers, such as agriculture, which is important in the fight against poverty.

The Tanzanian economy is constrained by low productivity, inadequate physical and economic infrastructures, dependence on the export of primary goods with very limited value addition through manufacturing and low product standards and standardization. These are key issues for reaping the full benefits from FDI. In its Vision 2025, the government has placed emphasis on the industrial sector to play the central role of transforming the economy from a low productivity agriculture to a semi-industrialized one led by modernized and highly productive agricultural activities, which are effectively integrated and buttressed by supportive industrial and service activities which are in turn, laid down in the Kilimo Kwanza framework. However, given the limited financial capabilities of the government, it is hoped that FDI will play a central role in this direction.

Tanzanian agriculture is dominated by smallholder farmers cultivating an average of 0.5- 2 hectares. Productivity has been especially low for smallholders compared to agricultural undertakings by estates or large commercial farms, which have been able to attract considerable FDI. Records from TIC show that more than 90 percent of FDI in agriculture went to the crop sub-sectors (e.g. sugarcane, jatropha, oil palm and sisal), whose smallholder farmers are well organized and sufficiently integrated to support foreign investments.

Although several factors (age, origin of the farmer, educational level, and farm area) have been observed to affect the technical efficiency of smallholders in the United Republic of Tanzania, the integration of smallholders with large enterprises was a major factor in some investments (e.g. Mtibwa Sugar Estate scheme). Furthermore, smallholders who are close to a processing plant or factory have been found to be more efficient compared to those who were farther away. This factor is closely associated with high transportation costs to smallholder farmers far from the factory, as in some cases, the large firms provide transportation for farmers close to the factory, while others are forced to use private transportation.

Appropriate reforms targeting the regulatory environment have been key factors influencing the attraction and harnessing of benefits of FDI in the United Republic of Tanzania. With respect to the regulation of FDI, the general trend over the past decade has been for the gradual liberalization of rules governing foreign investors and their investments. Furthermore, privatization programmes from the early 1990s have expanded the opportunities for foreign investors. For example, the intent behind the ongoing land reforms is to facilitate the use of land as collateral in bank borrowing and to spur private investment in agriculture. Investment promotion has concurrently become an important policy tool for attracting FDI. Policies aimed at attracting FDI have ranged from relatively passive and general promotion schemes to much more aggressive targeting of foreign investors combined with the use of investment incentives.

Despite these efforts and the recent growth of the sector, together with observed productivity and efficiency increasing capabilities, FDI flow into the agricultural sector has remained very small in the United Republic of Tanzania (Table 1). It is widely accepted that investments in agricultural and livestock projects are most efficient in creating employment and addressing poverty related issues. However, poor infrastructure combined with high energy and transportation costs, has rendered the United Republic of Tanzania's commodities non-competitive. A low level of domestic entrepreneurship coupled with poor quality products has resulted in loss of market share. Limited financial capital and an unfavourable regulatory environment deter the growth of medium and large-scale agricultural production, resulting in high dependency on poor quality, high cost products from small-scale producers.

On the other hand as the Tanzanian agricultural sector continues to depend on smallholder producers, the characteristics and institutional setup of smallholders will have an impact on the performance of the sector and thus its ability to attract FDI. Tanzanian smallholder farmers have limited education and experience, are frequently exposed to shock and have to deal with weak institutional arrangements for production. This has led to only slight increases in agricultural productivity and insufficient improvement in the quality of production. This is especially true when the productivity of smallholders is compared to that of estates or large commercial farms or even comparative smallholder production in other countries in the East African region. As discussed above, this difference in productivity led to more than 90 percent of FDI in agriculture being directed to the crop subsectors (e.g. sugarcane, sisal) whose smallholder farmers have proved sufficiently well organized to support the foreign investments. These findings justify the consideration of alternative institutional arrangements for smallholder farmers that will attract more FDI inflow and improve smallholder productivity.

12. Conclusions and recommendations

The United Republic of Tanzania's performance in the area of agricultural investments over the last decade is one with a mixed record. The earliest deals reflect decisions based on the assumption that investors would somehow link local smallholders into their investments and the latter would benefit automatically through employment, access to technology and market linkages. There was no conscious effort to determine how this would happen or to provide for it in contracts between the government and the investors. Further, the involvement of local communities in the deals was primarily limited to superficial consultations involving a lot of verbal promises with few obligatory commitments.

Today there is a much clearer understanding of the pitfalls involved, as evidenced in the initiatives on the drawing-board including the Biofuels Guidelines, the SAGCOT project and other interventions taking place under the umbrella of the Agricultural Sector Development Programme. It is in this context that it is felt that specific policy recommendations responding to the findings from the literature review can add value to the government's initiative to respond positively to emerging opportunities while mitigating against the inherent risks.

Findings and recommendations are drawn from the following issues: access to land and security of tenure; food security concerns; access to water rights and rights of way; business environment reforms; strategic development of infrastructure services; adoption of first best government policy intervention instruments; adoption of the principles of responsible agricultural investment and related business models; and effective M & E (monitoring and evaluation).

Food Security: the United Republic of Tanzania's challenge in addressing its food insecurity problems revolves around access to food – whether produced within the country or imported from neighbouring countries at times of need. The United Republic of Tanzania can meet its own food security requirements, even in times of drought and shortages. However, this is subject to improving accessibility to surplus production in the rich agricultural regions, most of which lie in the Southern Highlands, through investment in transport, rural infrastructure including post-harvesting facilities, and deeper integration of domestic and regional markets. Higher productivity could easily double grain production, if the appropriate policy intervention instruments were in place. For instance, a large programme for subsidizing food production through the delivery of subsidized fertilizer, based on voucher systems currently in operation is proving very difficult to sustain due to moral hazard problems. One question that comes to mind is whether the use of instruments like support for contract farming could prove a better alternative. Access to markets through improved transportation and removal of intra-district and export bans are also necessary to motivate farmers to invest more in their land and raise the level of surplus production.

The policy recommendation on this issue is to implement policy instruments to stimulate higher productivity by smallholder farmers, and to remove marketing bans. Improved transportation systems while lowering transport costs by reintroducing railway transport would also increase farmers' margins and motivate higher investment in smallholder agriculture.

Land Administration and Security of Tenure: the United Republic of Tanzania has an excellent land policy and equally good instruments for implementation in the Land Act 1999 and the Village Land Act 1999. Their effectiveness lies in actual implementation and in the efficacy of secondary implementation instruments, including secondary legislation and regulations in the areas of land use planning, surveying and mapping, land valuation and estate agency services, land acquisition and compensation, and land-based mortgages. The basic source of information on communal landholding patterns is embedded in the Village Land Act. This information is available in real terms for villages that have undertaken participatory land-use planning and have been issued with certificates of village land (CVLs).

The challenge is to extend the land use planning process from approximately 1 000 villages that have received this service to more than 10 000 villages that are on the waiting list and rolling out the service of land surveying, mapping and adjudication that is necessary to create a national land registry (which will guarantee security of tenure for village communities and smallholders). It is, therefore, recommended to expedite the rolling out of village land planning and certification as a means of securing tenure for land holding by local communities which will also significantly improve security of tenure for individuals within the villages.

Access to Water Rights and Rights of Way: Parallel to security of tenure is the issue of access to water rights in a world where consciousness of water shortages has become acute due to the climate change phenomenon. Further, rite of passage has become an issue in the rural setting, due to the tendency for large farmers to create a buffer between their land holding and the surrounding villages, leading to closure of public routes traversing through a large farm. Diversion of existing public routes and limitations to access to water – resulting from isolation of land transferred to large investors – tend to be a major issue in direct relation to the size of the land being acquired. Deals already concluded to-date ignore the future of local communities' access to water rights and, in some cases, this has been a source of conflict and tension between commercial farmers and local communities.

It is questionable whether such deals are sustainable in the long term without addressing this problem. It is recommended that future deals consider making provisions for acceptable alternatives for rights of way and equitable sharing in access to water rights between local communities and large investors. Further restriction of land-leasing contracts to shorter durations – say 33 years rather than 99 – would create the flexibility necessary to renegotiate contracts in the medium term, while extension of the biofuel guidelines to agricultural commodities and food products can redress the issue of speculative land acquisition. Finally, future contracts should seek to balance access to water rights where this becomes necessary and ensure that agricultural investors are obliged to pay for the water rights granted to them.

Business Environment and Investment Climate: One of the major challenges facing the United Republic of Tanzania in the course of bringing about economic transformation is the state of the business environment and the wider investment climate. With respect to the business environment, it is imperative to enhance policy and regulatory reforms that are already underway, starting with prioritization of sectors that have a major impact on the cost of doing business: registration to support intra-regional trade; land registration for improved security of tenure and use of land as a business asset; trade facilitation to promote regional market integration; taxation regimes and dispute resolutions. An even more daunting challenge is that of improving the delivery of infrastructure-based services, particularly in the transport, energy and water sectors, as well as the development of critical productive infrastructure in the agricultural and industrial sectors.

It is recommended to strengthen and expedite regulatory reforms and hasten the mobilization of private sector resources for development and management of infrastructure for the delivery of social and economic services through the PPP (public private partnership) approach. Identification of clear priorities in terms of specific sectors that provide a major initial contribution to economic transformation should be the primary yardstick in implementation. It is particularly critical to bridge major gaps between the supply and demand for power and transportation services that have become the binding constraints against private sector efficiency and the achievement of more rapid growth. The prerogative is to ensure the reliability and affordability of these services.

Agricultural and Industrial Infrastructure Development: Raising productivity and achieving sufficiency in food security as well as harnessing the opportunities emerging from increasing food demand and limited arable land resources require major investments in agricultural infrastructure such as irrigation infrastructure, as well as the development of industrial infrastructure such as industrial parks, special economic zones and export processing zones. Prioritization of investment in soft infrastructure, i.e. ICT and financial services, is critical for achieving more rapid economic growth. Current initiatives for improving access to finance and development of ICT infrastructure in key government institutions such as civil registries to support efficient services for private sector development should be enhanced and expedited.

Adoption of Relevant Areas in the Draft RAI Principles: The principles for responsible agricultural investment, business models and funding instruments provide a best practices framework for negotiation and conclusion of land-lease contracts as an alternative tool for acquisition of land that does not lead to dispossession on the part of local communities and individuals. Consequently it is recommended to undertake the following measures in handling agricultural investments:

i. Building up capacities for adoption of the existing business models and financing instruments as the primary tools for handling analysis, and development of decision-making options for consideration of requests and proposals for land-leasing contracts for agricultural investors.

ii. Extend application of the biofuel guidelines to cover crop production and address the issue of water rights as well as guarantees for rights-of-way and be backed by legal mandate. Ensuring equitable access to water rights in the future is one of the key factors for social sustainability.

Effective Implementation, Monitoring and Evaluation of Programmes and Projects: The failure of policy implementation in many sub-Saharan African countries including the United Republic of Tanzania is based largely on the poor track record of effective monitoring and performance evaluation. Even where this does occur, there is a tendency, amongst officialdom, to hide real developments in the field, starting from the planning stage through failure to establish realistic benchmarks. Africa has also been unwilling to adopt best practices emerging from other economies as a norm, preferring in many instances homebaked policy instruments that are known to suffer from failure to create change. Further, SSA governments have to adopt best practices in the development of economic strategies and strategic plans for their implementation. There is little meaning in redesigning the wheel as an excuse for adopting sub-standard policy measures that compound existing problems.

The agricultural sector in the United Republic of Tanzania offers potential investors opportunities not only in general commodity trading, but also investment in technology for supporting sectors such as irrigation works and refrigerated facilities; farm implements and agricultural inputs; fishing equipment and processing plants, and agro-processing businesses. As well as its huge potential in terms of national endowments, the government has attempted to increase investments mainly through fiscal incentives. The United Republic of Tanzania also has other features attractive to foreign investors, such as potential access to regional markets like those under EAC and COMESA (Common Market for Eastern and Southern Africa) arrangements.

While much of the regulatory framework impacting on the desirability to invest in the United Republic of Tanzania was reformed with the climate of liberalization and privatization in the 1990s, many of the regulations are now antiquated and require revision.

Many legislative revisions have been in the pipeline for a considerable period, but have yet to come into fruition. The government is in the process of re-examining its trade and investment legislation with mechanisms to involve stakeholders in the discussions on draft bills. These reform initiatives need to be speeded up and deepened.

The taxation regime is one example of a significant constraint to production with multiple duties in place at local and national level. It is recommended that the system is harmonized across the different crops and commodities to prevent price distortion, with a lowering of taxes and spreading of the tax base to remove the disincentives to production.

As well as enhancing and strengthening its existing attractive investment features, the United Republic of Tanzania should also work towards developing its weaker aspects. Despite its large human resource pool, the dearth of skilled workers and those with adequate technical capacity represents an area in which the government can promote private sector participation for capacity building and training schemes. The emphasis on technology transfer could be shifted somewhat to the provision of information on new technologies, and training the relevant stakeholders on their use, costs and appropriateness. Another problematic factor, particularly as regards agribusiness, is its infrastructure. Private sector (including foreign) participation is particularly useful in this regard, for example, in the development of its road development strategy. Thus, while the United Republic of Tanzania is strategically placed to continue with its success in attracting FDI into the country, many areas of the agricultural and allied sectors are in need of reform, revision and improvement in order to draw in a greater percentage of that same FDI to the agricultural sector specifically.

ANNEX
Investment opportunities in the crop sector

Crops	Investment opportunity
Coffee	Opening up new, large-scale coffee estates in Ruvuma, Mbeya, Iringa, Kigoma and Arusha regions. Creation of coffee processing plants.
Cotton	Establishment of large-scale cotton production farms, particularly in Morogoro, Coast, Singida, Tanga and Iringa regions; Establishment of spinning and textile industries.
Tobacco	Establishment of large scale woodlots for tobacco curing in Mbeya, Singida, Shinyanga, Rukwa, and Tabora regions; Purchase of tobacco and construction of processing factories.
Sisal	Establishment of large-scale sisal plantations in Dodoma, Shinyanga, Singida Kigoma, Tanga, Coast and Morogoro regions; Investment in new plantations and joint venture in the privatized sisal estates; Sisal spinning and weaving; Production of by-products: alcohol, particle boards, biogas and electricity, citric acid, pharmaceuticals, animal feeds, organic fertilizer, handicrafts. Sisal mattresses and padding for furniture and car seats; Sisal polishing cloth – a preferred material for polishing metals in industrial settings; Sisal composites in automotive, boats, furniture, etc. to replace fibre-glass. Establishment of pulp factories.
Tea	Establishment of large-scale tea production through opening up new plantations in Mbeya, Iringa, Mara and Tanga regions; Establishment of tea processing factories.
Pyrethrum	Establishment of contract and large-scale farming of pyrethrum in high altitude regions of Iringa, Mbeya and Arusha; Establishment of Pyrethrum crude extracts refineries.
Cashew nut	Cashew processing industries. Investment in large-scale cashew production; Investment in cashew marketing.
Sugar	Establishment of new sugarcane estates in Coast, Ruvuma, Kagera, Mara, Mbeya and Kigoma regions; Sugarcane processing factories.
Spices	Establishment of spice production, processing and marketing infrastructure in the coastal and high altitude areas of Tanga, Coast, Mtwara, Lindi, Morogoro, Mbeya, Kilimanjaro, Kagera and Kigoma regions; Establishment of spice processing and marketing infrastructure.
Floriculture	Opening up flower farms in Tanga-Usambara, Iringa, Mbeya, Kagera, Arusha. Kilimanjaro and Morogoro regions; Investing in lowland flower farming in Tanga, Dar-es-Salaam, Mtwara and Lindi regions; Flower seed production in Arusha, Mbeya, Iringa and Kilimanjaro regions.
Fruit & Vegetables	Opening up fruit and vegetable plantations in the potential areas for horticultural crops, Arusha, Kilimanjaro, Tanga, Morogoro, Dar-es-Salaam, Dodoma, Iringa, Mbeya, Mwanza and Kagera. Processing and canning for domestic and export markets.
Bananas	Expansion of banana production in Kagera, Kilimanjaro, Morogoro and Mbeya regions. Investment in production and marketing of banana seedlings like Williams, Lacatan, Pazz, Chinese, Cavendish, Grandmine
Oilseed	(Sesame, Sunflower, Palm oil and Soya); Production and Processing Sectors.
Other crops	(cassava, Irish potatoes, sorghum, millet and various legumes) Production in large quantities for food and animal feed for domestic and export markets.

Source: A summary of investment opportunities available in Tanzania's agricultural sector, Ministry of Agriculture and Food Security, available at http://www.agriculture.go.tz/

References

Abbassian A. Global food chain stretched to the limit, in *The Citizen (Tanzania)* Thursday, 20 January 2011. United Republic of Tanzania, Dar-Es-Salaam.

AgriSol Energy. 2011. *Report to the Prime Minister of the United Republic of Tanzania regarding proposed development of Katumba, Mishamo and Lugufu former refugee hosting areas*. 7 January, 2011. Power Point Presentation. Dar-es-Salaam.

Deininger, K., & Byerlee, D. 2011. *The rise of large farms in land abundant countries: do they have a future?* Washington, DC. March 2011. World Bank Policy Research Working Paper No. 5588.

Deininger, K. & Byerlee, D. *with* Lindsay J., Norton A., Selod H. & Stickler M. 2011. *Rising global interest in farmland: can it yield sustainable and equitable benefits?* Washington, DC. World Bank.

FAO. 2009. *Rapid assessment of aid flows for agricultural development in sub-Saharan Africa*. Rome, FAO.

FAO, IFAD (International Fund for Agricultural Development), UNCTAD (United Nations Conference on Trade and Development) and the World Bank Group. 2010. *Principles for responsible agricultural investments that respect rights, livelihoods and resources*. Rome, FAO.

FAO, IIED (International Institute for Environment and Development) & IFAD. 2009. *Land grab or development opportunity? Agricultural investment and international land deals in Africa*, by L. Cotula, S. Vermeulen, R. Leonard & J. Keeley. Rome, FAO and IFAD. London, IIED (available at www.fao.org/docrep/011/ak241e/ak241e00.htm).

Guardian The (Tanzania). Monday, 31 January 2011. *US to give $2million for farm project.* Dar-es-Salaam.

IIED, FAO, IFAD & SDC (Swiss Agency for Development and Cooperation). 2010. London/Rome/Bern. 2010. *Making the most of agricultural investment: a survey of business models that provide opportunities for smallholders*, by Vermuelen, S. and Cotula, L. London, IIED. Rome, FAO and IFAD. Berne, Switzerland, SDC.

Land Report, The. 2011. 2010 *Land Report 100: America's Top 100 Landowners.* Saturday 13 August 2011. The Land Report: Magazine of the American Landowner.

Ministry of Energy and Minerals. November 2010. *Guidelines for sustainable liquid biofuels development in Tanzania*. Dar-es-Salaam.

Ministry of Lands, Housing and Human Settlements Development (MLHHSD). 2005. *Strategic plan for implementation of land laws.* Dar-es-Salaam.

Munissi et al. *Study for the MOFEA and the planning process on climate change*. Sokoine University of Agriculture, Morogoro, Tanzania.

Ness B., Brogaard S., Anderberg S. & Olsson L. *The African land grab: equitable governance strategies through codes of conduct and certification schemes.* 2009 Amsterdam Conference on the Human Dimensions of Global Environmental Change, Amsterdam, 2-4 December 2009 (available at www.earthsystemgovernance.org/ac2009/?page=papers).

Oakland Institute. 2011. *Understanding land investment deals in Africa, AgriSol Energy and Pharos Global Agriculture Fund's land deal in Tanzania.* June 2011 (available at www.oaklandinstitute.org)

Oakland Institute. 11 December 2009. Memorandum of Understanding. *MOU for Conducting feasibility study between Mpanda District Council and AgriSol Energy Tanzania Limited*. June 2011. Oakland, California, USA, The Oakland Institute (available at: www.oaklandinstitute.org).

Sulle, E. & Nelson, F. 2009. Biofuels, land access and rural livelihoods in Tanzania, in *Theting & Brekke, Land investments or land grab? A critical view from Tanzania and Mozambique*.

Tanzania Investment Centre. 2008. *Tanzania Investment Guide 2008: Opportunities and Conditions*. Dar-es-Salaam.

Ministry of Agriculture, Food Security and Cooperatives. 2003. *Agricultural sector development program: support through basket fund.* Dar-es-Salaam.

UN General Assembly Human Rights Council. 2009. 13th Session, Agenda item 3. *Large-scale land acquisition and leases: a set of minimum principles and measures to address the human rights challenge*. December, 2009. New York.

UNCTAD & International Chamber of Commerce. 2005. *An investment guide to Tanzania: opportunities and conditions*. June 2005. New York and Geneva.

UNCTAD. 2009. *World Investment Report for 2009 and 2011. Transnational corporations, agricultural production and development*. New York and Geneva.

UNCTAD. 2009. Trade and Development Report 2009. *Responding to the global crisis: climate change mitigation and development.* New York and Geneva.

UNCTAD. 2010. The Least Developed Country Report 2010. *Towards a new international development architecture for LDCs*. New York and Geneva.

United Nations Press Release. 2009. World Population to exceed 9 billion by 2050 in Miller, Calvin; and Richter, Sylvia; 2009. *Agricultural Investment Funds For Developing Countries*. ConCap Connective Capital, Frankfurt.

United Republic of Tanzania. 1999. *The Land Act* (No. 4 of 1999). Dar-es-Salaam, Government Press.

United Republic of Tanzania. 2001. *The Land Regulations 2001 Subsidiary Legislation (Supplement No. 16 of 4 May 2001)*. Dar-es-Salaam, Government Press.

United Republic of Tanzania. 1999. The Village Land Act (No. 5 of 1999). Dar-es-Salaam, Government Press.

United Republic of Tanzania. 2008. President's Office, Planning Commission. *Economic Survey 2008*. Dar-es-Salaam, Government Press.

United Republic of Tanzania. 2005. President's Office. *Property and Business Formalization Project (PBFP or MUKUTABITA): Diagnostic Study*. Dar-es-Salaam, Government Press.

United Republic of Tanzania. 1999. *The Village Land Regulations 2002 Subsidiary Legislation*. Tabora, Tanzania, Peramiho Printing Press.

United Republic of Tanzania. 2011. President's Office, Planning Commission. *Five Year Plan 2011/12 to 2015/16*. June 2011. Dar-es-Salaam, Government Press.

Thailand:

Foreign investment and agricultural development in Thailand[1]

1. Introduction

Foreign direct investment (FDI) has played a pivotal role in the economic development of Thailand. In Thailand, FDI has grown rapidly with a clear shift in investment flows from import-substitution towards export-orientation, concentrating mainly in the manufacturing sector. Empirical studies have been largely concentrated on the role of FDI in this sector. Although Thailand is an agriculture-based economy and foreign investment in agricultural production has existed for a long time, the value of international investment in the agricultural sector is very small and the number of studies investigating the role of FDI in this sector is limited (Netayarak, 2008; Sattaphon, 2006). This chapter has two main objectives: first, to analyse the extent, nature and impact of international investment in the agricultural sector, and second, to analyse the policies, legislation and institutions affecting the international investment.

This chapter is divided into six sections including this introduction. The second section briefly reviews the background of Thai agriculture and explains the definitions of FDI statistics employed in this study. The third section describes policies, legislations and institutions affecting FDI in Thailand. The fourth section covers the analysis of FDI in Thai agriculture, with an emphasis on the extent and nature of FDI. A fifth section provides an analysis of the impacts of FDI with emphasis on the agricultural sector. A final section offers conclusions and policy recommendations.

[1] This chapter is based on an original research report produced for FAO by Waleerat Suphannachart, Faculty of Economics, Kasetsart University and Nipawan Thirawat, Independent Researcher.

2. Background of Thai agriculture and FDI data in Thailand

2.1 Overview of agricultural development in Thailand

Thailand has always had an agriculture-based economy in which the agricultural sector has played a crucial role in overall economic development. The agricultural sector was the economy's "engine of growth" in the 1960s and 1970s.[2] This leading role was superseded by the manufacturing sector in the 1980s. Since then the agricultural shares in overall GDP have declined. The decline in agricultural growth was in line with structural change toward an industrialized economy as well as many external factors, particularly a worldwide depression in major agricultural product prices (Poapongsakorn, 2006). Despite the declining shares of agricultural GDP, the agricultural sector continues to contribute to overall economic development by being an important source of rural income and export earnings.[3] It also provides raw materials for agribusiness and ensures household food security. The agricultural sector still managed to grow at an average growth rate of about 3 percent per year over the entire period of 1970-2008.

Within the agricultural sector, crop production has long occupied the largest share of total agricultural output, followed by fisheries,

[2] The main driving force was attributable to expansion of the land frontier and heavy public investment in roads and irrigation (Poapongsakorn, 2006).

[3] Thailand is a major net agricultural exporter, particularly of rice, rubber, cassava, sugar and poultry products (Warr, 2008). The majority of poor people in Thailand reside in rural areas and are directly involved in agricultural production (Warr, 2004).

livestock, forestry and agricultural services, respectively. However, in terms of the average annual growth rate, livestock GDP growth is largest during 1970–2008, followed by fisheries and crops. The expansion in livestock is mostly attributed to the higher demand for poultry exports, particularly from European markets (Poapongsakorn, 2006). Crop production has been dominated by staple crops such as rice, rubber, cassava, sugar cane, maize and kenaf.

Nonetheless, there has been a changing production structure in Thai agriculture in tandem with the changing comparative advantage and changing demand pattern toward high value-added and safe products. There has been a shift from traditional crops such as rice, maize and cassava to high value crops, particularly in horticulture (Poapongsakorn, 2006). Agricultural commodities and exports have also been diversified from major crops to processed agricultural products, such as frozen chicken, shrimp and canned pineapple, and high value products such as coffee, pepper, cut flowers, orchids, fruits and vegetables. While rice is still the dominant crop occupying the majority of land area and labour force, its export value has ranked after rubber since the 1990s, and after shrimp in the years 1991-1995 and 2001-2002. On average, the food processing sector[4] had greater growth rates than the agricultural sector. The food processing sector performed very well in 1986-1990, achieving the highest rate of growth of 8.95 percent while the agricultural sector's growth rate was only 3.17 percent. After that period, the growth rates of both sectors fell gradually over time. All in all, both sectors continued to be robust and remain among Thailand's most competitive and major sectors.

A large proportion of Thailand's food exports are processed foods, accounting for about 20 percent of total food exports. (6.45 percent in 2007 and 19.16 percent in 2008). Processed food exports, including canned seafood and processed fruits and vegetables, comprise critical components of Thailand's export structure. Moreover, higher value-adding products—like ready-to-eat food—are the fastest growing, even though they involve more complicated production processes than the others. This indicates the competitive advantage of Thailand's food processing industry in terms of its production capability and competitiveness. Thailand has achieved a significant reputation in exporting processed food, especially in the categories of processed tuna products (47 percent) and shrimp (20 percent of global market share, world's largest exporter in 2008), processed pineapple, world's largest exporter in 2008 and processed chicken products (25 percent of global market share)—world's largest exporter in 2008.

The sector is considered as reflecting one of Thailand's competitive strengths, and is judged important in the national economic development strategy. Thailand is one of the most important food exporters in Asia and the world. It is geared towards trade and investment liberalization (through free trade agreements and international investment agreements), and also tries to attract higher levels of FDI via its investment promotion programmes as well as export-led industrialization policies.

2.2 Foreign Direct Investment (FDI) data in Thailand

There are two main sources of foreign direct investment (FDI) statistics in Thailand: the Bank of Thailand (BOT) and the Board of Investment (BOI). Data from both are employed in this chapter. The BOT's FDI statistics cover overall FDI flowing into the Thai economy, while those from the BOI partially cover the FDI that receive the BOI's promotion packages. It is important to note that not all FDI projects apply for BOI promotion, and

[4] Regarding the food-processing industry, this research uses the same definition of the food industry as that of the Thai Ministry of Industry (2002), which defines "the food industry" in the national master plan for Thailand's food industry as: "Food industry means an industry that uses agricultural products such as plants, livestock and fisheries as main raw material in productions. The productions are based on technologies in order to get products for consumption uses or for other uses in further production processes. It is a method of preservation of agricultural products by primary manufacturing processes or intermediate manufacturing processes or final manufacturing processes."

the two data sources are compiled on a different basis.

FDI data collected by the Bank of Thailand follow the International Monetary Fund (IMF) Balance of Payments Manual, which is an international standard for collecting FDI statistics. The BOT's FDI statistics comprise three components: equity capital, with at least 10 percent of foreign shareholding, loans from affiliates, and reinvested earnings (Bank of Thailand, 2010). Since the data definitions are in accordance with the international standard, they are comparable among countries and widely used in the analysis of FDI. The BOT's statistics represent the entire streams of investment and are often reported as net FDI flows. Net FDI flows are defined as FDI inflows minus FDI outflows.

Foreign direct investment data collected by BOI refer to projects with foreign capital of at least 10 percent. The BOI's FDI definition does not strictly comply with the IMF's direct investment standard; therefore the data is often called foreign investment instead of foreign direct investment. The BOI's foreign investment data cover only projects which have applied for – or received approval from – BOI promotion. There are seven sectors under the BOI promotion: i) agriculture and agricultural products; ii) mining, ceramics and basic metals; iii) light industry; iv) metal products, machinery and transport equipment; v) electronic industry and electrical appliances; vi) chemicals, paper and plastics; and vii) services and public utilities. This study focuses only the first sector.

Since the two sources of FDI data are compiled on a different basis they are not comparable. Nonetheless, both data sets complement each other. BOT's FDI data represent actual flows of FDI into Thailand while BOI's data indicate trends of FDI. The BOT's FDI depict the overall picture of FDI at an aggregate level while BOI's FDI allows us to investigate the role of foreign companies at the project level.

3. Policies, legislations, institutions affecting FDI in Thai agriculture

3.1 Overview

Investment barriers

High Transaction Costs

Thailand has evolved towards an open economy. This is reflected in its declining tariff and non-tariff barriers over time. During the 1960s and 1970s, import tariffs were set at high levels, especially for those that were infant industries at the time (e.g. the automotive industry), when the import substitution policy was put in place to protect domestic industries (The Board of Investment of Thailand, www.boi.go.th). In the late 1990s, import duties on machinery and capital goods (61 categories) were removed for export oriented firms. Additionally, import taxes imposed on raw materials of exported products were exempted for both the Board of Investment of Thailand (BOI) and non-BOI promoted firms. Firms could obtain import tax refunds from Thailand's customs department.

High transaction costs still remain, due to inefficient public services, ambiguous regulations and duplicate/complex administration processes amidst the liberalization of trade and investment in Thailand. The Asian financial crisis in 1997 was a wake-up call for Thailand's wide range of reforms, including government transparency and economic reforms. Many Thai Government agencies like the Thai export promotion department and the BOI launched their One-Stop-Service centres in order to facilitate exporters and investors. To date, only some of these centres have proved to be efficient in providing services in a short period of time (i.e. visa and work permit approved within three hours as well as single window for submission of required customs/ business permits and standard certification documents). Nevertheless, processing time in the clarification and interpretation of the Harmonised System (HS) code, customs clearance and import tax refunds (maximum of 30 days with high possibility of delays), and value-added tax refunds

(15-90 days or more), is quite lengthy as a result of non-transparent rules and regulations as well as bureaucratic red tape. Last but not least, business permits, registrations and standard certificates involve many government agencies whose procedures and requirements are distinct to certain extent. This, in effect, requires significant time and increases in transaction costs which are among the factors influencing FDI inflows.

As a result of the issues described above, many firms (both new and established), have to acquire more information on, among other things, business permits and registrations, standard certification, product classification, customs and taxation procedures as well as relevant regulations. For example, a well known and established food processing firm (Company J), aiming to export its products to Australia would have to contact the Thai Government agency, the Department of Export Promotion (DEP) for detailed information on the bilateral FTA between Thailand and Australia. At the time, that company had not yet gained any benefit from the FTA, due to the fact that there was some confusion over the product categories entitled to enjoy lower tariffs.

The report of the World Bank on Thailand's investment climate assessment update (2008), is based on the analysis of 1 043 firms in manufacturing sectors which comprise automobile parts, food processing, furniture/wood, electronic parts, electrical appliances, garments, machinery, rubber/plastics, and textiles. These firms participated in the Thailand Productivity and Investment Climate Surveys (PICS), conducted in 2007. The report describes with great precision the difficulties that firms experience in doing business in Thailand. Complication and confusion over administration as well as procedures for getting business permits and standard certificates cost these firms both time and money:

In a nutshell, while the reductions in tariff, non-tariff barriers and taxes help induce FDI, Thailand still needs to simplify its taxation, customs and other public administration procedures and regulations in order to gain its position as one of the region's most attractive FDI recipient countries.

Political instability

Since 2006, Thailand has faced severe political uncertainty issues. There was a military coup in 2006 and political unrest and violence in 2010. Changing governments and prime ministers (seven prime ministers during the period 2006-2010), mean a possible modification of existing policies. In the worst case, some economic policies may be discontinued. For example, in 2006, right after the coup, changes in capital mobility policy were made via stricter currency and capital controls (30 percent reserve requirement on capital inflows). In addition, the government at the time tried to amend the Foreign Business Act 1999, causing an increasingly negative reaction on the part of investors. As expected, uncertainties caused many foreign investors to delay their decisions or search for alternative investment destinations. This has produced a continuously negative impact on FDI inflow (see Section 4).

The government announced that there would be no change to the Foreign Business Act 1999. Foreign firms could own up to 49 percent of shares in the service sector. The percentage of ownership was greater in case of foreign firms investing in Thailand's manufacturing sector. With regard to land ownership, foreigners and foreign firms could continue to purchase limited plots of land (mostly in industrial estates), but on condition that prior approval was obtained from the government. Clearly, amidst the political turmoil and instability of the period, Thailand's FDI inflows were declining. The Government perceived that a remedy could, nonetheless, be found through the creation of a stable and favourable macroeconomic climate as well as the development of clear, long-term policies.

The relationship between political turmoils and FDI prevailed in the case of demonstrations held during the first half of 2010; these undoubtedly adversely influenced Japanese investors' decisions and confidence. The Japanese Chamber of Commerce (JCC) in Bangkok conducted a survey to gauge business sentiment among JCC member companies in Thailand. A total of 375 firms out of 1 299 responded to the questionnaires (28.9 percent response rate). It was reported that the majority of firms participating in the survey (accounting to approximately 67 percent) recognized the demonstrations as a factor affecting their future investment in Thailand, while 7 percent of these firms increased their investment criteria in response to such political uncertainty (Japanese Chamber of Commerce, 2010). Remarkably, 99 percent of firms believed that the political unrest could cause possible negative effects on the domestic economy. Therefore, the impact on Thailand's FDI inflows is probably greater in cases of market seeking Japanese firms (primarily focusing on selling their products in Thailand) than those firms using Thailand as their production bases for exported products.

Limited government support on research and development and human resource development programmes

One of Thailand's weaknesses lies in research and development (R&D); another in its human resource development (HRD). There is a great need for public investments in these areas in order to enhance the attractiveness for FDI in the agricultural sector, and also increase agricultural productivity which has been included as the key area for development since the First National Economic and Social Development plan. This emphasizes the vital roles and importance of the agricultural sector as an engine for Thailand's economic growth. Agricultural products are exported to the world market; at the same time, they constitute raw materials and intermediate products for other industries including food processing. Thailand aims to be "the kitchen of the world" and global food exporters. In order to achieve this aim, the food processing industry has been included as a major priority sector in the Ninth National Economic and Social Development Plan. Agricultural development (both through R&D and HRD) requires concerted efforts by various government agencies, for example, the Ministry of Agriculture and Cooperatives and the Ministry of Science and Technology.

In the 1960s, government policy focused predominantly on increasing agricultural productivity and diversifying the production of major agricultural products that were in high demand in both domestic and international markets. Protection from epidemics and the development of fine livestock breeds were promoted during this period. Forest and natural resource conservation was also the key developmental goal aiming to utilize approximately 50 percent of land. However, research and development in the agricultural sector was limited to only some economic crops such as rice, rubber and corn. Additionally, regarding the fishery subsector, the Thai Government began to support research and training programmes for fishermen to increase their capabilities for deeper-sea fisheries.

Later, the Fourth National Economic and Social Development Plan (NESDP) reinforced the Thai Government's efforts to improving agricultural productivity and development by promoting advanced technologies, for example, fertilizer, pesticide, and agricultural machines, but most Thai agribusinesses and farmers still lacked the technological capabilities to create their own state-of-the-art technologies. As a result, most of these technologies were imported and adopted by Thai users in the agricultural sector. By so doing, they helped reduce costs of production and time consumption while increasing output. During the same period (mid-late 1970s), Thailand's Board of Investment (BOI) offered privileges to export-oriented manufacturers who employed capital-intensive production according to the Thai Investment Promotion Acts. This helped influence foreign investors to make investments in Thailand's agricultural sector including food processing as shown by positive figures for the first time (See Section 4 for details).

Agribusiness firms (both Thai and foreign) have played significant roles in the development of the agricultural sector. They become innovators and dominant players because they have better access to sources of funds, technology and expertise than farmers and other players in the value chain. Research and development requires a large amount of long-term investment; large firms are capable of mobilizing funds either via domestic channels or joint ventures with foreign firms, or internal capital support from international headquarters. These generate benefits to agricultural development in crops, livestock, aquaculture, and plantations as well as food processing. In addition, big firms possess technological skills and capabilities which can increase the success probability of their research projects. They build strong linkages with farmers via contract farming systems, allowing farmers to have access to newly developed technologies and thus to enhance their agricultural production skills.

The Ministry of Science and Technology also plays an important role in increasing Thailand's agricultural competitiveness and improving agricultural performance. This is clearly demonstrated in, for example, one of its agency's strategic plans. The National Science and Technology Development Agency (NSTDA)'s strategic plan (2007-2011) aims to promote research and development; implement activities related to technological transfer and human resource development; and develop science and technology infrastructures in order to achieve the main goal of the Tenth National Economic and Social Development Plan, to transform Thailand into a "knowledge based and creative economy". The NSTDA of the Thai Ministry of Science and Technology ranks the food and agricultural sector as one of its top priorities in line with the Ninth National Economic and Social Development Plan. A separate food and agriculture cluster is responsible for seed development, animal breeding technology, cost reduction and productivity enhancement technologies, improving production quality, food safety and risk assessment of seafood products.

Key indicators of the successful transformation towards a "knowledge based and creative economy" are the amount of investment dedicated to research and development as well as human resource development. Thailand's sustainable development depends on production capabilities, which can in turn be enhanced by utilizing technological capabilities; the latter can be promoted via research and development investment. The NSTDA is the main engine driving improvements in the industrial and agricultural sectors because it promotes new innovation and cooperation with partners. However, it is noteworthy that Thailand's research and development budget has remained unchanged at 0.5 percent of the GDP. Actual government spending on R & D is even less – only about half since the Fifth National Economic and Social Development Plan (1982-1986) up to the current national plan (2007–2011). Additionally, only 6 percent of Ministry of Agriculture and Cooperatives' spending is on research and development.

With regard to human resource development, the Thai Government acknowledges the low quality and poor access to education among Thai people. Labour quality has been the key issue affecting levels of gross FDI inflow and economic growth. As a result, education policy and its development have been set as the government's priority and included in the Tenth National Economic and Social Development Plan. Better-educated labour accelerates the rates of technological absorption, leading to higher productivity. At present, there is a mismatch between the skills offered by Thai labour and the skills needed by foreign firms. Approximately 40 percent of manufacturing firms indicated that labour shortages and mismatches is a major hindrance to doing business in Thailand (World Bank, 2008). The newly developed education policies and systems have now been put in place. The formation of strategic alliances between education and economic sectors can help solve the issue (close the skill mismatch gap) as well as generate research and knowledge suitable for sectoral development.

Singapore is a good example of successful human resource development programme in the Southeast Asian region. Singapore's government has spent a significant amount on education which has helped to build up knowledge and disseminate technology (Hobday, 1994). This may be the reason why Singapore is the most developed country of this group, attracting a huge amount of FDI. Although this is not yet the case for Thailand, the Thai Government is committed to achieve its long term human resource development goals through active education reform, encompassing a free, high-quality education policy. So far, the current Thai Government has provided full support for a 15 year free basic education programme.

Students are entitled to tuition fees, textbooks, learning materials, school uniforms, as well as other pertinent educational activities (free of charge). The reforms do not only focus on the quantity of education made available, but also on improving the quality. However, the government has not achieved much progress to date due to insufficient infrastructure (e.g. ICT systems), coordination and centralization issues arising from various agencies (e.g. Ministry of Education, Ministry of Science and Technology and Ministry of Agriculture and Cooperatives) involved in the human resource development as well as research and development programmes.

Investment policy climate

Macro-level policies

Export-led industrialization policy

Thailand is one of the most popular destinations in ASEAN (Association of South East Asian Nations) in which foreign investors choose to locate their operations since it is among the fastest growing economies in the Southeast Asian region. Obviously, many countries and their respective firms would want to enjoy and take advantage of its high rates of growth. Thailand has achieved remarkable economic growth since 1981, reaching a two-digit growth rate in late 1980s. Thailand's economic growth maintained a positive rate while that of Malaysia and Singapore declined in 1985. However, after the 1997 Asian financial crisis Thailand and Malaysia experienced the lowest economic growth in 1998, at -10.5 and -7.4 respectively, while Singapore's growth rate was -0.9 (Statistics Division of the United Nations, http://unstats.un.org). In the 2000s, Thailand's growth rate rebounded and reached 4.07 percent in 2006, in spite of political upheavals.

Thailand's development strategies have played important roles in accelerating economic growth. The development of Thailand's industrialization policy began with the formulation and implementation of an import substitution policy, initiated in 1958. The policy had been incorporated in Thailand's National Economic and Social Development Plan as well as the Thai Board of Investment's policy. The Thai Government selected certain industries to be entitled for benefits of such a shelter policy based on their direct linkages to domestic industries, as well as usage of domestic raw materials and contribution to Thailand's aggregate foreign exchange saving. This was achieved via tariffs, import restrictions and preferential treatment including special taxation for investment in the priority sectors. In the 1970s, the Thai Government started employing an export promotion policy. However, import substitution measures were used at the same time as protection tools for intermediate and capital goods producers as well as exporters. This is supported by evidence from food processing statistics with a very high effective tariff rate in 1975, estimated at 65.8 percent, and a nominal tariff rate of 22.6 percent (Urata and Yokota, 1994).

During the 1980s-1990s, Thailand progressed towards a more open and liberal economy by implementing its openness policy. In the early 1980s, the use of import substitution industrialization tools was minimized, as shown by a considerable decrease in tariff rates and other non-tariff barriers. Since 1987 (the Sixth National Economic and Social Development Plan), the Thai Government has implemented a full-scale, export-led industrialization policy focusing more on technology intensive sectors. This includes preferential measures through taxation and the provision of low cost funds, as well as the development of export processing zones. The success of the policy was marked by high economic growth rates from 1988 (13.29 percent) until the mid-1990s (9.24 percent). The changes made contributed to increased FDI much more than relying on the obsolete import-substitution policy, and resulted in an increase of Thailand's inward FDI to GDP ratio from 1.03 percent in the 1970s to 3.38 percent in the 1990s (see also Section 4). Additionally, Kohpaiboon (2003) found an empirical result of the increase in FDI generating higher economic growth in favour of an export promotion trade regime in the period of 1970-1999. This is not surprising as the nature of most FDI is export oriented. For example, Japanese MNEs and firms from the newly industrialized countries (NICs) like

Singapore, Hong Kong, the Republic of Korea and Taiwan Province of China established their subsidiaries in Thailand as production facility bases for manufacturing export products (Urata and Yokota, 1994). Clearly, appropriate and effective economic development policy help create a sound macroeconomic environment suitable for attracting FDI.

The economic implications of export-oriented policy for FDI growth of agricultural and food processing sectors succeeded in the 1980s and 1990s. In the past, the agricultural sector was a leading export sector for Thailand with little support from FDI. It seemed that the sector also did not receive much benefit from the import-substitution policy, given its nature of operations (natural resources intensive). Later, the export promotion policy partly expedited Thailand's agricultural and food processing exports. Food product export was the largest among other manufacturing sectors until 1990 (Julian, 2001). Such an open-door policy also helped attract foreign investors and companies to invest and take advantage of the low production and operating costs in these competitive sectors (see Section 4).

Crucial engines facilitating structural changes in Thailand were strong relationships and good cooperation among technocratic advisers, politicians, and industrial groups (Rock, 1995). The author also argued that Thailand's industrial policy has been well planned and consistent. In addition, Thailand successfully implemented an investment-incentive policy (Drabble, 2000; see also Section 3.2.2). Building up a sound investment environment and government initiatives and interventions are vital for economic and foreign investment growths. These government policies create advantages that can partially explain Thailand's internationalization success. The advantages are additional and complementary to conventional comparative advantages, such as low labour costs and other country-specific factors, which initially attract FDI.

Trade and investment liberalization

Thailand's government policy is geared towards a higher degree of economic integration and trade liberalization. Thailand is a member of trade organizations at both regional and global levels and is actively involved in the development of trade agreements at bilateral level. Apart from being a member of the Asia Pacific Economic Cooperation (APEC) forum and the World Trade Organization (WTO), Thailand aims to develop better bilateral trade and economic relationships with its trade partner countries. It is thought that these free trade measures and policies will help to expedite trade in goods and international investment and generate a sound environment for firms involved in international business activities. These are in accordance with the goals of the Ninth National Economic and Social Development Plan of Thailand (2002-2006) in obtaining bargaining power in international trade and investment (Thai National Economic and Social Development Board, www.nesdb.go.th). The Thai Government employs a bilateral FTA policy that partially helps them to achieve international trade and investment goals. In addition, the Tenth National Economic and Social Development Plan (2007-2011) continues to focus on a proactive trade strategy. This includes seeking new markets and enhancing competitiveness of Thai producers based on knowledge and abundant natural resources. Free labour mobility across countries through economic integration and liberalization is supported by the Thai Government as a means to attract foreign workers, businessmen and investment.

The Thai Government has undertaken free trade initiatives as a critical part of its overall international trade strategy. The policy was initiated in 2001, following the example of Singapore, which was the first ASEAN (Association of South East Asian Nations) country to implement a bilateral free-trade agreement regime. There are different stages of development and success in each free trade agreement negotiation process. In Thailand, many active free trade negotiations have been in progress for some time, for example, Thailand-United States. Others are already in effect: Thailand-Australia, Thailand-New Zealand, and Thailand-Japan (Thai Department of Trade Negotiation, Thai Ministry of Commerce, www.thaifta.com). Among these, Thailand's first bilateral, free-trade agreement with a developed country, the Thailand–Australia

Free Trade Agreement (TAFTA), was successfully agreed on 5 July 2004.

Apart from comprehensive FTAs, interim agreements, like the Early Harvest Schemes (EHSs) or the Early Harvest Programmes (EHPs) have also been reached. The interim trade agreements help to accelerate trade liberalization between the parties before bilateral FTAs are fully negotiated. In general, they comprise only one part of broader framework agreements. While the framework agreements cover trade in goods, services and investment embracing comprehensive economic cooperation, EHPs or EHSs focus on just one sector (mainly trade in goods). The interim trade agreements, like the Thailand–China EHP and the Thailand–India EHS, came into force in 2003 and 2004 respectively. At the regional level, Thailand is a member country of the ASEAN Free Trade Area (AFTA) which became effective in 1993. Moreover, ASEAN established many bilateral agreements with countries such as Japan, China, India and Republic of Korea.

The development of free trade agreements between Thailand and its trading partners has brought about a wider market opening for trade in goods. Tariff reductions are considered to be high in all these bilateral agreements. JTEPA, for example, eliminates tariffs from 95 percent of Thai goods. TAFTA and TNZCEP reduce tariffs for Thai products – including agricultural products, processed food, processed seafood and ready-to-eat food – by 83 percent and 79 percent respectively. Goods under the Thailand-China EHP are mainly fresh fruits and vegetables, while the Thailand-India EHS covers 84 items of agricultural and industrial products such as fruit and processed food products. Additionally, AFTA helps decrease tariffs by more than 60 percent including the removal of non-tariff barriers. The aforementioned FTAs have some exceptions with regard to the implementation of tariff elimination of agricultural products on the sensitive list –such as dairy products under TNZCEP – stating that complete tariff elimination is extended until 2015. But these constitute only a small minority of products, for which Thailand needs to enhance competitiveness by lowering their production costs.

Thailand's food exports, however, show a declining growth rate of -3.1 percent in 2009 (National Food Institute of Thailand, 2010b). This emphasizes the need to deepen current markets and, at the same time, expand into new markets. It is anticipated that the established FTAs will facilitate this process (National Food Institute of Thailand, http://nfi.foodfromthailand.com). The food industry is one of the key sectors in Thailand's free trade agreement strategy (Thai Department of Trade Negotiation, Ministry of Commerce, www.thaifta.com). As a result of successful negotiations, tariffs for some food products are subject to eliminations over time, while some others are immediately reduced to zero. This may well encourage international firms to take FTAs into account and to gain benefit from the favourable trade policy.

Clearly, the FTAs provide firms with competitive advantages (via tariff reduction) over those competitors whose governments have not yet liberalized their trade regime. There is also a provision for technical assistance and close cooperation, especially in agricultural technology (i.e. under TAFTA, TNZCEP and AFTA). It is postulated here that this cooperation will enhance productivity and the quality of Thai agricultural products used as inputs in processed food production. In essence, the established FTAs offer many benefits from trade liberalization, from wider business opportunities to larger and more easily accessible markets to technological development. However, there is one query concerning the major beneficiaries from trade liberalization. Although the FTA directly expand trade opportunities by widening market access for agricultural products and processed food products, the benefits to players such as agrobusinesses, exporters, distributors and foreign investors outweigh the benefits to Thai farmers at large. The annual income from agriculture for the Thai farmer household averaged US$3 821 in 2007, and increased slightly to US$4 406 in 2009. Similarly, net agricultural income was US$1 679 (per year) in 2007, and US$1 916 in 2009 (Office of Agricultural Economics, 2007, 2009). Most of the farming households remain poor. Although the existing contract farming system helps integrate Thai farmers into the agricultural and food industry value chain, most of them still cannot move up the value chain with their limited knowledge and technological know-how.

With regard to investment liberalization, there are two main types of international investment agreements (IIAs) that are increasing in their importance and popularity, namely FTAs (as described earlier) and bilateral investment treaties. The role of FTAs in driving FDI should not be neglected as they help promote and liberalize investment across countries. Dunning et al. (1998) argued that the internationalization of firms might be partly due to globalization and regionalization of markets and the pursuit of value-adding activities. Buckley et al. (2001) argued that the North American Free Trade Agreement (NAFTA) increased the possibility of non-member country firms' undertaking reorganization and rationalization. There would be higher foreign direct investment from European MNEs in the USA (Buckley et al., 2001). Rugman and Verbeke (1990) analysed the impact of Europe in 1992 on corporate strategy. They found that European firms would integrate related production and marketing activities across Europe. More generally, it seems that FTAs cause both insider firms (of countries party to the agreements) and outsider firms to increase investments.

While most interim agreements do not cover liberalization of investment or movement of people, the comprehensive bilateral agreements expedite investment by including investment promotion and liberalization provisions as part of investment chapters. This provides foreign firms with greater opportunities for investment in both service and non-service sectors in Thailand and vice versa. Liberalization in services and investment included in the FTAs is good for international firms in the food industry, since almost all value-adding activities are open to foreign investment. Higher levels of investment are encouraged by liberalization of the production and service sectors, as well as facilitation of natural person mobility. With regard to the movement of people, the most relevant feature is that Thailand agrees to facilitate temporary business entry for citizens from countries party to the bilateral FTAs, since the bilateral FTAs cover a chapter on the movement of natural persons. In addition, simplified and transparent immigration formalities for business people are employed and encouraged. The deregulation of movement for people helps foreign firms to relocate their human resources when they invest in Thailand; for example, in sales and distribution offices, or in setting up factories. Further, investment cooperation on research and development and capacity building of priority sectors including agroprocessing, is incorporated into many FTAs such as the bilateral FTA between Thailand and New Zealand.

Another category of IIAs falls to the bilateral investment treaties (BITs). The significance of bilateral investment treaties (BITs) between Thailand and its partner countries is to protect and facilitate foreign investors as well as increase inflows of FDI (Neumayer and Spess, 2005; Kerner, 2009). Since the multilateral investment agreement has not been established yet, the BITs are used as critical and universal tools to attract FDI. They gain popularity from their modest complexity and narrower scope/coverage, involving shorter time spent on the development process than other types of international investment agreements (IIAs) like double taxation treaties and FTAs. These BITs in effect help promote and, at the same time, protect FDI via provisions of national treatment, contractual right protection and investor-state dispute settlement as well as the relaxation of minority ownership restriction.

Up until 1 June 2010, Thailand signed off 40 BITs in total according to reports submitted by Thailand to the United Nations Conference on Trade and Development (www.unctad.org). The first BIT between Thailand and a developed country (Germany) was concluded successfully in 1961, followed by Thailand- Netherlands investment agreement concluded on 6 June 1972, and the Thailand-United Kingdom bilateral investment agreement signed on 28 November 1978. There was a tremendous growth in terms of numbers of Thailand's engagements in BITs. In the 1970s and 1980s, there were only four agreements signed, while 21 BITs were concluded during 2000-2010. These agreements have been reached with both developed countries (i.e.

Germany, Switzerland, United Kingdom) and developing countries (i.e. China, NICs, Indonesia). To date, Germany, China and Switzerland are among the most active countries engaging in the negotiation and development of BITs as shown by the numbers of signed BITs with these countries (www.unctad.org/iia). The Thai Government realizes the importance of FDI in economic development resulting in a rapid expansion of BITs and a change of policy towards a greater degree of investment liberalization after the Asian financial crisis in 1997.

Although Thailand is one of the most attractive FDI destinations, it has to compete with other countries in the same region and elsewhere for foreign capital. In particular, competition among developing countries is very stiff. Recent political unrest heightened concern about Thailand's competitiveness and its sound macroenvironment. Foreign firms may have to think more than twice before making the decision to invest, by taking divers variables into account, for example, market size, culture, legal systems, and political risks. Reduced level of political stability greatly affects uncertainty levels. These firms have to monitor possible changes in rules and regulations, particularly with regard to ownership, expropriation and profit remittance.

The establishment of Thailand's bilateral investment treaties help to build up confidence on the part of foreign investors, and reduce both political and commercial risks by providing protection for foreign investors against expropriation or nationalization. For instance, the BIT between the Russian Federation and Thailand clearly stated that investments of investors from countries party to the agreement shall not be nationalized, nor will ownership be transferred to the state (with some exceptions, such as public welfare protection requiring government intervention). In addition, several BITs between Thailand and partner countries include the provision of "prompt, effective and adequate" compensation in cases where expropriation occurs. This is in line with Thailand's Investment Promotion Act B.E. 2520 (1977) stating that the Thai Government will not transfer business ownership of promoted investors to the state. This reflects a high standard of Thai law in this aspect, although the Investment Promotion Act B.E. 2520 (1977) only provides safeguards for investors whose projects received approval from Thailand's Office of the Board of Investment.

In addition, these BITs grant foreign firms national treatment. In effect, foreign investors from different countries investing in Thailand will be treated equally without any discrimination or special preference toward any particular country. Foreign investors can sue the state when they receive reputedly unfair treatment. BITs also exempt foreign investors from minority ownership restrictions and, as a result, encourage firms to make direct investments. Foreign investors may find it faster and easier to utilize the benefits of BITs since they do not need approval from the BOI and can bypass all administration time and costs involved in the approval process. However, they still need to apply for industrial and commercial licenses as required by Thai rules and regulations during their establishment processes.

With regard to transfer of funds, many BITs between Thailand and partner countries guarantee "freedom of transfer" subject to domestic exchange regulations and practices which comply with international standards, such as that of the International Monetary Fund (IMF). However, most BITs do not include provisions for balance of payment safeguards, prudential measures and stability articles. Nuannim and Kaewpornsawan (2010) argued that Thailand should include these provisions in BITs to allow the state to implement emergency and appropriate measures to maintain financial system stability and to prevent any damages to the balance of payments as well as public interest as a whole. These are deemed sensible, especially when financial crises occur, because some negative aspects of free transfer and openness may be more vulnerable to external shocks.

There were many external shocks, e.g. increases in oil prices and the financial crisis, during the past two decades. An analysis of the Thai Government's response to external shocks in the short run helps us to understand the importance and role of economic policy on growth. After the financial crisis emerging in the Southeast Asian region, Thailand dealt with this problem by following the IMF rescue plan and maintaining high capital mobility. The Thai Government tried to induce foreign capital by raising domestic interest rates. This undoubtedly caused a reduction in domestic investment, while the huge influxes of FDI into Thailand increased from 99 733 million Baht in 1996 to 284 938 million Baht in 1998. Even with such a boost, Thailand's economic growth in 1998 was the lowest among Southeast Asian countries and continued growing at a lower rate than that of Malaysia and Singapore during 1999-2000 (Statistics Division of the United Nations, http://unstats.un.org). Malaysia, in contrast, responded to the crisis which occurred in 1997 by rejecting the rescue plan. Malaysia did implement a stricter capital control policy than Thailand, which led to a relatively lower domestic interest rate compared to that of Thailand in the same period (IMF, 2001). Malaysia successfully recovered within a year after the crisis. Thus, it may be concluded that the ability of the governments to effectively formulate and implement policies when external shocks occur is crucial for continuous and sustainable economic stability. Additionally, the government should build a good balance between domestic and foreign investments, as high fluctuations in FDI could cause macroeconomic turbulence. This should be taken into account and heavy reliance on FDI should be avoided.

Micro-level policies: BOI policies

The Office of the Board of Investment was established on 21 July 1966, commonly known as Thailand Board of Investment (BOI). The BOI is the core government agency responsible for promoting investments, both local and foreign, mainly in the manufacturing sector. Since 1966, the Board of Investment has played an important role in shaping Thailand's direct investment policies including the policies affecting FDI in the agricultural sector. Although there are several Thai agencies affecting investment policy climate, the BOI is uniquely positioned to provide policy feedbacks from direct access to foreign and domestic enterprises.

To maintain a favourable investment climate, the Thailand Board of Investment has adjusted its policies over time in accordance with economic conditions and the National Economic and Social Development Plans. The BOI (2006) summarizes the investment promotion policies as shown in Figure 1. There are three main policies: import-substitution, export-orientation, and the dispersion of direct investment to regional areas.

- Investment policy to promote import-substitution took place during 1958-1971, which is in line with the first and the second national development plans. This policy aims at encouraging firms to use local raw materials, developing infrastructures, and encouraging FDI in the form of joint ventures. The target industries during this policy include sugar, paper, automobile tyres, and plywood.
- Investment policy to promote export-oriented industries began in 1972 and continued through 1992 in accordance with the third to the sixth national development plans. This policy shifted emphasis towards promoting export-oriented activities as well as promoting small-scale and regional industries. A thrust was also given to agroprocessing industries such as canned food, fertilizers, and food processing.
- Policy to disperse investment activities to regional areas has been emphasized since 1993, as stated in the seventh national development plan and continues to the present. To maintain the country's competitiveness and promote more balanced growth, increased emphasis has been placed on the dispersion of industrial activities to regional areas. The agro-industry has been set as one of the target industries serving as a basis for long-run industrial development and linkages. The BOI has relaxed its conditions and offered more incentives in order to encourage investors to improve their production efficiency and technology. For example, the BOI encourages food-processing factories to enhance their operations to the level of international standards ensuring food safety (e.g. GMP, HACCP), and traceability.

FIGURE 1
Investment promotion policies

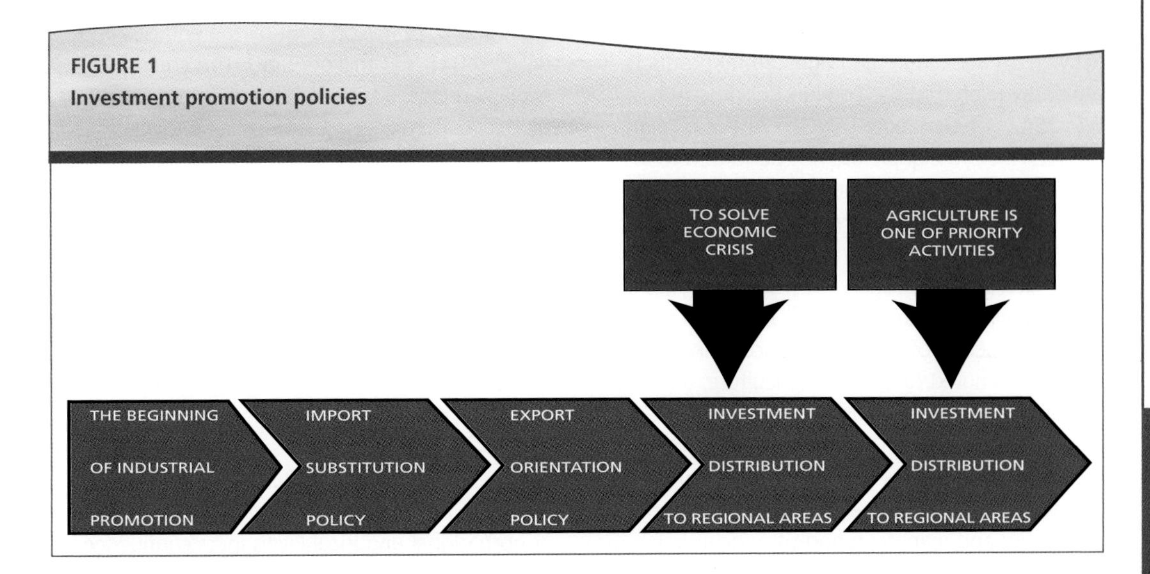

With regard to the Board of Investment's promotional packages, there is no discrimination; meaning that all approved projects receive the same privileges. Regarding foreign direct investment, BOI policies aim to promote and attract foreign investment into the country, particularly in activities deemed beneficial to the economy, using tax and non-tax incentives. The Board's tax privileges aim at reducing costs of doing business in Thailand by granting exemptions on corporate income tax (for a maximum of eight years), and import tariffs on machinery, equipment and raw materials. Rights and benefits vary according to factory location.[5] A promoted company is also allowed to own land under the approved project. These privileges are available to all investment projects, both local and foreign, approved by the BOI. In addition, the BOI provides the necessary information and assistance to facilitate investors' businesses. For example, the office helps investors to obtain official permits and documents required for conducting business, including visas, work permits and permanent residency permits. The Board also encourages industrial linkages between foreign firms and local supporting industries by bringing and matching those who want to find local business partners, subcontractors or specific raw materials.

The Board of Investment has granted promotional packages to investors or companies on a project-level basis. The promoted projects must comply with the BOI's criteria specified under the Investment Promotion Act B.E. 2520 (1977), which are transparent and periodically updated in response to current economic and investment conditions. The BOI has classified activities eligible for promotion into seven groups or sectors. They comprise agriculture and agricultural products; mining, ceramics and basic metals; light industry; metal products, machinery and transport equipment; electronic industry and electrical appliance; chemicals, paper and plastics; services and public utilities.

The BOI has accorded investment projects in the agriculture and agricultural products the status of priority activities. Priority activities are deemed important and beneficial for the Thai economy; they are granted maximum rights and benefits regardless of factory location. In general, an approved project is granted corporate income tax exemption subject to cap. That is, the tax break cannot exceed its project's investment value. This tax exemption limit is lifted for projects investing in agriculture and agricultural products. There is also no limit on machinery and equipment import duty exemptions.

[5] See details in 'A Guide to The Board of Investment' outlining the BOI's requirements for project approval, available at www.boi.go.th under BOI publications.

The criteria of foreign shareholding for activities in agriculture and agricultural products are partly related to the Foreign Business Act B.E. 2542 (1999). Under List One of the Foreign Business Act, foreigners are not permitted to operate the majority of agricultural activities (including rice farming, farming or gardening, animal farming, forestry and wood fabrication from natural forest, fishery for marine animals in Thai waters and within Thailand specific economic zones, extraction of Thai herbs). Accordingly, for BOI-promoted projects in agriculture, animal husbandry and fisheries under List One of the Foreign Business Act, Thai nationals must hold shares totalling not less than 51 percent of the registered capital. Other activities, such as food processing and manufacturing of agricultural products, are free from this shareholding criterion.

4. Analysis of international investments in the agricultural sector

The analysis of international investments in the agricultural sector of Thailand is divided into two subsections. First is the analysis of the overall international investment in the agricultural sector. The foreign investment data used in this analysis are mainly drawn from the Bank of Thailand (BOT). The second analysis focuses on the foreign investment promoted by the Board of Investment (BOI). Both BOT and BOI data have been commonly used to analyse international investment in Thailand.

4.1 Overall FDI analysis

Both GDP and total inflows of foreign direct investment portrayed a rising trend during 1997-2009. Although there are arguments over cause and effect issues between the two variables, it is obvious here that they move in the same direction. While GDP increased steadily over time, FDI fluctuated to some extent. In 1970, FDI accounted for 1 014.10 million Baht (GDP: 148 279.76 million Baht). Later, FDI reached its peak of 1 274 046.54 million Baht in the year 2006 (GDP: 7 850 193 million Baht) and declined to 459 938.44 million Baht by the end of 2009 (GDP: 9 041 551 million Baht). This could be explained by the United States subprime and global economic crises. Domestic factors like Thailand's political crisis also plays an important role in inducing sharp falls of FDI inflows from 2006 onwards. Although the fluctuation of FDI has not affected GDP that much in value, it is noticeable that GDP growth rates declined from 5.15 percent in 2006 to 2.46 percent in 2008, and the growth slowed to its lowest rate in a decade, reaching negative growth at -2.25 percent in 2009 (Thai National Economic and Social Development Board, www.nesdb.go.th). The authors of this chapter are of the view that macroeconomic and political stabilities at both global and local levels induce/influence FDI and vice versa. The analysis of FDI economic impacts on exports, output and employment of agricultural and food processing sectors will be discussed in Section 5 of this chapter.

During 1970-2009, FDI inflow is 192 710.32 million Baht on average (US$5 356.58 million); amounting to 3.66 percent of the GDP. It is noticeable that FDI to GDP ratios were very small before 1986 when there was a development of economic policy progressing toward a more export-oriented policy. Another observation is that, not surprisingly, the average FDI to GDP ratio of the industrial sector is the highest (1.37 percent), followed by FDI to GDP ratio of the service sector (0.25 percent). Agriculture FDI to GDP ratio is only 0.01 percent. This is consistent with structural adjustments that occurred in Thailand. It highlights the importance of effective shifting of resources away from the agricultural sector, while at the same time, shifting more towards the increasingly attractive, strong and competitive industrial sector.

In an early period (1970-1990), FDI inflow was quite low, ranging from only 1-2.08 percent of GDP and 4.69-6.19 percent of total investment (Gross Fixed Capital Formation) during 1981-1990. This may be due to the fact that global FDI inflow was at its lowest level, and Thailand had not developed much, both in economical and political terms. After the financial liberalization in the 1990s, Thailand's FDI increased considerably, from 2.83 percent to 8.72

percent of GDP in 2001considerably 2009 and, at the same time, increased from 7.03 percent to 35.98 percent of Gross Fixed Capital Formation (See Table 8). Interestingly, FDI increased up to 50.97 percent in 1996-2000. This helps explain the possible effects of the Asian financial crisis in 1997 on FDI inflow data. It was reported that parent companies (MNEs) injected capital into their subsidiaries in Thailand coping with Thai Baht devaluation and serious liquidity problems (www.bot.or.th).

In the 1990s, countries that contributed greatly to Thailand's economy via FDI, apart from the United States and the European Union, were Japan (one of the most advanced internationalizing economies in the region) and Asia's newly industrialized countries (NICs) like Singapore, Hong Kong, Republic of Korea and Taiwan Province of China. This was caused by the appreciation of their currencies after the 1985 Plaza Accord. In addition, their MNEs had located their value-adding activities in developing countries like Thailand where costs of operations and resources had been low since the late 1970s. Most Asian countries' international investment was made in countries less developed than their own, typically with lower wage rates and less sophisticated development (Lecraw 1992). After the Asian financial crisis in 1997, there were the recent surges in FDI inflows as shown by figures for the 2000s. For instance, Japan's FDI reached US$4 303.07 million (more than seven times the value in the 1990s), while Singapore's FDI was US$3 896.95 million (more than four times that of the 1990s). Such influxes of FDI into Thailand were reactions of these countries' MNEs to take advantage of economic opportunities in making investments at cheaper costs (i.e. buying up local firms in difficulty). Nevertheless, some were forced by the situation to inject more money into their own subsidiaries in difficult times.

Figure 2 exhibits FDI inflows of food processing and agricultural sectors during 1970-2009. On average, FDI value of food processing is substantially higher than that of agricultural sectors, that is, US$111.29 and 8.17 million respectively (Table 1). Food processing FDI rose significantly over the period, going from US$4.045 million in the 1970s to US$329.954

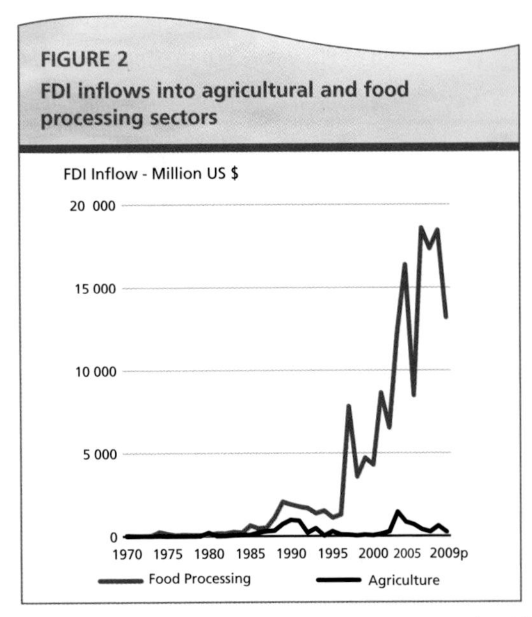

Sources: Bank of Thailand and Thai National Economic and Social Development Board

million in the 2000s. On the contrary, FDI of the agricultural sector evidently flew into Thailand in 1972, amounting to US$0.245 million. In the 1980s, there was a big jump of agricultural FDI, which increased by 4,389 percent over the amount during the 1970s. This is consistent with the movement of AgriFDI to GDP ratios and AgriFDI share figures of the same period. However, both AgriFDI to GDP ratio and AgriFDI share of total FDI dropped continuously since 1990s onwards. This was caused by the perceived high risk of investment and limited business opportunities in comparison to other sectors (Netayarak, 2008).

Furthermore, FDI inflow gaps between food processing and the agricultural sector grew larger over time in terms of values, FDI to GDP ratio and FDI share. Both Figure 4 and Table 10 clearly illustrate this fact. Clearly, Thailand is doing quite well in attracting FDI in the food industry and will possibly achieve its goal as a major world food exporter and producer in the longer term. However, low FDI in the agricultural sector is quite alarming since it is an indicator of the attractiveness and openness of the sector. Productivity and GDP growth of Thailand's agricultural sector could be enhanced

TABLE 1
Comparison of FDI value, FDI to GDP ratio and FDI share between food processing and agricultural sectors

Year	FP FDI (US$ million)	AgriFDI (US$ million)	FP FDI to GDP ratio (%)	AgriFDI to GDP ratio (%)	FP share (%)	Agri share (%)
1970s	4.045	0.178	0.028	0.001	2.606	0.110
1980s	22.654	7.990	0.043	0.016	2.998	1.076
1990s	88.527	12.036	0.070	0.012	2.120	0.365
2000s	329.954	12.487	0.174	0.007	2.065	0.085
1970–2009	111.295	8.173	0.079	0.009	2.447	0.409

Sources: Bank of Thailand and Thai National Economic and Social Development Board

through, among others, agricultural technologies and knowledge, market access and marketing capabilities from foreign partners. The agricultural sector is very critical as a part of the value chain producing inputs for the food processing industry. Ideally, the two sectors should prosper together; this is unlikely, however, as long as Thailand's policy vigorously promotes and opens up a particular sector (i.e. the food industry) for MNEs to invest in, while the other (i.e. agricultural sector) is quite restricted as shown by the Foreign Business Act B.E. 2542 (1999) – not allowing foreign investors to make their investments in largely primary agricultural production. Another example: Thailand offers a great deal of export promotion incentives and privileges for the food industry while imposing export taxes on rice and other agricultural products[6] – until 1986 for rice and until 1990 for rubber (Warr, 2008). This kind of policy has resulted in large discrepancies in terms of FDI inflows and sector growth rates.

Table 2 shows FDI inflows in both agricultural and food processing sectors by countries. Japan and the United States of America invested in the agricultural sector more than other countries from 1987-1999 on average. In the 2000s, Hong Kong ranked first in its FDI, totalling US$5.49 million. However, most of the countries reported here have a tendency towards decreasing their investment in the agricultural sector of Thailand through time. This may be related to the transparency and complexity of rules and regulations on land ownership, as well as limitations on minority business ownership and poor administration on complicated taxation when compared to other sectors. Structural changes also help to explain this phenomenon in Thailand as the country is trying to boost up competitiveness in manufacturing and high value-added sector by relocating both domestic and foreign resources from the primitive sector with the highest productivity to manufacturing and services sectors.[7]

Turning to FDI in the food processing industry, Japan contributed the most to this sector from 1987 onwards. The United States continued to hold second place (US$37.05 million) but in the 2000s it was overtaken by the Philippines (US$53.94 million). Ohmae (1985) emphasized the significance of the "Triad" consisting of the United States, Western Europe and Japan. Developed country firms have high market shares in the Triad countries, which are strategically important to the firms' growth and success. Additionally, these MNEs, in particular, from the "Triad" become key players in developing countries including Thailand. The empirical evidence of this study supports this stylized fact, illustrated by growing FDI from Japan, the United States, and European countries in the food processing industry over time. Moreover, figures from ASEAN countries such as Singapore and the

[6] Taxation on these agricultural products has decreased over time. For example, export tax on rice was about 40 percent in the 1960s and there has been no taxation on rice since the mid-1980s.

[7] Detailed discussion in Warr (2006) and Paopongsakorn (2006).

TABLE 2
Inflow of foreign direct investment in agricultural and food processing sectors of Thailand (US$ million)

Agri-sector	1987–89	1990s	2000s	FP Sector	1987-89	1990s	2000
Japan	8.74	5.98	1.99	Japan	12.34	22.31	70.50
USA	2.25	2.76	1.59	USA	9.13	16.92	37.05
Malaysia	0.10	0.01	0.00	Malaysia	0.21	0.15	7.50
Singapore	0.45	0.33	0.06	Philippines	0.00	0.04	53.94
Hong Kong	0.56	0.06	5.49	Singapore	3.20	10.87	22.34
Taiwan	4.44	1.70	0.27	Hong Kong	3.27	4.74	9.93
China	0.05	0.05	0.01	Taiwan	3.90	9.09	4.40
Canada	0.05	0.63	0.02	Canada	0.03	0.03	1.28
Australia	0.16	0.02	0.04	Australia	0.10	0.60	3.00
UK	0.13	0.07	0.48	UK	0.93	15.07	19.89
Netherlands	0.30	0.10	0.59	Netherlands	4.93	0.99	12.87
Germany	0.12	0.01	0.47	France	1.23	0.11	2.58
France	0.00	0.03	0.02	Belgium	0.02	0.11	10.58
EU	0.55	0.24	1.58	EU	7.83	17.75	48.11

Source: Bank of Thailand database

Philippines indicate their significance in Thailand. These reflect resource and market seeking behaviour of MNEs from the aforementioned investing countries. They may try to capitalize on their technological capabilities in the future, and take advantage of AFTA as well as favourable investment incentives provided by the Thai Government.

Foreign direct investment is divided into two major forms, namely, wholly owned subsidiaries and joint ventures. Total foreign investment in the manufacturing sector accounts for 11.3 percent of 23 677 firms included in the 1997 industrial census, and 0.7 percent of 457 968 firms included in the 2007 industrial census. Foreign investment in the food processing sector numbers 286 enterprises which is equal to 8.1 percent of total foreign investment in 1996, and 0.2 percent (217 enterprises) of total foreign investment in 2006 (See Table 3). Most foreign investors employ joint-venture as the major mode of entry. Firms with less than and equal to 50 percent of foreign ownership were 66.5 percent in 1996, and 54.8 in 2006. The percentage of minority foreign ownership of firms in the food processing sector is even greater than the average (of overall industries) accounting for 78.3 percent in 1996 and 77.9 in 2006. Data collected on wholly owned subsidiaries is only available for the year 1996. It was reported that 422 firms or 15.8 percent of total surveyed firms were 100 percent foreign owned firms, of which only 7.7 percent fell to firms in the food industry.

4.2 BOI's promoted foreign investment in the agricultural sector

Historical development

Since the establishment of the Office of the Board of Investment on 21 July 1966, agriculture and the agro-industry have been among the eligible activities for which the Thai Government tries to induce more investment from both local and foreign companies. At the beginning there was no foreign investment in agriculture and the agricultural products sector. Later, in the mid-1970s, foreign investors showed interest in this sector and brought in technology to invest in food ingredients projects. The projects used local agricultural outputs such as palm, cassava, and rubber as raw materials and added value to

TABLE 3
Foreign investment in the food processing sector classified by shareholders

	1996	Share in total	2006	Share in total
Total Foreign Investment (no. of establishments)	**2 672**		**3 160**	
> 50% Foreign (no. of establishments)	894	33.5	1 428	45.2
≤50% Foreign (no. of establishments)	1 778	66.5	1 732	54.8
Total Foreign Investment in food processing sector (no. of establishments)	286		217	
> 50% Foreign (no. of establishments)	62	21.7	48	22.1
≤50% Foreign (no. of establishments)	224	78.3	169	77.9

Source: Report Of the 1997 and 2007 Industrial Censuses, Whole Kingdom, Thailand's National Statistical Office, Office of the Prime Minister

their products (BOI, 2006). Since then, foreign investors' confidence has improved as shown by their continuous increased investments in this sector up to the present.

Although foreign investment in the agricultural sector promoted by the BOI has increased markedly to date, it has a relatively small share in total foreign investment compared with other sectors. Foreign direct investment in agriculture and agricultural products has concentrated in export-oriented activities, particularly in food processing and the agro-industry. Investors have largely operated in the form of joint ventures. Major investing countries have come from Asia, notably Japan. A more detailed discussion of the extent and nature of foreign investment in the agricultural sector is provided below.

Facts and figures

Over the period of 1970-2006, the value of foreign investment in agriculture and agricultural products was 291 901.7 million Baht; accounting for 5.3 percent of the total BOI's promoted foreign investment. The number of approved projects in this sector is 1 625 projects, accounting for 11.4 percent of the total number of approved foreign projects. The proportion of numbers of agricultural projects (11.4 percent) is not markedly different from other sectors but its investment value is quite small (5.3 percent). Most of the projects are small-scale with less than 50 million Baht of investment. As a result, the sector's share in total foreign investment is relatively small, ranking sixth out of seven BOI-promoted sectors (Table 4).

The value of foreign investment in the agriculture and agricultural products sector has generally increased over time despite some fluctuations, as shown by the bar chart in Figure 5. Although the sector's share in total foreign investment is relatively small, the average annual growth rate of its real investment value during 1974-2009 was 69.57 percent. Similarly, the number of approved projects has also risen with a sharp peak in 1988 (as shown by the solid line in Figure 3) which coincided with the overall FDI inflows and Thailand's economic boom (Warr, 2005). The average growth rate of the number of projects was 30.71 percent per annum, much less than its investment value. Thai employment generated by these foreign investments also shared an upward trend with an average growth rate of 79.74 percent per annum. Note that there was no foreign investment in the agricultural sector prior to 1974[8].

When considering foreign investment in the agricultural sector as a percentage share of total foreign investment, Figure 4 shows that its share (both in terms of investment value and number of project) has declined markedly since 1975. During 1974-1976, the agricultural sector has dominated with more than 60 percentage share

[8] This is perhaps due to fact that with regard to the agricultural sector during the early 1970s it was not in the interests of FDI to apply for BOI's privileges.

TABLE 4
Foreign investment approved by BOI classified by sectors, 1970-2009

Sector	No. of *Projects*	Share in *Total (%)*	Investment *(mill Baht)*	Share in *Total (%)*
Agriculture and agricultural Products	1 625	11.4	291 901.7	5.3
Minerals and ceramics	558	3.9	516 657.5	9.4
Light industries/textiles	2 015	14.1	266 847.8	4.8
Metal products and machinery	3 143	22.0	897 721.4	16.3
Electric and electronic products	3 096	21.7	1 102 796.4	20.0
Chemicals and paper	2 049	14.4	1 400 128.1	25.4
Services	1 784	12.5	1 031 745.0	18.7
Total	14 270	100	5 507 797.9	100

Source: International Affairs Bureau, BOI. Note: 1) Foreign Investment projects refer to projects with foreign capital of at least 10 percent. 2) Agriculture and agricultural products sector include eligible activities in primary production, food processing, manufacturing and services relating to agriculture and agricultural products.

FIGURE 3
Foreign investment in the agriculture and agricultural products sector approved by BOI during 1970–2009

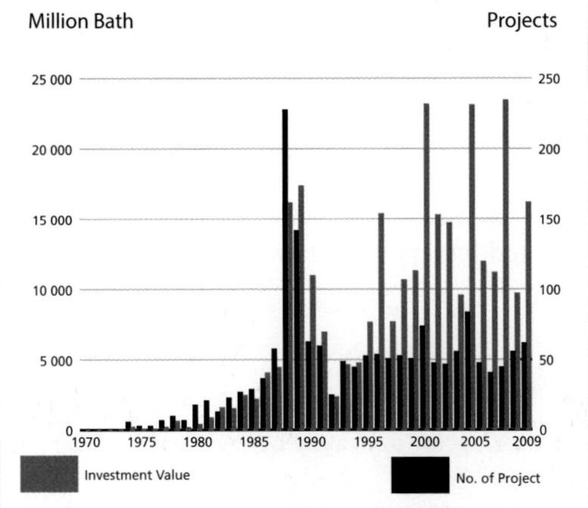

Million Bath
Projects
Investment Value
No. of Project

Source: International Affairs Bureau, BOI. Note: There is no investment in this sector prior to 1974. The investment value shown in this figure is in real terms, the nominal value was converted into real using GDP deflator.

FIGURE 4
Shares of foreign investment in agriculture and agricultural products in total foreign investment during 1970–2009

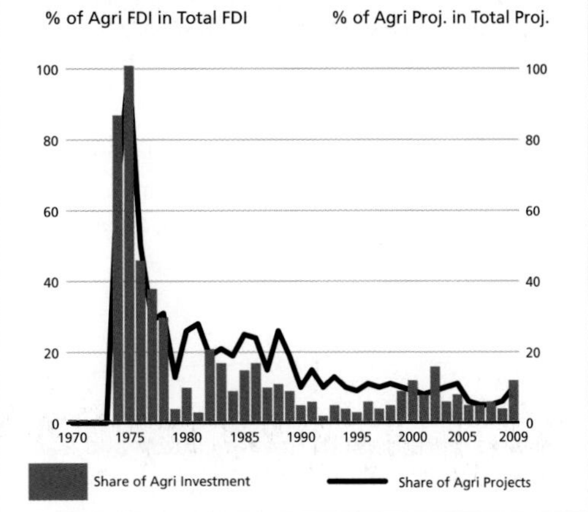

% of Agri FDI in Total FDI
% of Agri Proj. in Total Proj.
Share of Agri Investment
Share of Agri Projects

Source: International Affairs Bureau, BOI.

in total foreign investment. This is consistent with the agricultural growth period – 1960s-1970s – driven mainly by expansion of land frontiers and heavy public investment in roads and irrigation (Poapongsakorn, 2006). After 1976, its share fell significantly during the early 1980s and has continued to decline until the present. This is also in accordance with the period of agricultural decline, from 1980 to mid-1990s, categorized by Poapongsakorn (2006, pp.5-18). In addition, the declining share of FDI corresponds with the decreasing agricultural GDP relative to those of non-agricultural sectors.[9] The decline in agricultural growth was in line with structural change towards an industrialized economy as well as many external factors, particularly a worldwide depression in major agricultural product prices.

Characteristics of BOI's Promoted Foreign Investment

The majority of foreign investments promoted by the BOI are in the form of joint venture between local Thai investors and foreign partners. Particularly with regard to projects in agriculture, animal husbandry and fisheries under List One of the Foreign Business Act B.E. 2542 (1999), Thai nationals must hold shares totalling not less than 51 percent of the registered capital. As shown in Table 5, in terms of number of projects, foreign investments in agriculture and agricultural products during 1970-2009 are joint ventures, accounting for about 82 percent of the total, while the rest are totally foreign owned projects, mostly in agroprocessing activities that are not restricted by the law. In terms of investment value, joint venture projects account for 78 percent, whereas wholly foreign owned projects account for 22 percent of the total foreign investment in this sector.

[9] The relative decline of the agricultural sector has been explained by several studies, for example, Siamwalla, 1996; Martin and Warr, 1994; Coxhead and Plangpraphan, 1999.

The majority of these foreign projects are export-oriented. More than 80 percent of their products are produced to serve export markets. Specifically, there are 1 064 projects out of 1 625 projects that produce for exports. This accounts for 65.5 percent of the total number of foreign approved projects in the agricultural sector. The total investment value of export-oriented projects is 169 045 million Baht, sharing 58 percent of the total foreign investment value in this sector. This is in line with the export-oriented industrial policy that Thailand has pursued since 1972. The majority of the export-oriented projects were concentrated in the manufacture of the natural rubber products, which are one of Thailand's top export products. Other activities that also attract a large number of export-oriented foreign investments include the manufacture or preservation of food or food ingredients, using modern technology. This is because rubber products and food processing are two major activities with large export opportunities. The Board of Investment's promotional packages, which includes an exemption of import tariffs on machinery and equipment, is perhaps deemed attractive to export-oriented rather than locally served projects.

Export-oriented foreign investment has generally increased through time, both in terms of number of projects and investment value (Table 6). During the 1970s the value of export-oriented foreign investment was still less than a

TABLE 5
Foreign investment in the agriculture and agricultural products sector approved by BOI classified by shareholders

	1970–2009	Share in total (%)
Total Foreign Investment (no. of projects)[1]	1 625	
- 100% Foreign (no. of projects)	304	18.71
- Joint venture (no. of projects)2)	1 321	81.29
Total Foreign Investment Value (million Baht)	291 901.7	
- 100% Foreign (million Baht)	64 785.9	22.19
- Joint venture (million Baht)	227 115.8	77.81

Source: International Affairs Bureau, BOI. Note: 1) Foreign Investment projects refer to projects with foreign capital of at least 10 percent. 2) Joint venture projects refer to joint projects between local Thai investors and foreign partners with foreign capital of at least 10 percent.

TABLE 6
Export-oriented FDI in the agriculture and agricultural products sector

	1970s		1980s		1990s		2000s	
	No. of *Project*	Investment *(m.Baht)*	No. of *Pproject*	Investment *(m.Baht)*	No. of *Pproject*	Investment *(m.Baht)*	No. of *Project*	Investment *(m.Baht)*
Export-oriented	13	317.8	417	35 404.0	313	50 675.1	321	82 648.2
Others	11	775.9	159	15 316.1	171	31 368.3	220	75 396.3
Total	24	1 093.7	576	50 720.1	484	82 043.4	541	158 044.5

Source: International Affairs Bureau, BOI.
Note: Export-oriented foreign investment projects refer to projects which export their products of at least 80 percent.

half of total foreign investment. It has begun to dominate the overall foreign investment in this agricultural sector since the 1980s. Nonetheless, in terms of number of projects, foreign investment was roughly the same during the 1970s and reached its peak in the 1980s, during which time Thailand had experienced an industrial boom. This is partly attributed to the fact that Thailand had relatively cheap labour and raw materials at that time. Export-oriented companies had used Thailand as their production base for simple food processing and agricultural products.

With respect to major investing countries, Japan has been the largest investing country in the agricultural sector over the entire period, followed by the United States, Malaysia, Taiwan Province of China and China. These top five countries account for 63.5 percent of the total foreign investment value in this sector (Table 7). In terms of number of projects, Japan is also ranked number one, followed by Taiwan Province of China, Malaysia, the United States and China. Their share in the total number of approved foreign projects in this sector is 68 percent. Besides these top five countries, other major investing countries include Singapore, Hong Kong, Netherlands, United Kingdom, Australia, France, Germany, Canada and Luxembourg.

Considering by subperiods (Table 8), Japan, Singapore, United Kingdom and Taiwan Province of China were major investors during the 1970s. In later sub-periods, Japan and Taiwan Province of China still played a dominant role while the United Kingdom and Singapore invested relatively less compared with other countries. From the 1970s to 2000s, most countries had increased their investment in the agricultural and agricultural products sector. However, some

TABLE 7
Top 5 investing countries in the agriculture and agricultural products

Country	No. of Projects	Investment value (million Baht)	Rank of No. Projects	Rank of Investment
Japan	328	83 084.10	1	1
USA	159	29 390.90	4	2
Malaysia	218	28 529.00	3	3
Taiwan	300	23 638.80	2	4
China	98	20 820.80	5	5

Source: International Affairs Bureau, BOI.

TABLE 8
Promoted FDI classified by major investing countries, 1970–2009 (million Baht)

	1970s	1980s	1990s	2000s
Japan	12.4	664.1	2 220.0	10 158.2
Taiwan	6.1	1 187.1	1 018.0	1 970.5
Malaysia	-	309.0	1 752.0	2 059.4
USA	2.2	644.4	1 932.3	1 401.7
Netherlands	-	351.4	1 174.8	184.1
Singapore	10.0	237.9	557.2	779.5
Hong Kong	-	658.9	154.9	589.5
Australia	-	224.9	749.7	107.3
China	1.2	344.3	309.2	322.5
Luxembourg	-	748.9	-	-
UK	7.3	155.9	195.3	281.2

Source: International Affairs Bureau, BOI

countries have slowed down their investment during 2000-2009; for example, the United States of America, the Netherlands and Australia. This is in line with the declining trend of FDI in the agricultural sector.[10] It is worth noting that Japanese FDI has increased remarkably over time; furthermore, Japan is not only the largest investor in the agricultural sector but also in other manufacturing sectors, notably automotive and electronic products.

Decomposition of BOI's Promoted Foreign Investment

Disaggregating the agricultural sector's investment, BOI statistics (Table 9) reveal that foreign investment in primary agricultural production (including crops, livestock, fisheries and forestry) accounts for only 8 percent of the sector's investment value whereas the share of food processing accounts for 36.4 percent. More than 50 percent of the foreign investment value is concentrated in the manufacturing of other agricultural products and agricultural services. In terms of number of projects, primary agriculture accounts for about 10 percent and those of food processing and other agricultural products and services are about 35 percent and 55 percent, respectively.

The above findings suggest that international investments in the agricultural sector have concentrated in food processing and the manufacture of agricultural products. This is in line with the fact that Thailand has become industrialized with more emphasis on agro-industry and that the BOI is the government agency that principally promotes FDI in the manufacturing and service sectors. However, BOI-offered incentives and privileges may not be directly relevant to primary agriculture; in particular, primary agricultural production is under List One of the Foreign Business Act B.E. 2542 (1999), in which Thai nationals must hold shares totalling not less than 51 percent of the registered capital. This regulation more or less prevents foreign involvement in the agricultural sector. Moreover, the majority of FDI in this sector is export-oriented thereby investing in value-added agricultural products, using primary agricultural output as raw materials, to serve the world market.

Within primary agriculture, crops occupy the largest share in terms of number of projects,

[10] Because of time and data constraints, this study was not able to identify the particular reasons for the decline in these countries' investments.

TABLE 9

Foreign investment in the agriculture and agricultural products sector approved by BOI classified by subsectors during 1970–2009

Subsectors*	Total		Share in total (%)	
	No. of project	Investment value (million Baht)	No. of project	Investment value (million Baht)
Crops	61	4 015.6	3.75	1.38
Livestock	40	13 994.0	2.46	4.79
Fisheries	53	5 309.5	3.26	1.82
Forestry	3	245.5	0.18	0.08
Food processing	571	106 231.2	35.14	36.39
Non-food agricultural products	797	130 580.5	49.05	44.73
Others	100	31 525.4	6.16	10.81
Total	1 625	291 901.7	100.00	100.00

Source: Authors' calculation based on data from the International Affairs Bureau, BOI. *Crops include activity 1.1 and 1.2, livestock includes activity 1.4 and 1.5.1, fisheries include activity 1.5.2 and 1.8, forestry is activity 1.24, food processing includes activity 1.11, and manufacture of agricultural products include activity 1, 1.3, 1.9, 1.10, 1.14-1.16, 1.20, 1.25. Others include post-harvesting and other supporting agricultural services, under activity 1.7, 1.13, 1.17-1.19, 1.21-1.23, 1.26-1.30

followed by fisheries, livestock and forestry. Nonetheless, in terms of investment value, the livestock subsector accounts for the largest share of foreign investment, followed by fisheries, crops, and forestry. This is because the majority of approved livestock projects exist on a relatively large scale compared with crop projects that do not require as much investment. As shown in Table 9, the total value of foreign investment in livestock during 1970-2009 is 13 994 million Baht and those of fisheries, crops and forestry is 5 309.5 million Baht, 4 015.6 million Baht, and 245.5 million Baht, respectively.

There has been a changing investment structure within the primary agricultural production activities, as shown in Table 10. During the early periods (1970-1979), crops were the major recipient of foreign investment. Livestock and fisheries received moderate investment while forestry received none at all. The crop projects that were approved in early days were fast growing tree cultivation and pineapple cultivation projects. In more recent years, investment has shifted to the production of hybrid corn seeds, mushroom, and hydroponic vegetables. This is in line with agricultural diversification. There has been a changing production structure in Thai agriculture in tandem with the changing comparative advantage and changing demand pattern toward high value-added and safe products (Poapongsakorn et al., 2006). Since the 1980s, crops have received less investment while livestock and fisheries have gained more foreign investment. This is perhaps due to the growing export demands for poultry and fisheries. The amount of investment required in the crop sector is also relatively smaller than that for livestock and fisheries. There was no investment in forestry plantation prior to 2004, which is consistent with the minor role of forestry in the Thai agricultural GDP; consequently, there has been no foreign interest in this activity. The plantation projects approved from 2004 are in line with the public awareness over the extinction of forests, which attracted foreign investment in this activity.

The livestock projects approved by the BOI comprise livestock breeding and husbandry, mainly in swine and broiler chicken production. Fishery projects involve aquatic husbandry and deep sea fisheries, mainly prawn aquaculture. Crop projects are under the BOI's eligible activities categorized as plant propagation and development, and hydroponics cultivation. They are predominated by vegetables, fruits and field crops production. Foreign investment in forestry came mainly from a few forest plantation

TABLE 10
Foreign investment in the agriculture and agricultural products sector approved by BOI, classified in subsectors

Subsectors	1970s		1980s		1990s		2000s	
	No. of project	Investment (million Baht)	No. of project	Investment (million Baht)	No. of project	Investment (million Baht)	No. of project	Investment (million Baht)
Crops	2	433.0	18	728.4	22	1 298.0	19	1 556.2
Livestock	1	10.4	-	-	16	2 329.1	23	11 654.5
Fisheries	2	36.0	35	2 867.2	12	1 957.3	4	449.0
Forestry	-	-	-	-	-	-	3	245.5
Food-processing	6	137.0	178	20 069.3	166	22 909.3	221	63 115.6
Agri Products	10	352.8	318	25 468.6	242	43 697.6	227	61 061.5
Others	3	124.5	27	1 586.6	26	9 852.1	44	19 962.2

Source: Authors' calculations based on the data from the International Affairs Bureau, BOI

projects (teakwood, sandalwood and argarwood). Approval of projects in crops, livestock and fisheries has taken place since the mid-1970s while that of forestry has just begun in recent years (2004-2006).

The food processing subsector has received a relatively large number of the BOI approved foreign investment compared with primary production. The promoted projects include a variety of food processing products such as rice crackers, noodles, fruit juices, canned seafood, frozen foods, dried fruits and vegetables, etc. The first and oldest project in the BOI record was in the food processing subsector: a project producing Chinese cake made from rice and flour, which was approved in 1974. This project no longer receives BOI tax privileges but it is still active and is located in Chonburi province. In recent years, a number of approved projects produce ready meals which are in line with the changing consumer demands for faster and easier lifestyle.

Other approved projects are the manufacture of agricultural products and supporting agricultural services, which include a large number of agro-industry products, post-harvesting activities and supporting services. For example, the manufacture of rubber products has received a number of foreign investments from the past up to the present. The manufacture of oil or fat from plants or animals also attracts many foreign investments. Agricultural services mainly include grading and packaging of agricultural products, silo and crop drying, and cold-storage.

5. Impacts of FDI in Thai agriculture

5.1 Overview of FDI Impact

This section presents empirical evidence and discusses the impacts of FDI on the food industry's employment, export, output and value added. Data used for the analysis are from the Thai National Statistical Office. The food industry is divided into the four-digit International Standard Industrial Classification of All Economic Activities (ISIC) in order to see the detailed impact on its subsector. Data on some subsectors are not provided as there is no evidence of foreign ownership. In addition, the Thai National Statistical Office cannot publish data of firms in 1551 ISIC code (distilling, rectifying and blending of spirits; ethyl-alcohol production from fermented materials), or the 1553 ISIC code (manufacture of malt liquors and malt) because of disclosure rules and regulations which are applicable when the number of firms is less than three.

FDI and employment

The impact of FDI on employment according to the 2007 industrial census was positive. Official data show that 3 160 firms with foreign shareholders employed in total 983 778 employees (25.76 percent of total employment), generating an income of 142 426.05 million Baht (33.05 percent of total remuneration). Although firms with foreign ownership were only 0.7 percent of the entire manufacturing sector, their aggregate impact on employment was one-fourth of total employment and one-third of total employees' income. Foreign Direct Investment impact on Thailand's food industry also prevailed: there were 82 361 employees (13.34 percent of the total industry) employed by these foreign firms. These employees earned 9 605.15 million Baht, accounting for 15.67 percent of the total industry. It is noticeable that the positive effect on the employment share of the food industry is quite modest compared to the average figure of all manufacturing industries. This may be due to the fact that these foreign firms rely on the technology intensive production rather than labour intensive one.

Among others, fish and fish products processing and preserving gained the highest employment share of foreign firms in the food industry (19 648 employees) followed by fruit and vegetables processing sector with 16 069 employees. Examples of subsectors receiving the least benefit on employment were dairy products manufacturing (607 employees) and malt liquors and malt manufacturing sectors. Most of the foreign firms seem to invest a great deal in subsectors that offer them competitive advantages in terms of abundant and low cost of inputs. These firms can achieve their low cost targets by exploiting Thailand's resources and, at the same time, utilizing their internal strength and capabilities such as marketing and technological capabilities. Notably, some foreign firms choose to invest in subsectors that have know-how, even though they are among Thailand's weakest sectors (technology-wise). For instance, 19 firms invest in the dairy product manufacturing subsector. As Thailand is neither a dairy product exporting country nor a producing country, it would seem that these foreign firms invest in the sector in order to reap the benefits of a huge, untapped domestic market. Despite the fact that positive employment gain is not much, the potential for technological transfer is great. This may help improve Thailand's food sector as a whole especially in the subsectors which lack expertise and know-how through technological transfer processes between these foreign firms and Thai partners as well as relevant parties (i.e. workers and farmers).

FDI and export

The majority of foreign firms (2 040 firms or 64.56 percent) set up businesses/plants in Thailand as production bases for export. Foreign Direct Investment contributed to approximately 56.44 percent of total export value, amounting to 1 398 794.83 million Baht. This is a large proportion considering that it is derived from only 0.45 percent of total establishments (both local and foreign firms). Approximately 24 percent of these firms (758 out of 3 160 firms) exported more than 80 percent of total output. In 2006, the export share of foreign firms in the food industry was about 21.84 percent of total industry, and amounted to 62 612.79 million Baht. Two most prominent subsectors were 1) Processing and preserving of fish and fish products; and 2) Manufacture of other food products accounting for export values of 17 916.85 and 17 438.38 million Baht respectively. At the other end of the spectrum, up to 36.87 percent of these foreign firms (80 firms in total) did not get involved in exporting their food products at all. Obviously, they mainly focused on the domestic market. For instance, the dairy product manufacturing sector export value was only 2.85 million Baht, as most of the final output was sold to customers in Thailand.The greatest number of foreign firms in Thailand – totalling 644 firms (31.57 percent) – exported their products to Japan. The United States, Singapore and European countries were also among the most popular/preferred export destinations of these firms. There were 257 (12.60 percent), 242 (11.86 percent) and 232 firms (11.37 percent) respectively putting their efforts on the aforementioned target markets.

Main export markets for foreign firms in the food sector comprised similar countries to the manufacturing sector, except for China which ranked third in its importance by numbers of firms' choices of export markets, followed by Singapore and European countries. This may be driven by the large size of the Chinese market, FTAs between Thailand and China as well as AFTA. Not surprisingly, these countries were also major sources of Thailand's FDI in agricultural and food processing sectors. Their respective foreign firms have strong business linkages and marketing channels in their homeland while exploiting low cost advantages and abundant resources of the host country. This is a typical combined characteristic of resource seeking and efficiency seeking FDI. As a result, the authors observe that a great number of firms export the final output back to their home countries.

FDI and output and value added

In 2006, the share of foreign firms in the manufacturing sector's output was 43 percent accounting for 3 140 965.11 million Baht, whereas foreign firms' contribution to manufacturing value-added was 42.27 percent or 743 405.62 million Baht. The impacts of Foreign Direct Investment on output and value added were greater than its impacts on employment as shown by lower employment share of foreign firms (only 25.76 percent). However, the positive effect of FDI was greatest for Thailand's export with the highest foreign share of 56.44 percent of total export value.

The same pattern of results is repeated in FDI impacts on the food industry's output and value added. The degree of FDI positive impact seemed high on export, with an export share of foreign firms of 21.84 percent in comparison. Foreign firms were responsible for producing 13.37 percent of total output (150 889.52 million Baht) while generating total food processing value added of 15.50 percent of total industry (38 030.87 million Baht). At subsector level, soft drinks and mineral water manufacturing sector generated the highest output valued of 29 561.79 million Baht but its value added was quite low amounting to only 6 140.82 million Baht. The motivation of foreign firms undertaking FDI in this subsector was to seek markets and to maintain access to local markets with promising economic growth like Thailand. This was supported by marginal export value of 2 014.77 million Baht since most of the outputs were produced for customers residing in Thailand. Interestingly, foreign manufacturers of other food products (1549 ISIC code)11 did well in terms of both their output share and value added share accounting for 25 833.09 million Baht (28.02 percent of total subsector) and 12 621.83 million Baht (41.79 percent) respectively. These figures were higher than those of the top export subsector such as processing and preserving of fish and fish products.

5.2 Contributions of BOI's promoted FDI

The international investments through the BOI promotion have contributed to the Thai economy in several ways. The most obvious gains are in terms of employment generation and export earnings. Overall, foreign investment in this sector has generated a total of 369 514 jobs for Thai workers during 1970-2009. As shown in Table 11, the foreign projects have generally raised local employment over time despite a small reduction during the last decade. Over the entire period, the

TABLE 11
Employment generated by foreign investment in the agriculture and agricultural products sector during 1970–2009

Year	Investment value (million Baht)	No. of project (project)	Thai employment (person)
1970s	1 093.7	24	6 306
1980s	50 720.1	576	111 396
1990s	82 043.4	484	130 554
2000s	158 044.5	541	121 258
Total	291 901.7	1 625	369 514

Source: International Affairs Bureau, BOI

[11] Manufacture of other food products *not elsewhere classified such as* manufacture of soups and broths; spices, sauces and condiments; foods for particular nutritional uses; frozen meat, poultry dishes; canned stews and vacuum-prepared meals; herb infusions; extracts and juices of meat, fish, crustaceans or molluscs.

TABLE 12
Employment generated by foreign investment in the agriculture and agricultural products sector classified by subsectors during 1970–2009

Subsectors*	Thai employment person	Share in total %
Crops	10 624	2.88
Livestock	10 391	2.81
Fisheries	6 094	1.65
Forestry	314	0.08
Food processing	173 220	46.88
Agricultural products	146 528	39.65
Others	22 340	6.05
Total	369 514	100.00

Source: Authors' calculations based on the data from the International Affairs Bureau, BOI. *Crops include activity 1.1 and 1.2, livestock includes activity 1.4 and 1.5.1, fisheries include activity 1.5.2 and 1.8, forestry is activity 1.24, food processing includes activity 1.11, and manufacture of agricultural products include activity 1, 1.3, 1.9, 1.10, 1.14-1.16, 1.20, 1.25. Others include post-harvesting and other supporting agricultural services, under activity 1.7, 1.13, 1.17-1.19, 1.21-1.23, 1.26-1.30.

average annual growth rate of local employment is almost 80 percent per year, which is quite remarkable. The growth rate was particularly high comparing the 1970s to the 1980s.

When considering at subsector level (Table 12), food processing activities have created the largest number of jobs for Thai workers, totalling 173 220 persons which accounts for 47 percent of total number of job generated. This is mainly due to the concentration of foreign investment in this subsector. The employment under the manufacture of agricultural products accounts for about 40 percent while that of primary agriculture (including crops, livestock, fisheries and forestry) accounts for 7.4 percent. The small share in the primary agriculture is consistent with the relatively small investment in these activities.

With regard to the primary agriculture, the employment generated by crops and livestock are similar despite the fact that the overall value of investment in the livestock subsector is much higher. This reflects the nature of livestock production that is less labour intensive compared

TABLE 13
Employment generated by export-oriented FDI in the agriculture and agricultural products sector (persons)

	1970s	1980s	1990s	2000s
Export-oriented	1 995	95 715	94 957	86 971
Others	4 311	15 681	35 597	34 287
Total	6 306	111 396	130 554	121 258

Source: International Affairs Bureau, BOI.

with crops. Foreign companies have generally employed modern technology as required by the BOI's regulations.

Another obvious contribution is export earnings. As pointed out in section 4.2, the majority of foreign investment in the agricultural sector under the BOI scheme was export-oriented. More than 80 percent of their products were shifted abroad thereby boosting Thailand's agricultural exports. Expanding market size through export helps achieve the economies of scale that bring about real cost reductions thereby increasing productivity (Harberger, 1996). Exports also enhance market competition in the sense that export-oriented firms have to adjust to remain competitive in world markets by adopting new technology, marketing know-how and improving production efficiency. In the case of processed foods for exports, FDI has played a major role in the successes of these export industries (Netayarak, 2008). At macro level, these export gains help raise the country's GDP and hence productivity and living standards. Export-oriented FDI is also the dominant source of local employment since the 1980s up to the present, as shown in Table 13. Regarding the impact of FDI on agricultural growth and productivity the empirical evidence is limited as the presence of FDI in the agricultural sector is small (Furtan and Holzman, 2004, Sattaphon, 2006). Sattaphon (2006) found evidence that Japanese FDI had a positive impact on stimulating the growth process in Thai agriculture but the effect was not large.

FDI and technology transfer

FDI has been widely recognized as an important channel bringing in capital, new technology and know-how that can enhance the technological capability of the host country firms. However, these benefits – especially the technology transfer effect of FDI – varied among empirical case studies. Kohpaiboon (2006) investigated linkages between FDI and technology spillover using Thai manufacturing as a case study, some of which include food products, beverages, rubber and wood products. He found that gains from FDI technology spillover are conditioned by the nature of the trade policy regime, meaning that to maximize gains from FDI technology spillover, a liberalizing investment policy has to go hand-in-hand with liberalizing the trade policy (Kohpaiboon, 2006). Although his study did not specifically measure the gains from FDI technology transfer it has important policy implications. The implication from his study is that agricultural trade policy in Thailand has to be liberalized to induce the type of FDI inflows that are most likely to introduce technology spillover. According to Warr (2008), agricultural trade policy in Thailand is relatively liberal. This implies the relatively liberal agricultural trade policy has somewhat induced FDI with technology transfer. Since the extent of FDI in the Thai agricultural sector is quite small it is likely that the technology transfer impact is not large.

In Thailand, technology transfer to agriculture occurs mostly through non-FDI channels (Kohpaiboon, 2006). Private companies, particularly the Charoen Pokphand (CP) Group, have played an important role in transferring technology to farmers.[12] However, Netayarak (2008) found evidence that FDI projects have brought about new knowledge and technologies which were diffused very well to Thai farmers, entrepreneurs and labourers. In particular, the Thai agro-industries have benefited greatly from the technology transfer during the past decades.

Moreover, Netayarak (2008) observed increasing trends of agricultural R&D and agricultural technology transfer during 1994-2005. Since the majority of FDI are in the form of joint venture and export-oriented, R&D funds were financed by parent companies or subsidiaries abroad (Netayarak, 2008). In particular, foreign partners played a major role in choosing processing techniques that suit foreign demand, notably in processed agricultural product like chicken, pineapples and tiger prawns. Foreign companies also brought in seeds and animal breeds that were adapted with local conditions and benefited Thai agriculture (Suphannachart and Warr, 2009).

6. Conclusions and policy recommendations

The extent of international investment or FDI in the agricultural sector of Thailand is relatively small compared with other sectors. The majority of agricultural FDI is in the food processing sector and takes the form of joint venture producing mainly for export markets. The extent of FDI in primary agriculture is particularly small. This is perhaps due to a mix of several reasons, notably the rule of land ownership that prevents foreigners from owning land; uncertainty in export markets due to controls and restrictions on primary agricultural exports; and the enforcement of the Foreign Business Act that constrains the participation of foreign investors in primary agricultural production. There are larger investment opportunities in food processing and the agro-industry. Despite the limited extent of FDI, evidence of both overall FDI inflows and BOI's promoted projects suggests that the past investments have contributed to agricultural development and the overall economic expansion.

There are many benefits of FDI to the Thai agricultural sector in terms of output, value added, export and employment expansions as well as technological transfer. All these lead to a more sustainable agricultural development. While the export-led industrialization policy generates more benefits to the industrial sector than to the agricultural sector, IIAs like FTAs and BITs including

[12] The Charoen Pokphand (CP) has been instrumental in the research and development of broiler and shrimp cultivation, seed technology and a new variety of freshwater fish (Poapongsakorn, 2006, p.35)

BOI investment promotion policy are good tools encouraging foreign investors to invest in the agricultural sector. However, the Thai Government should effectively disseminate information and arrange in-depth consultation sessions with relevant parties including Thai firms and farmers prior to any changes or new development of policy. By so doing, it would help reduce short term shock and also prepare them for adjustment. There are large market and investment opportunities still to be tapped by Thai firms doing businesses in the agricultural sector. Hence the importance of appropriate internationalization strategies and the development of internal company and human resource strengths to enable Thai firms, labourers and farmers to capitalize on the increasing demand for food and to survive in very tight competition for FDI in the world market.

The Thai Government should try harder than before to facilitate FDI inflows and eliminate FDI's barriers to entry through deregulation and liberalization measures. This can be done by developing a greater number of international investment treaties such as FTAs. In terms of quality and coverage/scope of these IITs, the Thai Government should concentrate on developing comprehensive BITs and FTAs by incorporating provisions of investment promotion, liberalization as well as protection in investment chapters. A further step to enhancing the image of Thailand as an attractive international investment destination would be if its investment policies were geared towards a greater degree of openness and transparency. Public sector reform is in great need of increased transparency and the reduction in administration processing and approval time and costs. The efficient and integrated management of agricultural, industrial, trade and investment policies should be supported as a way to reduce production and operation costs and increase profitability of investment in Thailand. Furthermore, the relevant Thai Government agencies should collaborate in developing strategic, attractive and responsive investment promotion packages including grants to foreign investors' requirements (i.e. in terms of financial and human resource development), especially those prospective investors aiming to make investments in the agricultural sector.

While partnerships between foreign firms and Thai firms in the agricultural sector (most obvious in the food processing sector) are strong and increasing in numbers via joint ventures, linkages between MNEs and Thai farmers are expanding via contract farming arrangements. Such linkages should be maintained and established as agricultural production is a very important part of the value chain. Thai farmers often lack financial resources, skills and high-level agricultural technology. The agricultural productivity could be enhanced through the provision of training, new technological innovation and financial assistance. As it is now, most MNEs employ contract farming systems by supplying seeds, fertilizers and know-how/new technology to farmers. Such relationships and cooperation should be broadened and strengthened via activities such as research and development. Therefore, the Thai Government should develop a holistic policy to promote a higher level of FDI in research and development, as well as agricultural human resource development requiring concerted efforts by various government agencies, for example the BOI, the Ministry of Agriculture and Cooperatives and the Ministry of Science and Technology. Additionally, a better profit-sharing system (e.g. profit and loss sharing loans) should be put in place to increase Thai farmers' income and improve their well-being. All these efforts would generate numerous benefits to agricultural development as a whole.

References

Aggarwal, R. & Agmon, T. 1990. The international success of developing country firms: role of government-directed comparati. ve advantage. *Management International Review*, 30(2): 163-80.

Athukorala, P. & Jayasuriya, S. 2003. Food safety issues, trade and WTO rules: a developing country perspective. *World Economy*, 26(9): 1395-416.

Athukorala, P. & Sen, K. 1998. Processed food exports from developing countries: patterns and determinants. *Food policy*, 23(1): 41-54.

Bank of Thailand 2010. Foreign Direct Investment Statistics, Balance of Payments Statistics Team, Information Management Bureau.

Board of Investment 2006. 40 Years of BOI (in Thai), Office of the Board of Investment.

Buckley, P. J., Clegg, J., Forsans, N. & K. T. Reilly. 2001. Increasing the size of the "country": regional economic integration and foreign direct investment in a globalised world economy. *Management International Review*, 41(3): 251-74.

Coxhead, I. & Plangpraphan, J. 1999. Economic Boom, Financial Bust, and the Decline of Thai Agriculture: Was Growth in the 1990s too fast? *Chulalongkorn Journal of Economics,* 11(1): 1-17.

Department of Commercial Registration 1999. Foreign Business Act B.E. 2542 (1999), Bureau of Business Registration, Department of Commercial Registration.

Dunning, J. H. 1986. The investment development cycle and third world multinationals. In Khan, K. M., ed., *Multinationals of the south: new actors in the international economy*. London: Pinter Publishers.

Dunning, J. H., Van Hoesel, R., & R. Narula. 1998. Third world multinationals revisited: new developments and theoretical implications. In Dunning, J. H., editor, *Globalisation, trade and investment*. Amsterdam: Elsevier.

Fong, P. E. & Komaran, R. V. 1985. Singapore multinationals. *Columbia Journal of World Business*, Summer: 35-43.

Furtan, W. & Holzman, J. J. 2004. The Effect of FDI on Agriculture and Food Trade: An Empirical Analysis. *Agriculture and Rural Working Paper Series No.68.* Ontario: Statistics Canada, Agriculture Division.

Harberger, A. C. 1996. Reflections on Economic Growth in Asia and the Pacific. *Journal of Asian Economics,* 7(3): 365-392.

Hobday, M. 1994. Technological Learning in Singapore: A Test Case of Leapfrogging. *Journal of Development Studies*, 30(4): 831-58.

IMF 2001. Malaysia: From Crisis to Recovery: International Monetary Fund.

Japanese Chamber of Commerce. 2010. Survey of business sentiment on Japanese corporations in Thailand: for the spring, the 1st half of 2010: Japanese Chamber of Commerce in Bangkok.

Johnston, B. F. & Mellor, J. W. 1961. The Role of Agriculture in Economic Development. *The American Economic Review,* 51 (4) 566-593.

Julian, C. C. 2001. Japanese foreign direct investment in Thailand. *The Mid-Atlantic Journal of Business*, 37(1): 7-18.

Kerner, A. 2009. Why Should I Believe You? The Costs and Consequences of Bilateral Investment Treaties. *International Studies Quarterly*, 53: 73-102.

Kohpaiboon, A. 2003. Foreign Trade Regimes and the FDI-Growth Nexus: A Case Study of Thailand. *Journal of Development Studies*, 40(2): 55-81.

Kohpaiboon, A. 2006. Foreign Direct Investment and Technology Spillover: A Cross-Industry Analysis of Thai Manufacturing. *World Development,* 34(3): 541-556.

Lecraw, D. J. 1977. Direct investment by firms from less developed countries. *Oxford Economic Papers*, 29(4): 442-57.

Lecraw, D. J. 1992. Third world MNEs once again: the case of Indonesia. In Buckley, P. J. & M. Casson, editors, *Multinational enterprises in the world economy: essays in honour of John Dunning*. Hants: Edward Elgar Publishing Limited.

Lee, J. W. 2004. A new paradigm for the Korean economy: advanced state development (ASD) model approach. In Harvie, C., Lee, H. & J. Oh, eds. *The Korean economy: post-crisis policies, issues and prospects*. Cheltenham: Edward Elgar.

Martin, W. & Warr, P. 1994. Determinants of Agriculture's Relative Decline: Thailand. *Agricultural Economics,* 11(2): 219-235.

National Economic and Social Development Board 2008. *National Income of Thailand,* Bangkok, Office of the National Economic and Social Development Board.

National Food Institute of Thailand 2008a. *A forecast of Thailand's food export trend in 2008*. Bangkok: National Food Institute of Thailand.

National Food Institute of Thailand 2008b. *Thai food industry situation report*. Bangkok: National Food Institute of Thailand.

National Food Institute of Thailand 2010a. *Thai food industry situation report in 2009 and trend in 2010*. Bangkok: National Food Institute of Thailand.

National Food Institute of Thailand 2010b. *Thai food industry situation report (first four months, half year trend and overview of 2010)*. Bangkok: National Food Institute of Thailand.

Netayalak, P. 2008. Characteristics and Impact of Foreign Direct Investment on Thai Agricultural Production. *University of the Thai Chamber of Commerce Journal,* 28(1): 1-18.

Neumayer, E. & Spess, L. 2005. Do bilateral investment treaties increase foreign direct investment to developing countries? *World Development*, **33(10):** **1567**-85.

Nuannim, M. & Kaewpornsawan, S. 2010. "Bilateral Investment Treaties (BITs): implications and policies", Bank of Thailand.

Office of Agricultural Economics 2007. Economic and social conditions of farmer households and agricultural labor 2006-2007. Bangkok: Office of Agricultural Economics.

Office of Agricultural Economics 2009. Outlook on agriculture 2009 and trend in 2010. Bangkok: OfficeSricultural Economics.

Ohmae, K. 1985. *Triad power*. New York: The Free Press.

Pananond, P. 2007. The changing dynamics of Thai multinationals after the Asian economic crisis. *Journal of International Management*, 13: 356-75.

Peng, M. W. 2000. *Business strategies in transition economies*. Thousand Oaks: Sage Publications.

Poapongsakorn, N. 2006. The Decline and Recovery of Thai Agriculture: Causes, Responses, Prospects and Challenges. *Rapid Growth of selected Asian economies: lessons and implications for agriculture and food security.* Bangkok: FAO Regional Office for Asia and the Pacific.

Poapongsakorn, N., Siamwalla, A., Titapiwatanakun, B., Netayalak, P., Suzuki, P., Pookpakdi, A. & P. Preedasak. (1995) *Agricultural Diversification / Restructuring of Agricultural Production Systems in Thailand,* A Paper prepared for The Food and Agricultural Organization of the United Nations. Bangkok, Thailand Development Research Institute.

Porter, M. E. 1990. *Competitive advantage of nations*. New York: Free Press.

Rock, M. 1995. Thai Industrial Policy: How Irrelevant was It to Export Success? *Journal of International Development*, 7(5): 759-73.

Rugman, A. M. & Verbeke, A. 1990. *Global corporate strategy and trade policy*. London: Routledge.

Sattaphon, W. 2006. Do Japanese Foreign Direct Investment and Trade Stimulate Agricultural Growth in East Asia? Panel Cointegration Analysis. *International Association of Agricultural Economists Conference.* Gold Coast, Australia.

Siamwalla, A. 1996. Thai Agriculture: From Engine of Growth to Sunset Status. *TDRI Quarterly Review,* 11(4): 3-10.

Sim, A. B. 2006. Internationalization strategies of emerging Asian MNEs-case study evidence on Singaporean and Malaysian firms. *Asia Pacific Business Review*, 12(4): 487-505.

Suphannachart, W. & Warr, P. 2009. Research and Productivity in Thai Agriculture. EconPapers 2009-11: Departmental Working Papers from Australian National University, Economics, RSPAS.

Thai Ministry of Industry, 2002. *National master plan for Thai food industry*. Bangkok: Office of Industrial Economics, Thai Ministry of Industry.

Thailand's National Statistical Office 1997. **Report Of the** 1997 Industrial Census, Whole Kingdom: Thailand's National Statistical Office, Office of the Prime Minister.

Thailand's National Statistical Office 2007. **Report of the** 2007 Industrial Census, Whole Kingdom: Thailand's National Statistical Office, Office of the Prime Minister.

Urata, S. & Yokota, K. 1994. Trade liberalization and productivity growth in Thailand, *The Developing Economies*, 32(4): 444-459.

Vernon, R. 1966. International investment and international trade in the product cycle. *Quarterly Journal of Economics*, 80(2): 190-207.

Vernon, R. 1979. The product cycle hypothesis in a new international environment. *Oxford Bulletin of Economics and Statistics*, 41(4): 255-67.

Warr, P. 2004. Globalization, Growth, and Poverty Reduction in Thailand. *ASEAN Economic Bulletin,* 21(1): 1.

Warr, P. 2005. Boom, Bust and Beyond. *In:* Warr, P. G. (ed.) *Thailand Beyond the Crisis.* New York: Routledge Curzon.

Warr, P. 2006. Productivity Growth in Thailand and Indonesia: How Agriculture Contributes to Economic Growth. *Working Paper in Economics and Development Studies*. Department of Economics. Padjadjaran University.

Warr, P. 2008. Trade Policy and the Structure of Incentives in Thai Agriculture. *ASEAN Economic Bulletin,* 25(3): 249-70.

Warr, P. 2009. Aggregate and Sectoral Productivity Growth in Thailand and Indonesia. *Working Papers in Trade and Development, 2009/10 Arndt-Corden Division of Economics.* The Australian National University.

Wells, L. T. 1983. Third world multinationals: the rise of foreign investment from developing countries. Cambridge: MIT Press.

World Bank. 2008. Thailand Investment Climate Assessment Update: Poverty Reduction and Economic Management Sector Unit, The World Bank.

Uganda:

Analysis of private investment in the coffee, flowers and fish sectors of Uganda[1]

1. Introduction

In Uganda, like other African countries, foreign investment in commercial agriculture though growing since 2000, is still relatively low. Most of the companies engaged in commercial agriculture – about 70 percent of the total – are domestic-owned. This is also illustrated by the small number of planned projects in the sector that were registered by the Uganda Investment Authority (UIA) between 1992 and 2008. A total of 124 projects were registered in the sector and they account for just 3.5 percent of all projects registered by the Authority. About half of the registered agricultural projects were in four subsectors: fish, general farming, flowers, and forestry. The majority of the planned foreign projects in commercial agriculture were from investors from three countries: India (21 percent), United Kingdom (16 percent) and Kenya (10 percent).

FDI flow to commercial agriculture are concentrated in: the supply of agricultural chemicals and fertilizers; coffee processing and export; floriculture; and fish processing and export. Nevertheless, the data on the largest taxpayers during 2005/2006 do not show a dominance of foreign-owned companies in the agricultural sector in general. The total number of companies ranked among Uganda's top 50 taxpayers is evenly split between domestic and foreign-owned. Foreign-owned companies in coffee and flowers had a lower value of assets but higher sales than their domestic counterparts in 2007.

This chapter focuses on private investment in three value chains based on their importance for the Ugandan economy (in terms of export earnings). These are (i) coffee – the main export commodity, (ii) fish – the main non-traditional export commodity; and (iii) flowers and cuttings – among the top three non-traditional export commodities in 2007. The first and second are fish and maize, respectively. Maize will not be analysed as an export commodity because there is no foreign-owned companies involved in this subsector. Most of the maize produced is sold by domestic enterprises to the World Food Programme (WFP).

2. Foreign direct investment flows in agriculture

Foreign direct investment inflows into Uganda have been on an upward trend since the 1990s, from US$25 million in 1991[2] to US$2.2 billion in 2007 (UIA, 2008). Similarly, the total value of planned foreign projects registered by the Uganda Investment Authority (UIA) increased by 14.6 percent per annum from US$270.5 million (1992) to US$2.38 billion (2008), in line with the increased number of registered planned projects. A total of 3 513 foreign-owned projects were registered by the UIA between 1992 and 2008. The UIA's mandate includes maintenance of a database of all foreign projects.

The value of FDI is highly correlated with the value of planned projects registered by the UIA between 2000 and 2007. The correlation coefficient between the value of FDI and the value of planned projects registered by the UIA in the period 2000 to 2007 is 0.79 (Figure 1).

[1] This chapter is based on a research report produced for FAO by Alice K. Gowa, Consultant.

[2] Obwona, Marios V., 1996: 8, as cited in the UIA Database from July 1991 to December 1995.

FIGURE 1
Value of FDI and projects registered by UIA in Uganda, 2000–2007

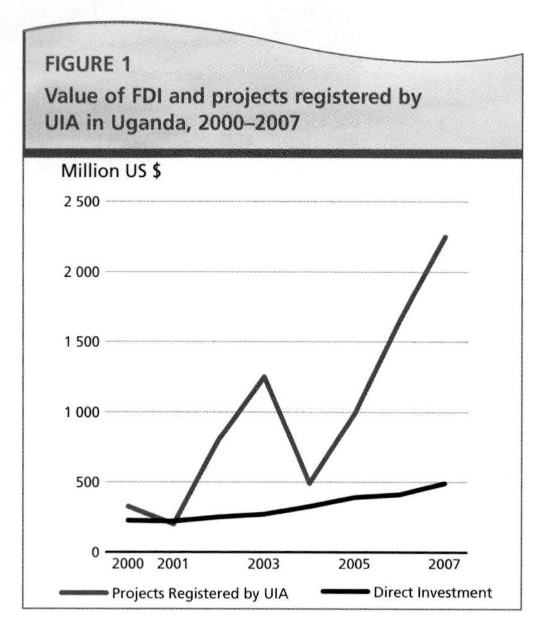

Source: 2008 Statistical Abstract; UIA Database.

FIGURE 2
Number of planned foreign investments in agriculture in Uganda, 1992–2008

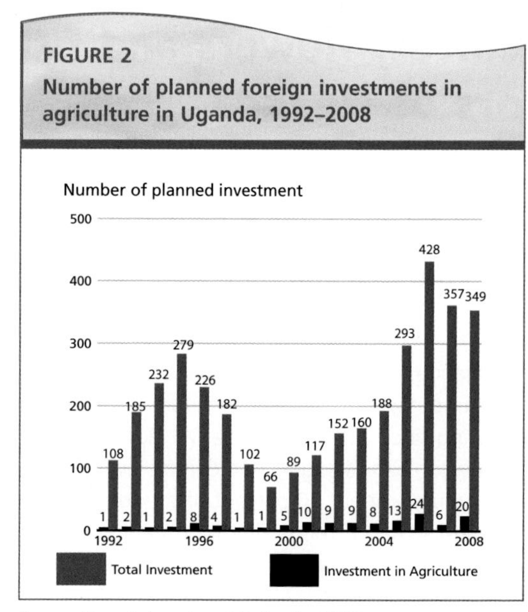

Source: Uganda Investment Authority, 2009

Nevertheless, it was noted that the values of actual FDI for a given year were much lower than that of the planned projects for that year. The range was 16 percent at the lower end, and 81 percent at the higher end. This is expected as not all the planned projects are implemented and, in some cases their value may be either under-estimated or over-estimated. Our analysis shows that the average value of FDI is 40 percent of the value of planned foreign investment. The median value is 32 percent.

The latest published data on foreign direct investment in agriculture were for the year 2000; they show that the market value of FDI in agriculture, forestry and fishing was very low at US$406 548 or 0.06 percent of total FDI stocks (Uganda: Bank of Uganda, 2002). The sector, however, has attracted increasing investment since 2000. The number of planned projects registered by the UIA between 2000 and 2008 was 104 (total value: US$238 154 846), compared with a total of just 20 projects (total value: US$39 039 500) registered in the sector between 1992 and 1999 (Figure 2). Based on the comparison between total FDI and the value of total planned foreign investments discussed in section I.2.1, we estimate the total value of foreign-owned investments in agriculture from 2000 to 2008 to be between US$77.3 million and US$100 million[3]. In the period 1991 to 2008, the value of planned investments in agriculture ranged from US$0.8 million to US$55 million, and was less than 20 percent of the value of all projects (Figure 3).

Over the period 1992 to 2008, the number of projects in the agriculture have accounted for less than 10 percent of all planned foreign-owned projects. Since its establishment in 1991, the Authority has registered a total of only 124 planned, foreign-owned projects in commercial agriculture, which had a total value of US$277 million.

About half of the registered projects were in four subsectors: fish (22 percent); general farming (14 percent); flowers (7.8 percent) and forestry (7.8 percent), (Figure 4). The majority

[3] The value of 32.4 percent (which is the median of value FDI as a percentage of value of planned investments for a given year between 2000 and 2008) was multiplied by the total value of planned investments from 2000 to 2008 to obtain the value of US$77.3 million. In order to obtain the value of US$100 million we used the average of value of FDI as a percentage of the value of planned investments for a given year between 2000 and 2008 of 42.1 percent.

FIGURE 3
Value of planned foreign investments in agriculture in Uganda, 1992–2008

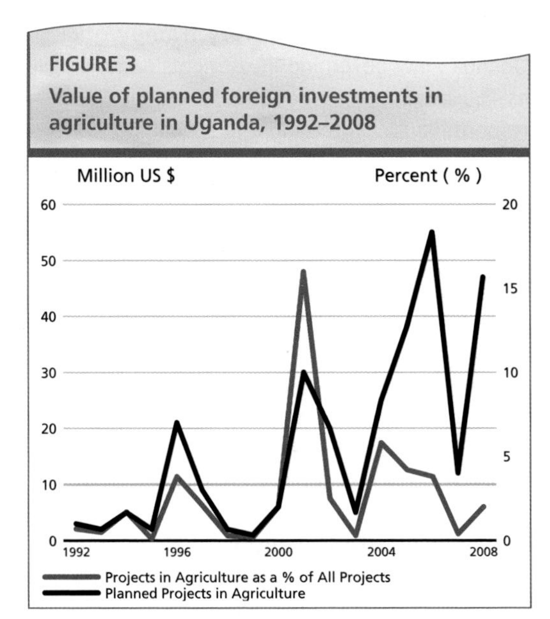

Source: Uganda Investment Authority, 2009

of the planned foreign projects were from three countries: India (21 percent), United Kingdom (16 percent) and Kenya (10 percent) as shown in Figure 5.

3. Policies to encourage private investment in the agricultural sector

Uganda's agricultural sector has undergone major policy reforms over the past two decades. The reforms centred on economic liberalization and privatization of public enterprises with the aim of promoting private sector participation in the development process. Previously, the government controlled the agricultural sector by setting prices and establishing marketing boards that were engaged in buying commodities from smallholder farmers, and selling them abroad. However, this system proved ineffective in running the agricultural sector, prompting the government to implement structural changes.

Currently, Uganda's overall development policy framework is the Poverty Eradication Action Plan (PEAP) that was introduced in 1997 and revised in 1999 and 2004, respectively. The PEAP has five pillars that were identified as the key areas to steer Uganda's development agenda. These are: 1) Economic management; 2) Enhancing production, competitiveness and incomes; 3) Security, conflict resolution and

FIGURE 4
Agricultural sectors of planned FDI in Uganda, 1992–2008

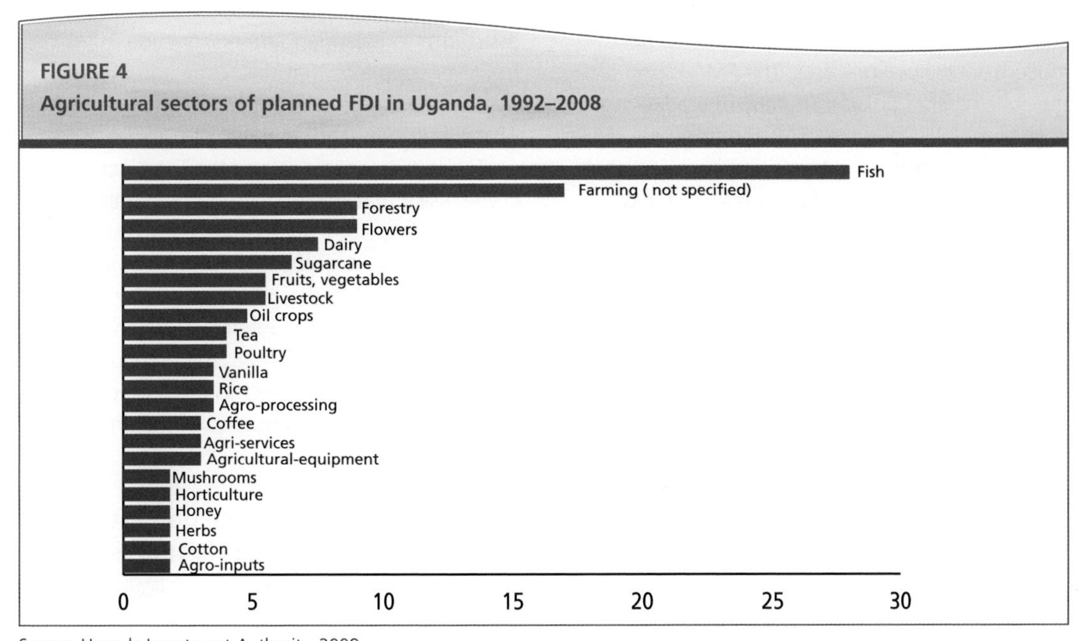

Source: Uganda Investment Authority, 2009

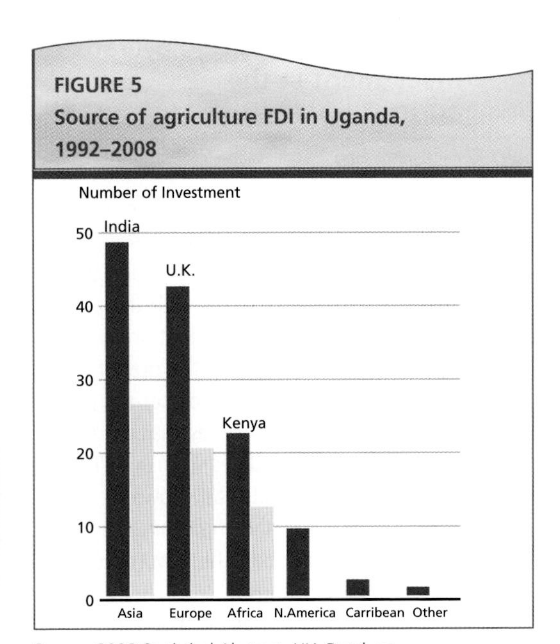

FIGURE 5
Source of agriculture FDI in Uganda, 1992–2008

Source: 2008 Statistical Abstract; UIA Database.

disaster management; 4) Good governance; and 5) Human development. The agricultural sector falls under pillar two, which focuses on improving the livelihood of farmers by supporting them, and increasing their incomes through agriculture.

In 2000, the Government of Uganda introduced the Plan for the Modernization of Agriculture (PMA) – a strategic policy seeking to transform the lives of poor farmers through introducing modern agricultural practices. The PMA is part of the government's broader strategy of eradicating poverty as outlined in the PEAP and its overall objective is to increase incomes, improve household food security, provide gainful employment and promote the sustainable use and management of natural resources (Uganda, MAAIF, 2006).

The government policy on agriculture is aimed at increasing household incomes to at least UShs 20 million per annum in the short and medium term (Uganda, MFPED, (2008). However, government expenditure on the agriculture sector is small, relative to expenditure on other sectors[4].

[4] The Government of Uganda allocated 3.2 percent of its total expenditure in the 2006/07 financial year to agriculture, compared to 13.4 percent and 10.7 percent to Security and Public Administration respectively. (Uganda, MFPED, 2007: 23)

Institutional and regulatory framework

Uganda's law governing investments is known as The Investment Code (2000)[1], and the body responsible for promoting and facilitating investment in the country is the Uganda Investment Authority (UIA). The Investment Code does not have provisions for TNC participation in Uganda but instead provides a broad regulatory framework for both local and foreign investors.

Foreign investors are required to have a minimum of US$500 000 in planned investments in order to secure an investment license from UIA, while local investors require US$50 000 (Uganda, UIA 2000). The TNCs operating in Uganda are regulated by the Companies Act (Cap 85).

The land tenure situation

The land tenure system in Uganda includes a mix of traditional practice, colonial regulations and post-colonial legislation. There are four main forms of land tenure systems in Uganda namely: customary, mailo, freehold and leasehold tenure (World Bank, 1993).

The Investment Code discourages foreign investors from owning land in Uganda. Part II, Section 10(2) of the Investment Code Act (2000) bars any foreign investor from engaging in crop or animal production, or to be granted lease land for the purpose of agricultural production. However, the code permits the investor to provide assistance to Ugandan farmers. Section 10(4) however stipulates that this restriction may be overlooked by the relevant Minister, on the advice of the Authority and approval of Cabinet.

The Investment Code and the Land Act (1998) state that foreign investors can only hold leasehold land titles; leases can be for up to 99 years (Uganda, UIA, 2001). According to section 42 of the Land Act, the government may acquire land from citizens in the interest of defence, public safety or public use. However, this compulsory acquisition of land is subject to the prompt payment and fair compensation of the affected people by the government.

3.1 Policies to enhance domestic capabilities and safeguards

Uganda does not have a specific policy targeting the participation of TNCs in agriculture but instead

TABLE 1
Investment incentives, Uganda

Incentive	Description
Investment Capital Allowance	Initial allowance on plant and machinery (50-75 percent)
	Start-up cost spread over 4 years (25 percent p.a.)
	Scientific research expenditure (100 percent)
	Training expenditure (100 percent)
	Mineral exploration expenditure (100 percent)
	Initial allowance on hotels, hospitals and industrial buildings
	Deductible annual allowances (depreciable assets)
	Depreciation rates of assets range (20-40 percent)
	Depreciation rates for hotels, industrial buildings and hospitals (5 percent)
VAT Refunds	Investors who register as investment traders are entitled to VAT refund on building materials for industrial/commercial buildings
Duty and tax free import of plant and machinery	
First Arrival Privileges (FAPs)	FAPs in the form of duty exemptions for personal effects and motor vehicles (previously owned for at least 12 months) to all investors and expatriates coming to Uganda
Export Promotion Incentives and facilities	Manufacturing under bond
	Duty exemption on plant and machinery and other inputs
	Stamp duty exemption
	Duty drawback – a refund of all or part of any duty paid on materials, inputs imported to produce for export
	Withholding tax exemption on plant and machinery, scholastic materials, human and animal drugs and raw materials
	Ten year tax holiday – duty remission scheme for exporters involved in value addition

Source: Uganda Investment Authority (UIA)

focuses on creating an enabling environment for all private sector players to engage in agricultural production. In view of this, the government set up measures to control the quality of agricultural products and the enforcement of quality standards, in order to ensure food safety and compliance with international standards.

Under the PMA Policy, the government identified seven priority areas that would create domestic capabilities and maximize benefits from the participation of TNCs. The areas include: agricultural research and technology development; agricultural advisory services; rural financial services; agricultural education; agricultural marketing and agroprocessing; sustainable natural resources management; and supportive physical infrastructure. These key areas are to be implemented through the coordination of the various relevant ministries.

Incentives offered to encourage private investment in agricultural sectors

Uganda has an open investment climate with regard to foreign investment. The Ugandan Government has in place a number of specific incentives for investors (both foreign and Ugandan). The main criterion for investors to benefit from these incentives is a minimum initial capital investment of US$500 000 for foreign investors and US$50 000 for Ugandan investors. The investment incentives that apply to all investment are provided in Table 1.

Uganda has a number of obligations under international law that are relevant to the fisheries

TABLE 2
Summary of Uganda's international obligations for the fishing sector

Name of obligation	Details/objectives
The Convention on Biological Diversity	Objective is to develop national strategies, plans or programmes for the conservation and sustainable use of biological diversity.
The Treaty for the Establishment of the East African Community	**Objectives**: • To jointly and efficiently manage the natural resources within the community • To adopt common regulations for the protection of shared aquatic and terrestrial resources.
The Ramsar Convention	• Signed in Ramsar, Iran in 1971. • The main objective is to provide the framework for national action and international cooperation for the conservation and wise use of wetlands and their resources.
The Convention on International Trade in Endangered Species	• Regulates the international wildlife trade worth billions of dollars annually. • The signatory countries act by banning commercial international trade in an agreed list of endangered species and by regulating and monitoring trade in others that might become endangered.
Technical Corporation for the Promotion of the Development and Environmental Protection of the Nile Basin (Tecconile) 1992	• Established by the Ministers of Water Affairs in the Nile Basin. There are ten signatory countries: Burundi, Democratic Republic of the Congo, Egypt, Ethiopia, Eritrea, Kenya, Rwanda, Sudan, United Republic of Tanzania and Uganda • Main objective is to provide for cooperation in the sustainable development, conservation and joint use of the River Nile's waters.
Convention for the Establishment of the Lake Victoria Fisheries Organization 1994	• Adopted by Kenya, Uganda and United Republic of Tanzania. • Objectives of the convention are to foster cooperation among the parties; harmonize national measures for the sustainable utilization of the living resources of Lake Victoria; and to develop and adopt conservation and management measures.
The FAO Code of Conduct for Responsible Fisheries 1995	• Adopted at the 28th Session of the FAO Conference in October 1995. • Provides principles and standards applicable to the conservation, management and development of fisheries. • Covers the capture, processing and trade in fish and fish products, fishing operations, aquaculture and fisheries research.

Source: Uganda Investment Authority (UIA)

sector and the National Fisheries Policy, which have been summarized in Table 2.

Uganda is a signatory to the International Treaty on Plant Genetic Resources for Food and Agriculture, commonly known as the International Seed Treaty. This treaty aims at guaranteeing food security through the conservation, exchange and sustainable use of the world's plant genetic resources for food and agriculture. Some of the issues that have made member countries raise concerns about the treaty include: 1) the extent to which farmers and communities will be allowed to freely use, exchange, sell and breed the seeds; and 2) what enforcement procedures will be used by national governments to ensure that principles of farmers' rights are respected. Some critics of the treaty state that its provisions are either still unresolved, or even open to interpretation.

4. Investments by transnational corporations in the agricultural sector

According to UNCTAD, a transnational corporation (TNC) is generally regarded as an enterprise comprising entities in more than one country, which operate under a system of decision-making that permits coherent policies and a common strategy. The entities are so linked, by ownership or otherwise, that one or more of them may be able to exercise a significant influence over the others and, in particular, to share knowledge, resources and responsibilities with the others.[5] According to data from the Uganda Bureau of Statistics, 168 enterprises

[5] http://www.unctad.org

TABLE 3
Summary of enterprises in commercial agriculture and related services, 2008

No.	Sub-sector	Foreign-owned/TNCs	Domestic	Total
A	Agricultural Chemicals, Fertilizers and Irrigation	3	7	10
B	Agricultural Engineering, Equipment and Services	19	29	48
C.	Agricultural Seeds	3	5	8
D.	Beekeeping, Equipment Manufacturing and Trainers	0	5	5
E.	Coffee Processing and Export	6	26	32
F.	Cotton Exporters	3	1	4
G.	Cotton Ginning Machinery	2	0	2
H.	Cotton Lint	1	10	9
I.	Fish Farms	0	1	1
J.	Fish Processors and Exporters	3	19	22
K.	Fishing Equipment and Supplies	0	2	2
L.	Floriculture and Flower Exporters	11	10	21
M.	Grain Millers	2	6	8
N.	Hides and Skins	0	2	2
O.	Milk and Dairy Products	2	4	6
P.	Poultry Hatcheries and Poultry Feeds	0	3	3
Q.	Rice Growers, Dealers and Exporters	1	1	2
R.	Sugar Manufacturers	1	3	4
S.	Tobacco Processing and Export	1	1	2
T.	Tea Processing	1	2	3
U.	Edible Oil Processing	1	2	3
	Total	**60**	**139**	**199**
	Percentage	**30**	**70**	**100**

Sources: company websites, Uganda Flower Exporters Association (UFEA); Uganda Fish Processors Association (UFPA); and Uganda Coffee Development Authority (UCDA), The *Monitor* Directory 2008.

employing 5 or more persons were engaged in commercial agriculture (excluding livestock agriculture) and fishing during 2006/7[6].

A review of the listing of the firms engaged in agriculture or related activities, summarized in Table 3, shows that three out of every ten companies are foreign-owned, including TNCs. Given the limited availability of data on asset holdings of companies engaged in agricultural production, there is a tendency to assume that most of the commercial agricultural enterprises are foreign-owned. However, some are owned by Ugandans of Asian ethnicity[7]. Information on the ownership of the enterprises engaged in agricultural production was collected from company websites and interviews with the respective associations: the Uganda Flower Exporters Association (UFEA); the Uganda Fish Processors Association (UFPA); and the Uganda Coffee Development Authority (UCDA). Enterprises whose existence or ownership could not be confirmed have been excluded. The listing estimates about 200 enterprises including

[6] Uganda: Uganda Bureau of Statistics (2007): Report on the Uganda Business Register 2006/7 :16, 19. The number of enterprises engaged in commercial agriculture (excluding livestock agriculture) has been calculated as 0.44*382 (total number of enterprises in commercial agriculture including livestock rearing). The number of enterprises engaged in fish processing and export has been calculated to including all private limited liability companies (3 percent of all companies); all partnerships (3 percent); and other companies (2 percent). The total number of enterprises engaged in fishing and related services was 124 during 2006/7.

[7] Uganda Investment Climate Report 2004.

66 enterprises providing agro-inputs (seeds, fertilizers, machinery) .

Most foreign-owned enterprises, including TNCs, have invested in agricultural engineering, equipment and services; floriculture and flower exports; and coffee processing and export. Investment by foreign-owned enterprises, including TNCs as a percentage of total investment, is concentrated in the following areas: cotton ginning and export (75 percent of enterprises); floriculture (52 percent); rice growing and export; and tobacco processing and export (1 of 2 enterprises in each subsector); agricultural seeds (38 percent); supply of agricultural chemicals and fertilizers (33 percent); and milk and dairy products (33 percent of enterprises). Information on the actual contribution of the TNCs in terms of percentage of total contribution in each subsector is not available for most sectors. We can therefore make a reasonable assumption from the information presented in Table 3, that ownership of enterprises in agriculture is predominantly in the hands of Ugandans.

5. Value chain of selected commodities: Coffee, flowers and fish

A value chain is a supply chain consisting of the input suppliers, producers, processors and buyers that bring a product from its conception to its end use.[8] The value chains of the three selected commodities are outlined in the following sections:

- **Coffee subsector**

In Uganda, coffee is a smallholder crop cultivated on small farms with an average size of 0.2 hectares. Two varieties of coffee are cultivated: Robusta accounts for about 85 percent of coffee cultivated and Arabica accounts for 15 percent. It is the main source of income for about 500 000 rural households (Sayer, 2002). Uganda's Robusta coffee is considered one of the best varieties in the world (Uganda, National Exports Strategy 2007). Arabica coffee is also an important variety, with good harvests in the Mt. Elgon area and fetching relatively high farmgate prices (DANIDA, Agriculture Sector Programme Support (ASPS II) Annual Progress Report 2005/2006).

Since the early 1990s, the coffee sector's main challenge has been the coffee wilt disease. This disease is estimated to have affected about 55 percent of the total area planted with Robusta coffee trees (Uganda, UCDA, 2005). A replanting programme, which is replacing affected coffee trees with wilt-resistant varieties, and ageing trees, has gradually reversed years of declining production. For example, between October 2007 and September 2008, 3.2 million bags of coffee worth US$362 million were produced compared to 2.5 million bags worth US$238 million in October 2006 – September 2007 (Uganda, UCDA 2009).

Coffee processing and export Uganda[9]

The ripe coffee fruits (cherries) go through a number of operations aimed at extracting the beans from their covering of pulp, mucilage, parchment and film, to improve their appearance. The resulting clean coffee bean of fairly average quality (FAQ) can then be roasted and ground to obtain the coffee powder fit for human consumption. Two main techniques are employed in Uganda to obtain clean coffee beans. Wet processing is applied to the choice Arabica coffees produced at high altitudes in the Mount Elgon areas in the East, the Highland areas of Nebbi in the North and the mountainous areas of Kisoro and Rukungiri in the Southwest. The coffees so produced are generally described as 'mild'. Dry processing produces coffees that are described as 'hard'. These are mainly the Robustas grown around the Lake Victoria basin. The wet processed (washed) coffees are generally superior to the dry processed in terms of physical appearance and cup taste.

[8] Dempsey Jim, Campbell Ruth. A Value Chain Approach to Coffee Production; Linking Ethiopian Coffee Producers to International Markets: http://www.acdivoca.org/852571DC00681414/Lookup/WRSpring06-Page5-7-ValueChainCoffee/$file/WRSpring06-Page5-7-ValueChainCoffee.pdf

[9] http://ugandacoffee.org: Primary and Secondary Processing.

Over 95 percent of the total annual coffee production is exported as green beans; just 5 percent of the coffee is processed locally. Secondary processing – also known as export grading – transforms the clean coffee (FAQ) into the various coffee grades that meet international standards. According to the UCDA, there are about 28 coffee exporting companies, 19 export grading factories, 251 primary processing mills and 9 roasters (Uganda: UCDA Annual Report, 2006/07: 7). Uganda has one only one coffee processing company.

During the 2006/2007 coffee season, the average farmgate price for dried Robusta coffee cherries was about US$0.59 per kilo, while the price of clean coffee (FAQ) was double, at about US$1.18 per kilo. The export price for Robusta coffee rose from an average of US$1.7 per kilo in October 2006 to US$2.0 per kilo in September 2007, and more than 50 percent of the price of FAQ.[10] On a price basis alone, locally, the greatest value is added during the processing of FAQ.

Uganda has a total seven foreign-owned enterprises and eighteen locally owned enterprises engaged in coffee buying, processing and export. Six TNCs are among the largest coffee processors and exporters, including: ED&F Man Holdings Limited, United Kingdom; Olam International Limited, Singapore; Ecom Agroindustrial Corporation Limited, Switzerland; Sucafina S.A., Switzerland; Neumann Gruppe GmbH, Germany; and Great Lakes Coffee Company Uganda Limited owned by two Greek nationals. These TNCs accounted for about 59 percent of coffee exports in the 2008/9 season.

Eleven of the 16 major coffee processing and export companies, including all the TNCs, started exporting in the 1990s, following the liberalization of the coffee sector in 1991. Four of the five companies that started exporting after 2000 were domestic-owned companies. In part, this suggests that there are no 'crowding out' effects by TNCs in the sector. Indeed, domestic companies that started exporting after 2000 were able to take a sizeable share of the market, and accounted for 25 percent of coffee exports in 2008/9.

Whereas TNCs that entered the coffee business early (following liberalization) had first-mover advantages with respect to market shares, in that they commanded a larger share of the total market for coffee than the later entrants, the domestic-owned companies that entered the coffee business more recently – after 2000 – had a larger market share than similar companies that started exporting in the 1990s. This could be attributable to the recent entrants having more resources and also having benefited from studying the operating practices of the TNCs and their domestic-owned counterparts who entered the market before them.

There is also evidence that TNCs are becoming more involved in the lower end of the value chain by establishing demonstration farms and providing training support to subsistence farmers, in order to have more reliable supply of coffee and to manage product quality. Examples of demonstration farms include Project Nakanyonyi by Kyagalanyi Coffee Ltd (established in 2007), Kaweri Coffee Plantation Ltd (established by Ibero (U) Ltd in 2001) and demonstration farms started by Ugacof Ltd.

- **Flowers and cuttings subsector**

Uganda's floriculture sector was established in 1992. At the time, the main commodity produced was rose flowers. By 1998, there were 18 companies engaged in production and export of roses; ten of these companies have since closed.[11]

[10] Uganda: UCDA Annual Report: 2006/07: 7, 11); The exchange rate applied was US$1: UGX 1 721, which was the bureau weighted average selling rate in 2007 (Uganda: 2008 Statistical Abstract: 223).

[11] In 1998, the following companies were producing roses: Equatorial Flowers, Harvest International, Horizon Roses, Jambo Roses, Kajjansi Roses, Mairye Estates, Melissa Flowers, MK Flora, NBA Roses, Nile Roses, Nsimbe Estates, Pearl Flowers, Royal Flowers, Scoul Roses, Tropical Flowers, UgaRose, Van Zanten (U), Victoria Flowers, and Zziwa Horticultural Exporters (Djikstra T, 2001). In 2008, the following companies had closed: Equatorial Flowers, Harvest International, Horizon Roses, MK Flora, NBA Roses, Nile Roses, Nsimbe Estates, Royal Flowers, Scoul Roses, Tropical Flowers, UgaRose, and Zziwa Horticultural Exporters.

BOX 1

TNCs in the coffee processing and export sector in Uganda, 2009 *

Kyagalanyi Coffee Ltd (www.volcafe.com), is a subsidiary of ED & F Man Holdings Limited, United Kingdom, which bought VOLCAFE in 2004. Kyagalanyi was the first coffee exporter to be certified under the ISO:9001:2000. The enterprise is also certified and verified under the OQS, a member of the Australian member of the International Certification Network - IQNET. In 2007, Kyagalanyi started Project Nakanyonyi to train farmers on improved farming practices in order for them to receive better prices for their produce.

ED & F Man is the market leader in procurement and preparation of green coffee. The company has operations in 21 countries worldwide.

Olam (U) Ltd (www.olamonline.com), a subsidiary of Olam International Ltd in Singapore. Olam International specializes in 17 agricultural products. The enterprise's strategy is to manage each activity in the supply chain, from origination to processing, logistics, marketing and distribution. This has allowed for operational efficiencies, and value addition. Olam Uganda has its head office in Kampala and its procurement/distribution units are spread over the entire country. The first product on Olam Uganda's portfolio was Robusta coffee. Subsequently Arabica coffee, cotton, sesame, rice and sugar were added on to its products. Olam has invested in a state-of- the-art coffee processing facility in Kampala.

Kawacom (U) Ltd (www.kawacom.com), a subsidiary of Swiss based ECOM Agroindustrial Corp. Ltd Kawacom was established in 1996. The enterprise spearheaded the development of the first organic coffee farm in the country, and was the first exporter of organic coffee. Four buying centres. Kawacom currently operates processing mills and one central processing mill for the preparation of export coffee. By procuring the coffee directly from the source, Kawacom can offer guaranteed quality and timely delivery to its buyers. Since 1998, Kawacom has expanded its trading business from the better known Ugandan Robusta into washed Arabicas.

At the time of writing, the company is developing three organic coffee projects in partnership with small farmers. Two of these projects focus on washed Arabicas, and the third on Robusta – the first organic Robusta from Africa.

ECOM is among the leading supply chain managers in the world and an integrated supplier of both raw and semi-processed agricultural commodities

UGACOF Ltd (www.ugacof.com) is a subsidiary of Sucafina, a Swiss based enterprise. UGACOF has been in the coffee business since 1994, exporting both coffee and cocoa. The enterprise is also engaged in transport and shipping services through its sister company UGATRANS. To improve on the quality and yield of the coffee, UGACOF installed demonstration plots and developed training sessions for farmers.

Ibero (U) Ltd (www.nkg.net) is a subsidiary of Neumann Kaffee Group. In addition to Ibero (U), Neumann Kaffee Group (NKG) operates Kaweri Coffee Plantation Limited in the Mubende district. This is a large-scale Robusta coffee farm, established in 2001 as part of the NKG farming strategy.

NKG operates an arm's length relationship with its 40 subsidiaries. Each subsidiary is run as its own profit centre within the Neumann Gruppe GmbH, the holding company of Neumann Kaffee Gruppe. Neumann Gruppe GmbH is located in Hamburg, and directs and coordinates all activities of the group.

Great Lakes Coffee Company Ltd was established in 1999 by two Greek nationals with a 50:50 shareholding. The company is engaged in coffee buying, processing and export. Great Lakes sells about half of its total production (57 percent in 2008) on the domestic market to Kawacom, Olam and Kyagalanyi Coffee Ltd and exports the rest, mostly to customers in Italy, the United Kingdom, Germany and Switzerland.

* This information was primarily obtained from the companies' websites

CHART 1
Uganda coffee value chain

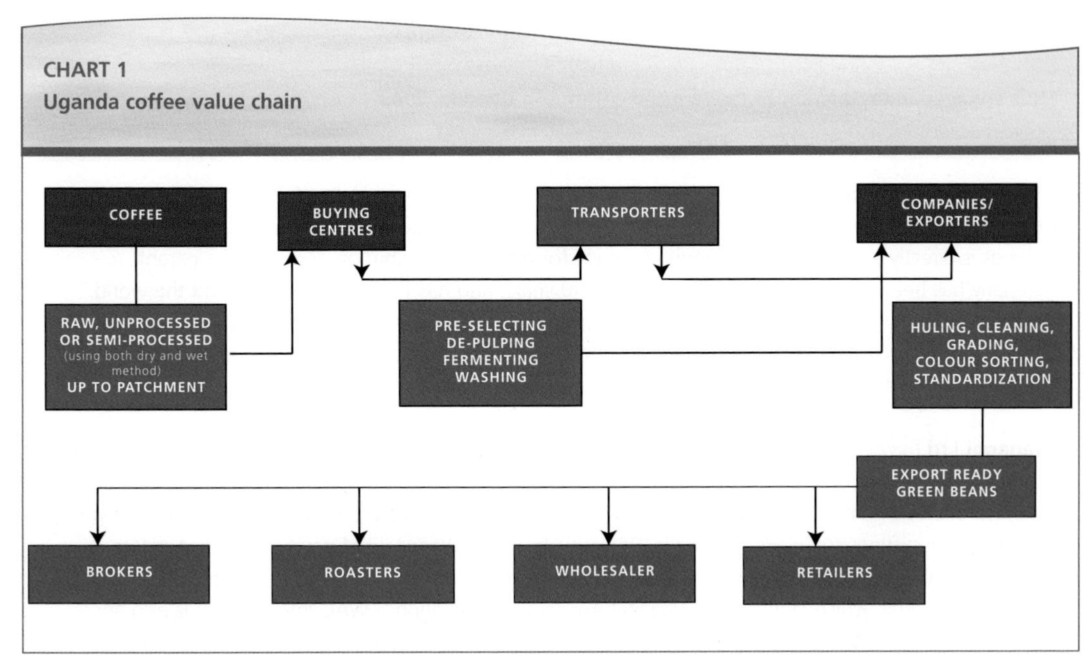

Source: Uganda National Export Strategy 2007.

Cultivation of rose flowers in Uganda is predominantly undertaken by majority Ugandan-owned companies. In all, 14 out of 20 companies (or 70 percent of the total companies) produce roses. These companies comprise Ugandan-owned and foreign-owned companies but do not include companies owned by Dutch investors. The three largest exporters of roses: Rosebud Limited, UgaRose Limited and Jambo Roses Limited, are Ugandan-owned companies.

All the five companies established by investors from Holland, which boasts expertise in flower production, produce plant varieties other than roses (mostly chrysanthemum cuttings). One Ugandan-owned company (Chrysanthemum Cuttings Ltd), which was established in 2007, has ventured into the production of chrysanthemum cuttings. This company has the same ownership as Kajjansi Roses Ltd, which was established earlier and cultivates roses.

Flowers and cuttings have emerged as major, non-traditional export commodities for Uganda, with an estimated value of US$22.8 million in 2007, making these products the fourth largest non-traditional export commodities after fish, gold and maize (Uganda: Statistical Abstract 2008). Floriculture exports are dominated by cut flowers (virtually all cut roses), and chrysanthemum cuttings (Uganda, UFEA 2007).

There are three types of roses currently grown in Uganda: T-hybrids (long stem, big flower heads), sweethearts (short stem, small flower heads) and floribundas (intermediate). The sweetheart rose variety is most suitable for Uganda's warm, humid climate. Trials with chrysanthemum cuttings started in 1995, through joint ventures with Dutch companies, and very high yields of cuttings under Ugandan conditions were indicated. Indeed chrysanthemums grow very well in Uganda's climatic conditions (Uganda, UFEA 2007).

Currently, the flower and cuttings sector comprises 20 enterprises covering more than 200 hectares of land and producing over 40 varieties of flowers (Uganda, UFEA 2008). The sector has grown considerably over the last eight years, at an average annual rate of 20 percent (Uganda, UEPB 2007). In 2007, total investment in the sector, both local and foreign, was estimated at over US$60 million (Uganda, UIA 2007). The flower sector has also emerged as an important non-traditional export earner and a major employer. About 6 500 persons (mostly women) are employed in the flower industry or 325

BOX 2

TNCs engaged in production of flowers and cuttings in Uganda, 2009

Fiduga (Uganda) Ltd (www.fides.nl) is a subsidiary of FIDES BV Group, Holland. It started as a trial farm on 2,500 m2 in 1996. Currently the farm stands on 20 hectares. Fiduga's production of chrysanthemum cuttings is directly exported to the parent company for growing, distribution and sale. The parent company has been in the floriculture business for 40 years, and has five subsidiaries around the world. Each year FIDES BV Group introduces new varieties of chrysanthemums, which after considerable testing, are introduced into subsidiary countries for propagating and production of cuttings. Fiduga is currently Uganda's largest exporter of chrysanthemum cuttings.

Wagagai Ltd (www.wac-international.com) is a Ugandan-based enterprise, and Dutch owned. It has been in the floriculture business since 2001. In the last five years, the enterprise diversified into the cuttings business. The farm is on 22 hectares and supplies chrysanthemum cuttings to Deliflor in the Netherlands and pot plant cuttings to Selecta Klemm in Germany. In 2002, Wagagai Ltd partnered with Agricom, a breeding company based in Holland, to allow for easy production and breeding. This partnership is called WAC International, which stands for "Wagagai Agricom Combination". WAC international is also an agent of Delforge in East Africa and Kenya. WAC's strategy is to introduce varieties of chrysanthemums slowly and on a small scale into the market. Wagagai Ltd is currently the second largest exporter of chrysanthemums from Uganda.

Royal Van Zanten Uganda (www.royalvanzanten.com) is a subsidiary of Royal Van Zanten, Holland. The enterprise has been operating in Uganda for the last 12 years, and is the third largest exporter of chrysanthemums. Royal Van Zanten, Holland operates nine subsidiaries worldwide, and has been in the floriculture business for the last 160 years. It has a modern and advanced department where the latest techniques are used to research and develop improvements to current plant types and varieties. Royal Van Zanten Uganda exports its production directly to the parent company for growing, distribution and sale. The enterprise's current arrangement with the parent company is that key decisions are made at the local level

employees per company, on average. Eighty-five percent of Uganda's companies have less than 50 employees (Uganda: Uganda Business Register 2006/07:88)

The value chain shows that the subsidiary (or partner) in Uganda principally propagates the chrysanthemum cuttings and exports the shoots or cuttings back to the parent company or partner for growing rooting material (or flowers). The processes of plant breeding and flower growing (from shoots) are performed outside Uganda.

Within the flower and cuttings sector there is evidence of a shift – driven by the TNCs – to cultivating flower varieties other than roses. Most of the flower companies established between 1994 and 1999 concentrate on rose production (9 out of 11 companies, or 81 percent of total companies established). Just two companies established within the period, both Dutch-owned, are engaged in the production of other flower varieties, specifically chrysanthemum cuttings. This trend, however, changed after 2000. Nine flower companies were established between 2000 and 2008. Four of these companies (or 44 percent of total companies established during the period), are producing chrysanthemums and other plant varieties, e.g. kalanchoe cuttings, bedding plants, pot plant cuttings and vegetables, among others. The proportion of the companies producing roses reduced to 56 percent of total companies established.

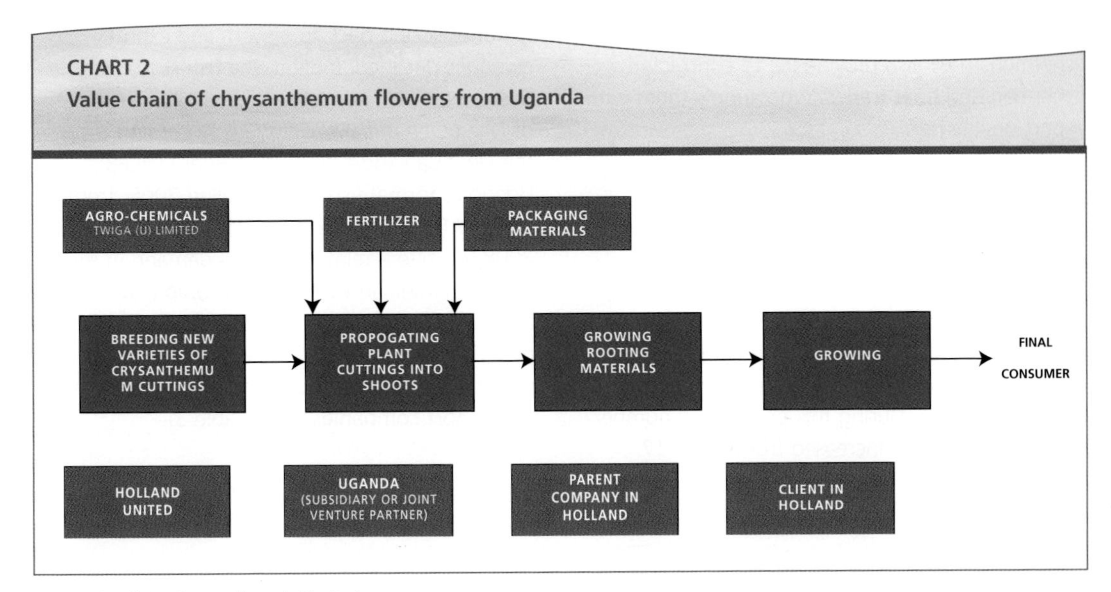

Source: Royal Van Zanten Uganda Limited.

The shift suggests that the failure of many rose growing companies during the 1990s provided an important lesson for later investors, who are now concentrating on producing plant varieties that are better suited to Uganda's climate. Uganda's warm, humid climate[12] is very favourable to the cultivation of chrysanthemum cuttings and other plant varieties. Most rose varieties, however, thrive better in cooler climates, e.g. in the highlands of Kenya and Ethiopia. There is no evidence of crowding effects in the flowers and cuttings sector.

- **Fish subsector**

The fish value chain consists of five players: the primary producers (the fishermen), the fish collection boats (wooden and motorized), the fish transporters (traders and factory agents), the local traders and processors, and the regional and international exporters (Diagram 3). Fish processors do not operate fishing boats but can purchase fish from fishermen or middle men.

The fishermen, who were estimated at about 136 000 in 1997 (National Fisheries Policy for Uganda, May 2004) – land their catch on various landing sites on three major lakes: Lake Victoria, Lake Kyoga and Lake Albert. Lake Victoria is estimated to have about 600 landing sites (Uganda, Fisheries Resources Research Institute, 2003). At this stage, the processors provide ice (from their own icemaking plants) to fish collection boats to preserve the fish until it reaches the landing sites. The number of fish collection boats was estimated at 960 in 2003 (National Fisheries Policy for Uganda, May 2004). At the landing sites, fish is sold to traders and suppliers, who in turn supply the fish processors or domestic traders. The number of fish traders was estimated at about 20 000 in 2003 (Uganda, Fisheries Resources Research Institute).

At the processing plants, frozen fish products for export are packed in corrugated carton boxes, which are sourced locally, while chilled fish products for export are packed in styrofoam boxes, which are also sourced locally. By-products are also processed and exported; they include the swim bladders, visceral fat and skins. It is expected that most of the value is created during fish processing.

Along the value chain, fish factories are involved in the provision of inputs (packaging materials), technical support and quality control (implementing the EU and USA Hazard Analysis Critical Control Point System), processing and

[12] Temperatures range from a maximum of 28 °C during daytime, down to around 18 °C at night (UFEA, 2007).

exporting of fish, marketing and training of fishermen. They also provide ice to contracted fishermen and boat traders who supply them with export-quality fish.

Fish processing and export

The fish processing and export sector comprises 17 factories and employs over 800 000 Ugandans, directly and indirectly (Uganda, UFPEA, 2008). Between 2001 and 2005, the sector registered its highest growth, with export earnings increasing from US$87 million to US$143.4 million. During this period, the number of fish factories also increased from 11 to 17. In 2007, however, the number of operational factories scaled back to eleven, and export revenues declined. Indeed, the fish sector has been negatively impacted by dwindling stocks of Nile perch in Lake Victoria.[13] According to the Uganda Export Promotion Board (UEPB), Uganda's formal fish exports fell in 2005, from US$143.6 million in 2005 to US$112.2 million in 2008, despite a relatively stable demand in the export markets in the European Union. This was attributed largely to a decline in stocks of Nile perch as a result of overfishing by the fishing industry. The current number of fish processing and export companies is 22 (Table 3).

[13] Interview with Greenfields (U) Ltd.

BOX 3
Major fish processing and export companies in Uganda, 2009

Marine & Agro Export Processing Ltd (www.marineandagro.com) is the leading fish processing and exporting company in Uganda. The enterprise is affiliated with Kendag Ltd, in Nairobi, Kenya, which operates six processing plants. Marine & Agro Export Processing Ltd has been in fish processing and exporting business for more than 20 years. Presently, the enterprise operates 5 processing plants in Uganda and exports to more than 20 countries worldwide.

Uganda Fish Packers (www.alphauganda.com) is a subsidiary of Alpha Group, a multinational company, which has been operating in sub-Saharan Africa and the Gulf countries for the last 50 years. The enterprise is the second largest fish processing and exporting company in Uganda, with an installed capacity of 6 000 metric tonnes of fish fillet. Uganda Uganda Fish Packers owns nine processing plants in Uganda, United Republic of Tanzania and Kenya.

Hwan Sung (U) Ltd (www.hwangsungbiz.com) is owned by Korean nationals and the third largest fish processing and exporting company in the country. The enterprise has been in the fish processing and exporting business for close to 20 years. It has invested heavily in technology, with a capacity to store up to 390 tonnes of frozen fillets. Hwan Sung is also the leading supplier of various sizes of Styrofoam boxes that are used for packing fish, fruits, vegetables and flowers.

Greenfields (U) Ltd was established in 1989 and is owned by two Belgian nationals (95 percent shareholding and one Ugandan (5 percent shareholding). The enterprise is strategically located in Entebbe, along the shores of Lake Victoria, which allows for easy access to fish from landing sites and water. Greenfields processes Nile perch and Tilapia and exports over 4 000 tonnes of fish annually. More recently, the company in partnership with the Lake Harvest Group, Luxembourg established the Source of the Nile Fish Farm Ltd (SON). The enterprise is pilot testing a commercial fish farm to meet the increasing demand for fish.

Source: company websites and field interviews

CHART 3
Value chain for fish processing in Uganda

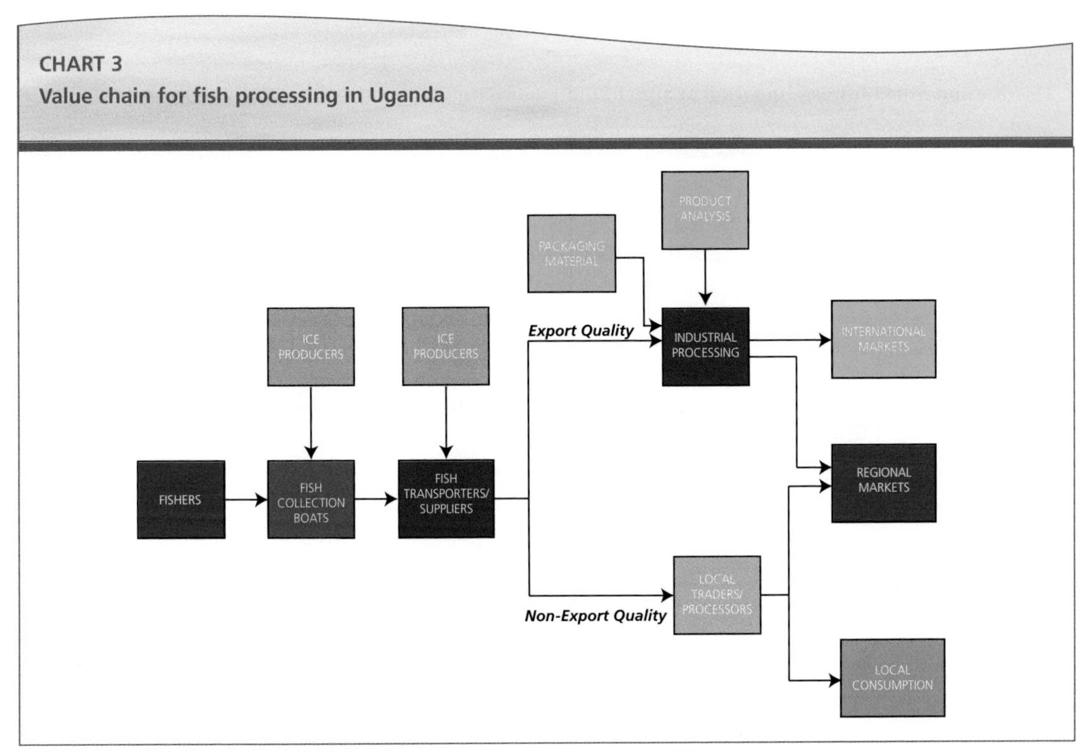

Sources: Nyombi K, Bolwig S (2003); Uganda: Uganda National Exports Strategy, 2007.

Of the three fish factories that started operating in the 1980s, the two domestic-owned factories (Gomba Fish Industries Ltd and Ngege Ltd) have recently closed due to poor financial management (Ngege Ltd) and the depleting fish stocks (Gomba Fish Industries Ltd).

Eleven of the operational fish factories were established after 2000. The new companies in the sector are established to maximize the fish catch. For example, seven fish processing factories were established in 2005, a year before the quantity of fish exported started declining. Five of these companies have one owner, who established each company on a different landing site to maximize the raw material catch.

Surprisingly, however, even with the dwindling fish stocks, the sector is attracting new investment: three new companies were established between 2007 and 2009. These companies are Wildcatch Fisheries Ltd, established in 2007 and fishing from Lake Albert; Lake Bounty Ltd, established in 2008 and using premises rented from Ngege Ltd and IFTRA (U) Ltd, established in 2009 and using premises rented from Gomba Fisheries Ltd Furthermore, it is notable that some foreign companies in this sector have diversified into other sectors that are not related to the fish value chain. For example, Hwan Sung (U) Ltd, a Korean TNC, also engages in the manufacturing of furniture, while the Alpha Group of Companies (Riyaz Kurji) produces meat and dairy products.

The latest entrant in this sector is the Source of the Nile Fish Farm Ltd (SON). The enterprise is pilot testing a commercial fish farm to meet the increasing demand for fish, which is partly attributable to the reducing fish stocks (Uganda, UIA, 2007). SON is jointly owned by the Lake Harvest Group, from Luxembourg and Greenfields, a Belgian owned enterprise, based in Uganda.

BOX 4
Distribution of enterprises engaged in agriculture among Uganda's 1 000 largest taxpayers in 2005/2006

The data on the largest taxpayers during 2005/2006 do not show dominance of foreign-owned companies in the agricultural sector. For example, the total number of companies ranked among Uganda's top 50 taxpayers shows almost double the amount of domestic companies compared to foreign-owned companies. Three foreign-owned companies were ranked among Uganda's top 50 taxpayers, compared with five domestic-owned companies. These companies were engaged in producing and processing of tobacco, sugar and edible oil.

Distribution of enterprises engaged in agriculture among Uganda's 1 000 largest taxpayers in 2005/2006

Ranking	Number of companies	
	Foreign-owned	Domestic-owned
1-10	1	0
11-20	1	1
21-30	0	0
31-50	1	4
51-100	1	0
101-500	12	6
501 -1000	9	9
Total	25	20

Source: Uganda Investment Authority

5.1 Transnational Corporations (TNCs) in agricultural production in Uganda

Uganda had 25 large foreign-owned companies[14] engaged in commercial agriculture. These companies paid total taxes of at least US$90 000 during 2005/2006[15]. The largest tax paying companies were engaged in tobacco processing (BAT (U) Ltd); sugar processing (Kinyara Sugar Works Limited) and edible oil processing (Bidco (U) Ltd). These three companies were ranked among Uganda's 50 largest taxpayers in 2005/2006 (Table 6). The data also show that 20 domestic-owned companies in agriculture and related activities were among Uganda's top taxpayers.

Most of the largest foreign-owned companies were concentrated in produce farming, processing and export of various products. The majority of TNCs were in coffee processing and export (five companies); fish processing and export (three companies); chrysanthemum growing and export (two companies) and the supermarket business (two companies). Foreign-owned companies also performed the following functions in the agricultural value chain: input supply; sale of agricultural produce on the domestic market; and testing of agroproducts, e.g. fish.

Activities of foreign affiliates

The foreign-owned companies in the coffee, flowers and fish sectors operate as limited liability

[14] Wagagai was recorded twice as Wagagai Chrysanthemum Ltd and Wagagai Ltd The company has been counted once for this study.

[15] Metro Cash and Carry Limited has since closed its operations in Uganda.

companies and fully owned subsidiaries of the parent companies. Field interviews also show that these companies were established using financing from the parent company (as a loan or equity). Most of the companies interviewed were unwilling to provide information on assets and sales. The asset value of foreign-owned companies is not always higher than for comparable domestic-owned companies (see example of the coffee and flower sectors). Nevertheless, foreign-owned companies showed more efficient asset utilization (about 80 percent higher), in both sectors; generating more sales from their assets (and indeed had higher sales values) in 2007. All the companies interviewed were export-oriented.

There is no clear information to identify the proportion of foreign affiliates established by different types of foreign parents, including sovereign wealth funds and private equity funds, for the commercial agriculture sector.

Main competitive advantages, motivations and strategies

Transnational corporations operating in Uganda have the following competitive advantages over their domestic-owned counterparts:

Access to affordable finance: Most of the TNCs interviewed were established using financing from the parent company, at affordable interest rates. For example, Fiduga obtained a loan from its parent company at an interest rate of 2 percent per annum, with no deadline for repayment. Royal Van Zanten (U) Ltd financed 60 percent of its start-up costs using a loan from the parent company. On the other hand, Melissa Flowers Ltd, a domestic-owned company, obtained a loan from a Ugandan-based bank at start-up, at an interest rate of 11 percent per annum and a repayment period of five years.

Access to management and technical expertise: The Dutch-owned companies obtain material for propagation from the plant breeding laboratories owned by their parent companies and have expatriate management. The TNCs in the coffee sector can readily source and hire international expertise in the sector. For example, Kyagalanyi Coffee Limited employs two specialists in washed coffee production from Colombia. The domestic-owned companies do not always have access in terms of contacts and financial resources, to hire similar expertise.

Ready market for commodities: The companies producing chrysanthemum cuttings directly supply their parent companies. The TNCs in coffee processing and export have the option to sell to their parent company or other buyers, if they are offering better terms than the parent company. Conversely, domestic-owned companies depend solely on international buyers.

High visibility: Uganda, through the UIA, has made tremendous efforts to attract FDI. A large, foreign-owned company planning to enter the market is therefore highly visible and could use this position to demand (and even receive) discretionary incentives. This situation has not occurred in the three sectors. However cases exist where foreign investors have seemingly received preferential treatment[16] (Kalema, W., Nsonzi, F. (2008) .

Motivation for investing in Uganda

Most TNCs chose to invest in Uganda primarily because of production factors including: ready availability of raw material (fish and coffee); excellent climate for production and availability of water (flowers and cuttings);. Other reasons included the liberalization of the coffee sector and availability of a low cost, easily trainable, English speaking workforce. The reasons provided for investing in Uganda are provided in Box 5.

[16] In 2004, the Government gave a comprehensive package of incentives, including a 25 –year holiday on income tax and a 17-year holiday on Value Added Tax, to encourage an investor, BIDCO (from Kenya) to establish a US$120 million palm oil project. Other edible oil producers complained, alleging unfair treatment. The BIDCO project has been very slow in its implementation.

Tri-Star Apparel, an investor in garment manufacturing targeting the United States market under the Africa Growth and Opportunity Act (AGOA), received US$15 million in Government guaranteed loans but closed with huge losses after five years, and failed to repay the loans.

BOX 5
Reasons for investing in coffee, flowers and fish sectors in Uganda, 2009

Coffee Processing and Export	Flowers and Cuttings	Fish Processing and Export
Uganda is a good volume producer of coffee and has a liberalized market. Multinationals were invited to invest n the coffee industry following liberalization in the 1990s.	Two novel features about Uganda are its climate and the availability of adequate water for farming. Uganda's climate is characterized by hot and humid conditions and all-year-round high temperatures, which are ideal for production of small budded (sweetheart) roses and chrysanthemums cuttings. *Preferential Market Access*: Uganda's floriculture exports benefit from preferential tariffs to the European markets. Products that are destined for the USA market, quota and duty free under the African Growth and Opportunity Act (AGOA) preferential trade arrangement.	Uganda has extensive fresh water resources. Half of Lake Victoria, the second largest fresh-water lake in the world, is located in Uganda. The country is also endowed with an additional 160 smaller lakes and a number of rivers, including the Nile, on which substantive fish harvesting and farming can be done (Uganda, UIA, 2007).
	In 1995, floriculture enterprises formed the Uganda Flower's Exporters Association (UFEA). Through this association, members supported the setting up of a cold storage facility, Fresh Handling Ltd, to efficiently handle horticultural products in cold storage and arrange for appropriate air transport. Currently, members pay a handling fee, which includes the use of cold stores, the professional fee and air freight charges. Fresh Handling is presently operating at full capacity, and plans are underway to expand it.	Lake Victoria is home to the Nile perch (Lates Niloticus), which is in high demand in Europe, and the wild tilapia (Oreochromis Niloticus). Uganda's bio-physical environment also favours warm water fish aquaculture. In 2002, the Food and Agriculture Organization (FAO) estimated that over 70 percent of districts in the country have the potential for aquaculture development. Uganda's other competitive advantages include; (i) the low cost of labour (lower than many other countries); (ii) the highly trained professionals in fisheries related fields; and (iii) low cost of raw fish; the price is lower in Uganda than in neighbouring countries.

Source: Field interviews.

FIGURE 6
Average number of employees in formal enterprises in Uganda, 2006/2007

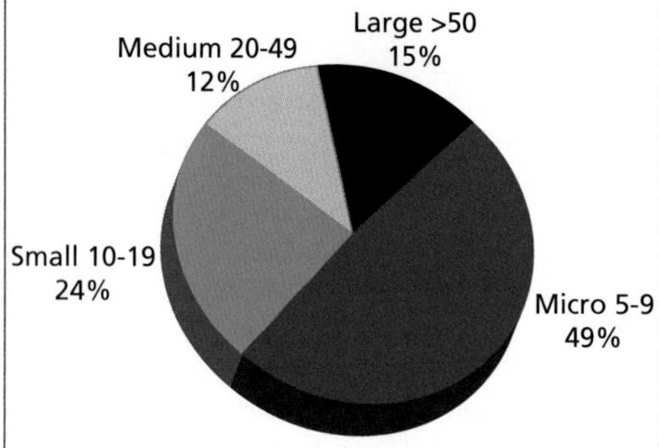

Large >50 15%
Medium 20-49 12%
Small 10-19 24%
Micro 5-9 49%

Source: Uganda Business Register, 2006/2007

Challenges

i. *Delays in operationalizing Investment Policies:* Uganda's investment incentive of a ten-year tax holiday that was introduced in the Budget Speech of 2007/8 is yet to be operationalized.

i. High production Costs: These are attributable to the high cost of electricity, the recent reintroduction of taxes on generator diesel and the high freight charges. Presently, air freight charges in Uganda range between US$1.9 per kg and US$2.2 per kg of product, compared with US$1.6 per kg of product in Kenya.

ii. In addition to the high freight charges, the fish processing sector faces risks from

TABLE 4
Employment in surveyed companies in commercial agriculture in Uganda, 2009

Commodity	No. of firms	Full-time		Part-time		Total		
		Average	Median	Average	Median	Average	Median	Sum
Coffee	3	52	25	243	279	295	304	884
Flowers and cuttings	5	147	180	217	240	324	350	1 620
Fish	3	96	37	92	85	187	122	562
Total	11							3 066

Source: Economic Survey, URT 2010

FIGURE 7
Trends in production of seven major crops in Uganda, 1991–2005

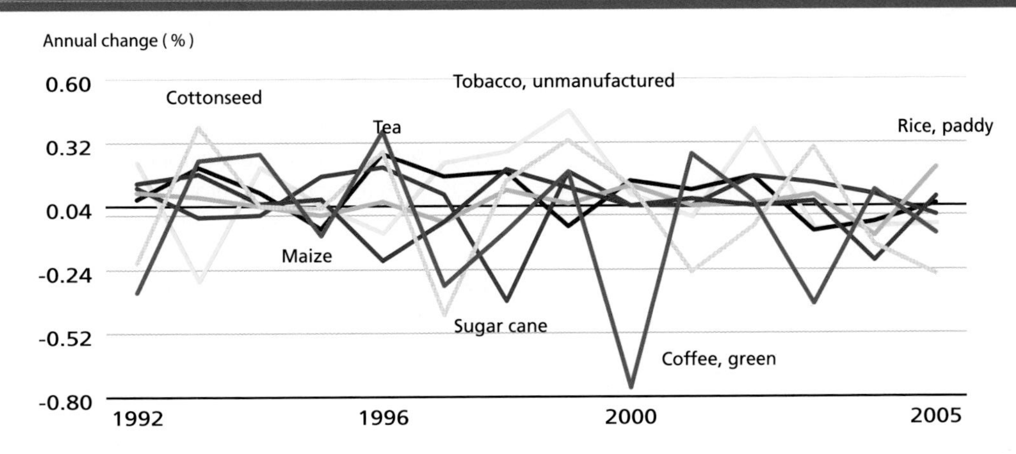

Source: FAOSTAT database, downloaded on 10 February 2009

(a) the decreasing fish stocks and (b) the inadequate budgetary support from the Government's Department of Fisheries, to enable it to effectively monitor landing sites.

6. Impact and implications of private investment in Ugandan agriculture

6.1 Impact on employment

Most of the companies in Uganda (73 percent) employ less than 20 people and are categorized as micro or small enterprises (Figure 6). The 11 surveyed firms employed a total of more than 3 000 employees, including part-time workers. Their average number of employees is more than 50 and therefore they are classified as large companies. These companies contribute considerably to employment in Uganda, especially those in the flower sector because of the high average number of employees per firm. There was no notable distinction in employment size between foreignowned and domestic companies.

6.2 Impacts on agricultural production in Uganda

Positive impacts

Transnational corporations such as Tilda Uganda Limited have contributed to increased food

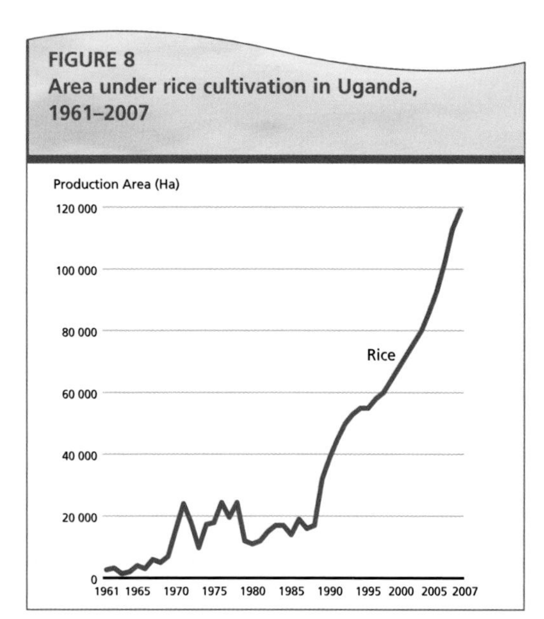

FIGURE 8
Area under rice cultivation in Uganda, 1961–2007

Source: FAOSTAT database, downloaded on 10 February 2009

production in Uganda. According to the FAOSTAT database, between 1991 and 2005, among seven major crops: coffee, maize, rice, cotton, sugarcane, tea and tobacco, rice was the only major crop to register a consistently positive increase in the production area (Figure 7). Tilda (U) Limited, a subsidiary of Tilda Limited, United Kingdom, contributed to sustaining this trend. After the company started operating in Uganda in 1997, the production area for rice increased from 60 000 hectares in 1997 to 119 000 hectares in 2007 (Figure 8).

Negative impacts

The considerable number of large fish processing companies has contributed to the high demand for Nile perch fish, which in turn has led to the depletion of Nile perch fish stocks in Lake Victoria. Although fish is a renewable resource, there should be mechanisms in place to ensure that this resource is continually replenished.

Uganda does not have export quotas for the fish sector. Fish is harvested mostly from Lake Victoria and fish stocks are replenished seasonally in line with natural fish breeding patterns. Therefore, overfishing (in that the quantity of fish harvested exceeds the new fish bred in a given season), or harvesting of immature fish are likely to result in a decrease in fish stocks.

In the fish export sector, between 2001 and 2004 the quantity of fish exported increased considerably to about 30 000 tonnes from an average volume of 15 000 tonnes in previous years, 1995–2000. The increased quantities of fish have exerted pressure on the existing fish exporters to either close operations or expand to other landing sites to maximize their catch, as discussed earlier. The volume of fish exports from Uganda declined from 36 614 tonnes in 2005 to 22 731 tonnes in 2008. It is important to note that fish companies are not directly engaged in fishing activities. Instead, they contract fishermen and other suppliers to supply fish to their processing plants.

6.3 Impact on agricultural exports

Uganda exported commodities valued at US$889.43 million in the financial year, 2005/2006 (Uganda: Background to the Budget 2007:18). Exports of the three selected commodities (coffee, flowers and fish), where TNCs are dominant, accounted for 39.1 percent of that total value.

FIGURE 9
Relationship between quantity of fish harvested and number of fish factories in Uganda, 1995–2008

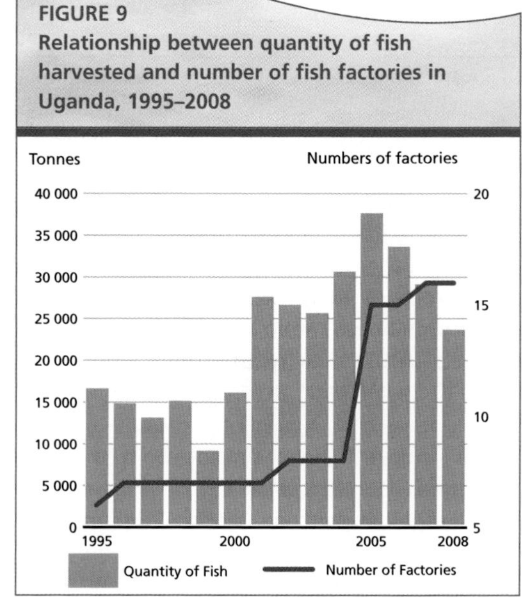

Source: Uganda Fish Processors and Exporters Association, 2009

TABLE 5
Sources of investment financing for selected TNCs operating in Uganda

Company	Source country	Percentage of funding sourced from :		
		Bank overseas/Parent company	Bank in Uganda	Other sources
			%	
IFTRA (U) Ltd	UAE	100		
Royal Van Zanten Ltd	Netherlands	60	40	
Great Lakes Coffee Ltd	Greece		95	5
Xclusive Cuttings Ltd	Netherlands	10		90
Kyagalanyi Coffee Ltd	Switzerland	100		
Fiduga Ltd	Netherlands	100		

Source: Field interviews, May 2009.

6.4 Impact on agricultural financing

Access to finance for smallholder farmers
Small farmers who work with TNCs usually have improved access to finance. As Box 7 illustrates, foreign-owned enterprises sometimes provide credit facilities to contract farmers or out-growers so that they do not need to obtain credit from financial institutions. The financing provided by the foreign-owned enterprises (including TNCs), is at a low interest rate and is usually tied to farmers' outputs. What the farmer borrows from the enterprise is deducted from earnings. Although some domestic-owned enterprises may also be providing credit financing to smallholder farmers, the authors could not find supporting information to this effect.

Impact on the domestic banking sector
Transnational corporations in the three sectors have limited impact on the domestic banking sector. Most of these companies source funding either from their parent company or from a bank overseas. Findings from the field interviews showed that only two companies, Royal Van Zanten and Great Lakes Coffee Ltd, obtained investment financing from a bank in Uganda (Table 6).

6.5 Impact on technology and knowledge sharing

Agricultural technologies include labour technologies: soil fertility management, crop protection, disease control, farm management, on-site storage, and non-labour technologies such as improved agricultural inputs (Uganda: Uganda National Household Survey 2005/06). The Uganda National Household Survey reports that the percentage of agricultural households that utilize labour technologies ranges from 7.1 percent to 23.2 percent (UBOS, 2005: 104). Further, only 7.3 percent of agricultural households reported that they were visited by an extension worker. Below, the authors highlight how the companies in the coffee and flowers sectors are utilizing agricultural technologies:

- The nature of the flower and cuttings sector requires a 100 percent utilization rate of labour and non-labour agricultural technologies to ensure profitability. Successful production of flowers and cuttings requires that the company ensures soil fertility management, crop protection, disease control, farm management, on-site storage, and utilizes agricultural inputs. For example, flower companies use fertilizers and agro-chemicals purchased from Balton (U) Ltd, a TNC, or Greenhouse Chemicals Ltd (agrochemicals only), a domestic company. They apply steam to the soil to ensure disease control, and are obliged to protect their crops by constructing greenhouses.

- Transnational corporations in the coffee sector are becoming involved in the lower

BOX 6
Introduction of upland rice varieties in Uganda

Prior to 2002, rice in Uganda was predominantly grown on paddy fields. This limited the production capacity of Uganda, an effect that was due to two reasons: low rice yields and a long maturing cycle of six months. Between 1997 and 1999, Tilda partnered with the West African Rice Development Association (WARDA)* in the field-testing of 30 upland rice varieties on Tilda farms. With further funding from USAID's Investment in Developing Export Agriculture (IDEA) project, Tilda trained field workers and farmers and established on-farm demonstrations in three additional districts in Uganda. In 2002, Uganda officially released two upland rice varieties from these activities - WAB 165 and WAB 460 (New Rice for Africa, Nerica 4), making the latter only the third NERICA variety to be released anywhere.

In the early years after its release, Tilda was the leading adopter of this new variety. However, the adoption of Nerica 4 by smallholder farmers also increased significantly. This increase was mainly due to two reasons: deliberate government promotion of upland rice to increase household income and food security, and the high rate by which Tilda was losing her highly trained employees who opted to become farmers themselves due to the high returns offered by upland rice production.

Source: WARDA. "The Africa Rice Centre – Recognizing WARDA's role in sub-Saharan Africa. WARDA Annual Report 2002/03 Features.http://warda.org/publicationsAR2002-03/recognizing%20warda%275%20role.pdf

* WARDA was renamed "WARDA – The African Rice Centre" in January 2003.

BOX 7
Interviews with farmers planting the Monsanto DEKLAB Hybrid Maize Variety

"I have been in the maize growing business for seven years and plant five acres of maize per season. I own a total of 12 acres. I plant the DEKLAB hybrid from Monsanto. Yields every season are between 2 and 3 tonnes per acre. I use about 20 kg of seed for an acre. A 5 kg bag of DEKLAB seed costs about US$2 (UGX 16 000).
In 2005, I tried to replant part of my harvest, because I did not have enough money to purchase seed from the stockist. The yields this season were lower by 30 percent. Consequently, I now buy the seed I need every season to maintain the high yields. The main challenge I have with this hybrid is that to have maximum yields, the rains have to be good and the soil well fertilized".

Richard Nusu, farmer in Jinja. Interviewed on 14 May 2009

"I have been in the maize growing business for eight years and plant about 20 acres. I buy seed every season from stockists, as efforts to replant my harvest (in previous years) produced no yield. I plant the DEKLAB Monsanto hybrid. To get yields as high as 3 tonnes per acre, the rains have to be very good as this breed is a heavy feeder. With this good weather, one maize cob can have up to 18 lines of seeds. I plant 20 kg of seed per acre. Another challenge I face is the counterfeit seeds on the market, which when planted yield nothing. Unfortunately, the counterfeit seeds are sold in the same packaging as the Monsanto, DEKLAB variety'.

John Wabwire, farmer in Mukono. Interviewed on 15 May 2009

BOX 8

Partnerships with EPOPA to introduce (a) Standards for Organic Coffee and (b) Sustainable Fisheries and Inspection Protocols

Introduction of Organic Coffee Production by Kawacom (U) Limited

Kawacom is a Ugandan based coffee exporting company, and member of the Ecom Agro Industrial Corporation, an international agribusiness enterprise. In 1998, Kawacom initiated organic coffee production in Uganda in conjunction with the Export Promotion of Organic Produce from Africa (EPOPA). The project was initiated in the coffee-growing district of Bushenyi in western Uganda. By 2002 after EPOPA had withdrawn their involvement, Kawacom independently started two other programmes in the Sipi and Paidha areas. These two areas have 13 000 coffee farmers. Kawacom trained farmers and field officers through the use of demonstration plots and nurseries (Tulip, 2005).

Source: Uganda, UCDA, 2006: 14

Greenfields (U) Limited – International Standards for Fisheries

In 2004, Greenfields constructed a fishing landing site in Nakasongola district on the shores of Lake Kyoga. The site was constructed in accordance with EU standards. Greenfields partnered with EPOPA in the training of fishermen on fish quality and standards. The objective of the construction project and the training was to comply with UgoCert standards of sustainable fisheries and inspection protocols.

Source: Beule (2008)

end of the value chain. They are increasing providing training to local farmers to ensure proper handling and storage of coffee. They also supply tarpaulins for improved coffee drying.

Research is also being undertaken currently by the private sector to produce new seed varieties. According to the Variety Release Committee of the Ministry of Agriculture, Animal Industry and Fisheries, a total of 41 seed varieties were released by the private sector between 2000 and 2008 (Mugoya, 2009). These varieties were for maize, rice, beans, soya beans, sorghum, barley, sunflower, cowpeas and sweet potatoes. They were developed to address specific production constraints including low yields, drought persistence, or pest and disease persistence. Two notable examples are: (i) the introduction of the upland rice variety (Nerica 4) into Uganda by Tilda (U) Limited, with support from USAID's Investment in Developing Export Agriculture (IDEA) project and the West African Rice Development Association (WARDA); and (ii) the introduction of the DEKLAB maize variety by Monsanto (U) Ltd.

Introduction of the NERICA 4 Rice Variety into Uganda: The introduction of the Nerica 4 upland rice variety, which was led by Tilda (U) Limited and the USAID IDEA project, has significantly increased the production of rice, even by smallholder farmers, and has contributed to Uganda's self-sufficiency in rice production (Boxes 6 and 7). Previously, most of the rice produced in Uganda was imported.

Introduction of the DEKLAB hybrid maize variety: Monsanto opened a branch in Uganda in March 2000. Currently, Monsanto (U) Ltd, deals in two products: DEKLAB Hybrid maize and vegetable seeds. The company mostly sells to distributors and suppliers and indeed, is the main supplier of maize hybrid throughout Uganda. The DEKLAB hybrid maize has higher yields than other maize seed brands on the Uganda market, e.g. Longe-5 (a maize hybrid developed by Uganda's Kawanda

CHART 4
Avenues for access to domestic and international markets

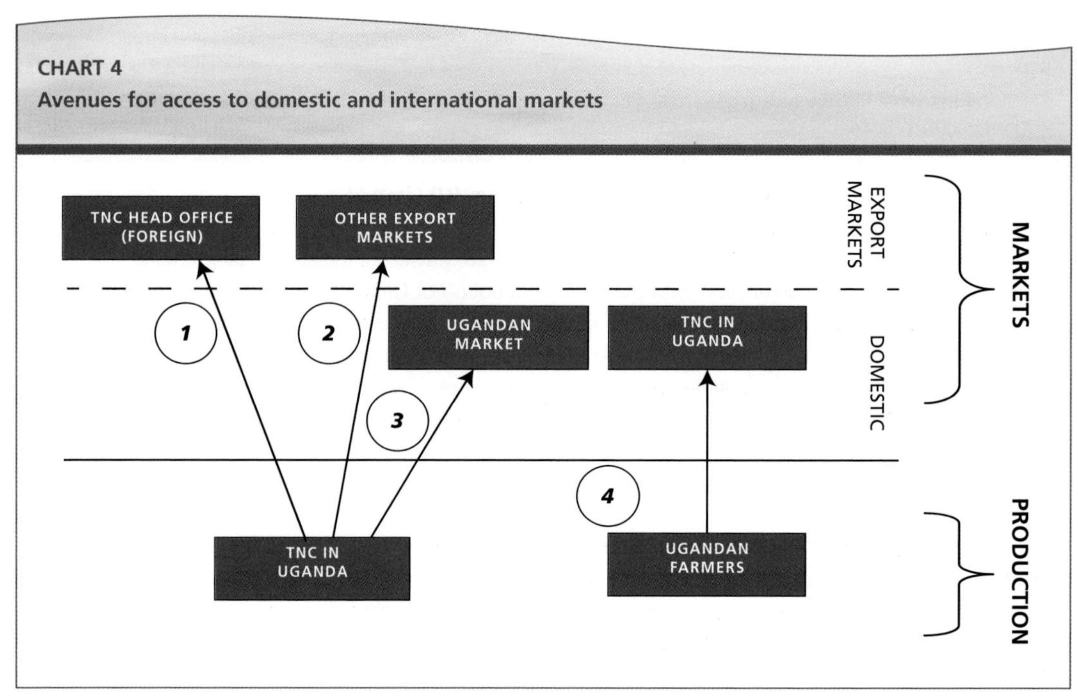

Source: Prepared by author

Research Institute).[17] Farmers reported that before the introduction of the DEKLAB hybrid maize, one acre of land could produce 200 kg of maize, from which farmers could reserve seeds for replanting for the next season. Presently, with the DEKLAB Hybrid and other hybrids on the market, one acre produces between 2 and 3 tonnes, an increase of up to 12-fold.

The introduction of the DEKLAB hybrid variety, however, has created dependence by farmers on Monsanto's seed. Farmers interviewed indicated that they need to purchase the Monsanto hybrid every season in order to have consistently high crop yields.

6.6 Enforcement of production and processing standards

Transnational corporations in the three sectors (coffee, flowers and fish) are export-oriented. Therefore, their activities must adhere to the various standards in their respective destination markets. Fish exporters adhere to EU fishing standards and protocols, flower firms adhere to MPS standards,[18] while coffee companies adhere to standards set by UCDA at the beginning of the coffee season. In addition, some of the exporting companies are ISO certified. Some TNCs, specifically Kawacom (U) Ltd, have also introduced production standards for specialized products (organic coffee) as presented in Box 8.

6.7 Investment in training

Training in agricultural production is critical to improving the existing skills of farmers. All the companies that were interviewed for this case study reported that they train their employees. The general training is conducted mainly for the low-level and mid-level workforce. It is provided on-the-job, and covers areas such as crop harvesting, general safety standards and cleanliness standards. Specialized training is provided to top level management in the different departments and includes modules such as ISO certification, production handling and quality control.

[17] Longe-5 yields about 16 bags (1.6 tonnes per acre), while Monsanto yields between 25 and 32 bags (2.5 - 3.2 tonnes) per acre.

[18] MPS is a Dutch audit company.

TABLE 6
Market destinations for selected exporters

Company	Sector	Avenue 1		Avenue 2	
		Destination	% of exports	Destination	% of exports
Wagagai Ltd	Flowers	Deliflor (Holland)	100		
Kyagalanyi Coffee Ltd	Coffee	Vocafe (Switzerland)	< 20	Various	
Royal Van Zanten Ltd	Flowers	Royal Van Zanten (Holland)	100		
Kawacom Ltd	Coffee	Ecom Industrial	100		
Xclusive Cuttings Ltd	Flowers	Floritech (Holland)	100		
Lake Bounty Ltd	Fish			EU, USA, UAE	100
Fiduga Ltd	Fish	FIDES BV	100		

Sources: Field interviews; "Uganda's Horticulture Veteran, Wagagai awarded" West African Business Week 25 February to 2 March 2008.

6.8 Market development

Transnational corporations are engaged in agricultural production in Uganda with the main objective of sourcing raw materials. This buyer-supplier arrangement is beneficial to both parties. For the supplier (TNC subsidiary in Uganda), the TNC is a ready market and a reliable source of income. For the buyer, there is a steady source of raw material.

6.9 Market access and exports

Positive impacts of market access

There are four different avenues through which markets for agricultural products can be accessed (Diagram 4). With all these avenues, TNCs either have a direct or indirect involvement in the production process. The TNCs in the flower sector are directly involved in agricultural production (FIDES BV, Royal Van Zanten). TNCs in the fish and coffee sectors mainly contract farmers/suppliers (Kawacom), or hire outgrowers.

Avenue 1: From Subsidiary in Uganda to Main Company

Several TNCs set up operations in Uganda in order to source raw materials for their operations abroad. In these cases, nearly 100 percent of the TNC's production output in Uganda is exported directly to the parent company. Examples of this avenue are in chrysanthemum cuttings: FIDUGA (U) Ltd exports chrysanthemum cuttings solely to FIDES BV Holland; Royal Van Zanten exports cuttings to Royal Van Zanten Holland; and coffee, where Kawacom exports most of its coffee to Ecom Agroindustrial Corporation Ltd.

Avenue 2: From TNCs in Uganda to other export markets

As the TNCs expand their production capacities in Uganda, they search for new markets to either absorb their increased output, or to increase their

FIGURE 10
Uganda coffee exports in 2006/2007 and 2007/2008

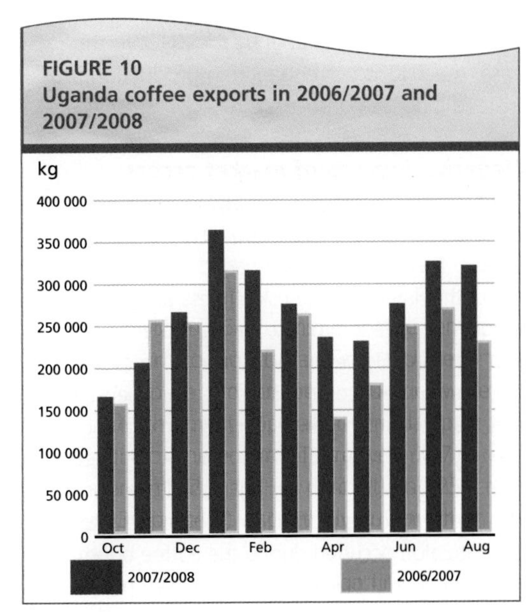

Source: UCDA Annual Report 2006/07

regional/international presence. This avenue is employed by Kyagalanyi Coffee Ltd and IFTRA (U) Ltd.

Avenue 3: From TNCs in Uganda to the Ugandan market

The third avenue pertains to TNCs that engage in agricultural production in Uganda and then sell their output on the Ugandan market. Most of the companies in the selected sectors are exclusively export-oriented. From the field interviews, only three out of eight companies sell on the domestic market. Not surprisingly, domestic sales are a very small portion of total sales (less than 1 percent for the flower companies, and less than 5 percent for the fish processing company). The TNCs in other sectors that sell on the Ugandan market are mainly motivated by the potential for import substitution and usually receive strong government support, for example, Tilda (U) Ltd in the rice sector.

Avenue 4: From local farmers direct to TNC in Uganda

Transnational corporations in Uganda serve as a market for farmers' agricultural produce. Farmers produce under contract and then supply to the TNC after harvest. Such arrangements usually require the farmers to undergo training to produce the specific type of product that is required as a raw material by the TNC. In this case, the TNC is not engaged in production, but instead induces production.

Negative impacts of market access

Over-reliance on one commodity: Markets induced by TNC involvement could also have negative effects. There are two main negative aspects that are both related to price risk. Access to markets induces farmers to produce the marketable commodities. However, there may be an over-reliance on these commodities, at the expense of other agricultural products. This over-reliance can be risky: when the price of the commodity drops, farmers are at risk of incurring heavy losses, and are discouraged from producing in subsequent seasons. This situation occurred during the coffee boom of 1994/95 during which farmers increased production, and the subsequent decline in prices resulted in

FIGURE 11
Relationship between number of coffee processing factories and FAQ price for coffee in Uganda, 1991–2007

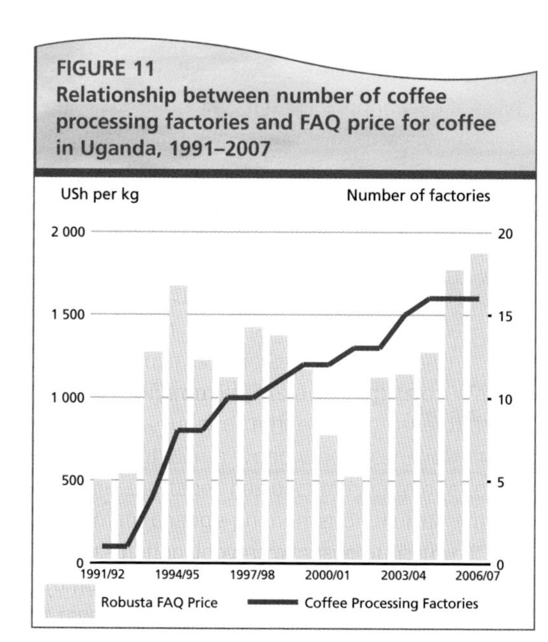

Source: UCDA (2003); UCDA (2007); UCDA (2008: 11)

the neglect of coffee farms and contributed to the spread of the coffee wilt disease[19].

Impact of the global economic crisis: Although the long-term impacts of the global financial crisis on commercial agriculture in Uganda are not yet known, there is evidence suggesting that some sectors are being affected even in the short term. For example, since 2008, sales in the flowers and cuttings sector have dropped by nearly 20 percent. However, it is important to recognize that the industry was already going through a series of economic problems such as increased production costs – especially freight costs (increasing by 15 per cent in three years) – and increased competition from neighbouring Kenya and Ethiopia. Flower exporters in both countries incur lower freight charges of US$1.67 per kg, compared to US$2.20 per kg for flowers from Uganda going to the same destination.

Conversely, according to the East African Fine Coffee Association (EAFCA), the coffee industry, thus far, appears to have escaped the effects of the economic crisis. Indeed, overall coffee exports in the 2007/2008 coffee season were

[19] Interview with Kyagalanyi Coffee Limited.

BOX 9
Community services provided by selected TNCs in Uganda, 2009

Sector	Number of firms	Community activities
Flowers	41*	Construction of schools
		Construction on water points like boreholes and wells
		HIV/AIDS prevention and counselling services
		Construction of football pitch
		Construction of clinic
		Erection of power lines
		General medical services, e.g. malaria treatment, family planning
		Construction of toilets
Coffee	1	Micro finance services

Sources: Field interviews; "Uganda's Horticulture Veteran, Wagagai awarded" East African Business Week 25 February to 2 March 2008
* Each flower company is engaged in at least four different community activities

higher than in the 2006/2007 season (Figure 10). There are two likely reasons for this occurrence: first, Uganda's coffee exports account for a small share of global coffee trade, only 2.3 percent in the 2006/07 season (Uganda, UCDA, 2008: 12). Second, there is increased emphasis on the export of organic and washed coffees to niche markets.

6.10 Increased competition

The participation of TNCs has increased the level of competition in the commodity market. Since the selected sector are export-oriented, the impact is at the supply level, and not consumer level. Increased competition has both positive and negative effects on the industry.

Positive impact: Increased demand leading to increased farmgate prices

Increased competition for raw materials has driven up the commodity prices, to the benefit of farmers. In the coffee industry, exporters compete for high quality coffee, which leads to increased prices for the farmers.

The fall in the world coffee price in the 2000/01 and 2001/02 coffee seasons led to a subsequent fall in the price paid to local farmers. In February 2002, the price at the local mills ranged from US$0.22 – US$0.24 per kg. The following month, the world price was on the road to recovery, and increased by 30 percent. The price paid in Uganda, however, increased appreciably more: by over 60 percent, to US$0.36 – US$0.39 per kg. This sharp increase was attributed to increased competition for two reasons: low supply from farmers and the need for the coffee exporting companies to fulfil their contractual obligations with their international buyers (Sayer, 2002:9).

The data also show that the increased competition by the coffee sector as a whole has enhanced the bargaining power of farmers. This effect, however, was more evident in the period starting 2001/2002, where there was a strong positive correlation between the number of coffee processing factories and the FAQ Price for coffee, of 0.81. Starting in 2001, three large domestic companies joined the coffee sector. However, it was difficult to isolate the contributions of domestic and foreign-owned companies on the impacts of the increased bargaining power of farmers. There was little correlation between the FAQ Price and quantity exported (correlation coefficient of 0.27), or the FAQ Price and the international price for Robusta coffee (correlation coefficient of 0.16)

CHART 5
Industry linkages in the flower, fish and coffee sectors

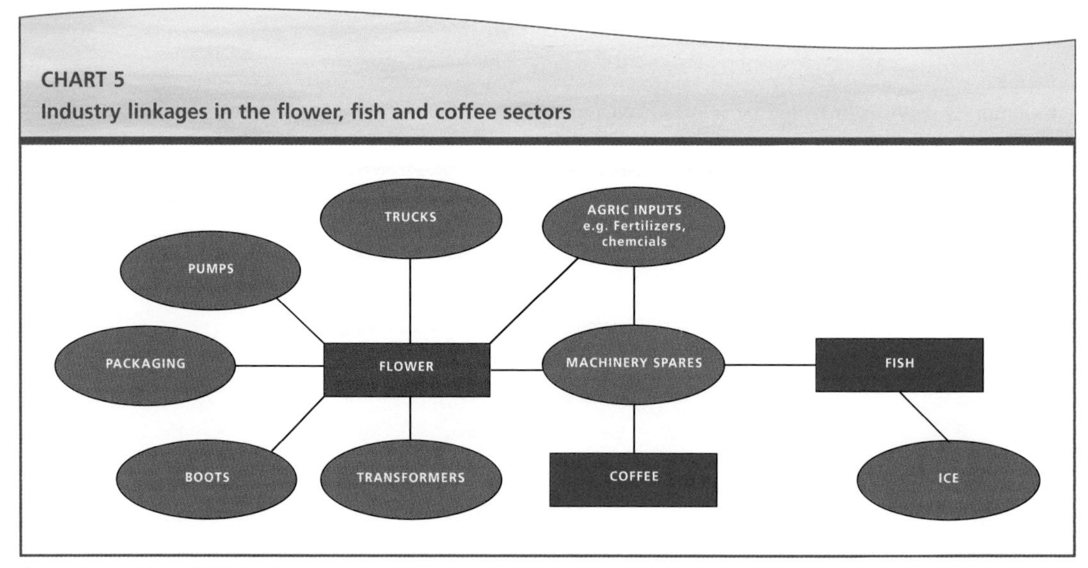

Source: Derived from field interviews

Negative impacts: Increasing demand has squeezed out (local) companies

Increased competition for limited supplies of fish from new players has led to the closure of most of the domestic-owned fish processing factors. Whereas the TNCs have used their ample resources to consolidate operations or to expand to new landing sites in search of fish, the domestic companies have not been able to sustain operations and have closed. Three of the four Ugandan-owned fish processing companies have closed down: Gomba Fish Industries Ltd, Ngege Ltd; and Byansi Fisheries Ltd.

6.11 Community impacts

Transnational corporations have contributed to the increased provision of social services and increasing demand for goods and services in the communities where they operate. With the exception of the new companies (those established after 2007), all of the firms interviewed reported that they contribute to their communities in various ways (Box 9).

Some domestic-owned companies also reported community programmes that they are directly engaged in, for example, construction of a local borehole, supply of fish to orphanages, providing scholastic materials, construction of community toilets and allowing the community to access medical services provided by the company for its workers. Most of these benefits are provided by companies in the flower sector.

The flower companies are more involved than the coffee and fish processing and exporting companies because of their direct involvement in the production chain. A survey of five flower farms and the communities in which they are located, conducted in 2003, revealed very positive findings in terms of socio-economic impact. The survey covered five flower farms, 25 retail shops, nine clinics/ drug stores, and over 100 employees of the flower farms (Donohue, 2003: vi). A summary of some of the relevant findings are outlined below:

- *Increased business:* Slightly less than half (44 percent) of the shopkeepers indicated that most of their customers were employed on the nearby farm. Similarly, five out of the nine drug stores indicated that most of their patients/customers were employed by the flower farm;
- *Increased medications:* Seven of the nine drugstores noted that the availability of medications had improved dramatically since the establishment of the flower farm;
- *Land purchase:* 18 percent of employees were able to buy themselves land;
- *Building a house:* 11 percent of employees were able to build their own house;

BOX 10
Examples of environmental impacts by TNCs in Uganda

Positive impacts in the floriculture industry
The Code of Practice audit report and a survey of five flower farms revealed that all the five utilize proper run-off control measures. Further, all the five use soak pits for the disposal of crop chemical rinseate (Donohue, 2003).

Comparison with Kenya's flower industry
This is not the case, however, in neighbouring Kenya, one of the leading flower exporting countries in Africa. The country's floricultural sector is dominated by large-scale flower farms around the Rift Valley area near Lake Naivasha, Kiambu and Thika. Much as the industry has grown steadily over the years; the environmental impacts are significant. Since most farms have neither soak pits nor wetland areas for the disposal of pesticides and chemical products, the waste ends up in the lake leading to water pollution. Further, as the industry expands, the land is continually being encroached upon, limiting human and animal access to the lake (Fedha, 2009).

Forest depletion to grow palm oil
Kalangala district is a collection of 84 islands on Lake Victoria. The total land area is 9 067 square kilometres (906 700 hectares), of which 26 783 hectares is forest cover (about 3 percent of the total land area in the district). Bidco, with headquarters in Kenya, is the leading marketer of edible oils, soaps and hygienic products in East and Central Africa. In 1998, the Uganda government gave the company's Ugandan subsidiary, BIDCO Uganda Ltd, ten hectares of land in Kalangala to grow palm oil. As of 2008, the land allocated to BIDCO had increased to 26 500 hectares, of which about 3 200 hectares is forest cover

Sources: "Uganda Districts Handbook, 2005-2006" 2005. Fountain Publishers: Kampala (86-87); Bidco. http://bidco-oil.com/regional/index.php?conid=2. Accessed on 12 May 2009; "BIDCO to undertake largest private Project", The New Vision, 10 November 2005, http://newvision.co.ug/D/8/220/464984. Accessed on 12 May 2009; "Government to limit land for foreigners" The Daily Monitor, 25 March 2009.; http://monitor.co.ug/artman/publish/news/ Govt_to_limit_land_for_foreigners_82092.shtml Accessed on 12 May 2009

CHART 6
Push and pull factors of TNC participation in commercial agriculture in Uganda

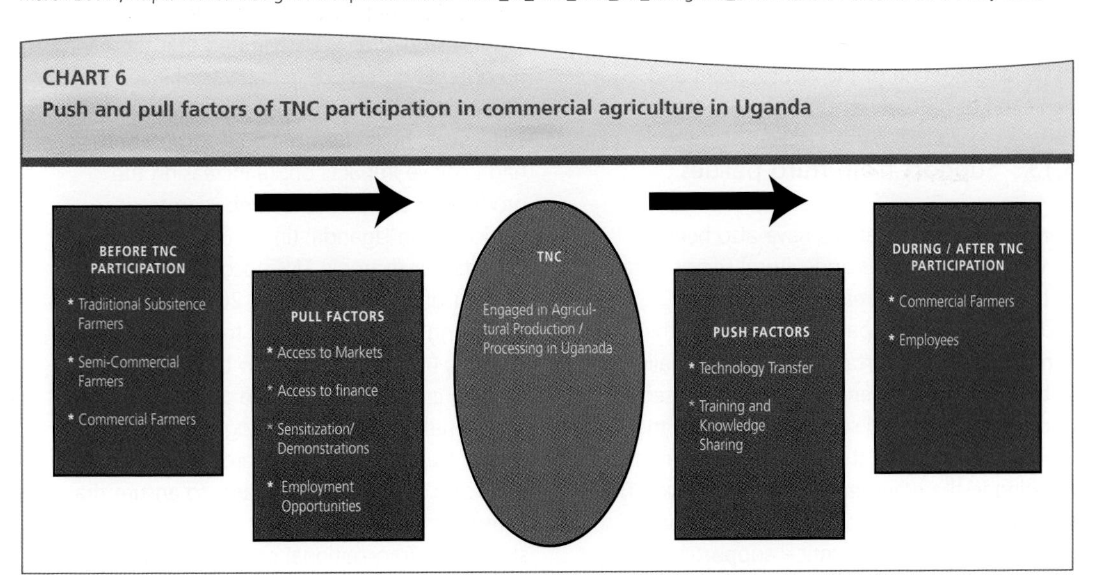

- *Savings:* 63 percent of employees reported that they save money each month.

Growth in the three selected commodity subsectors has led to growth in other sectors and industries to which they are linked. Such industries include packaging, vehicles (trucks), machinery (generators), pumps, footwear (boots) and motor vehicle spare parts. The flower sector has the most widespread linkages because flower firms are directly engaged in production, as illustrated in Diagram 5. All the flower companies interviewed reported that they purchase packaging materials from Riley Industries Ltd, a domestic company. Prior to 2007, the companies used to import boxes for packaging from Kenya. However, in 2007, Riley Industries purchased new machinery that meets international standards, and started producing the type of boxes required to package cut flowers for export. The three companies that released their cost information spend up to US$200 000 annually on purchasing boxes from Riley Industries Ltd.

6.12 Impact on the environment

The involvement of TNCs in agricultural production in Uganda has had both positive and negative environmental impacts. On the positive side, the companies have set, implemented and encouraged the use of environmentally friendly production techniques and practices. On the negative side, some activities of TNCs have led to the degradation of the environment and depletion of natural resources.

6.13 Support from third parties

Transnational corporations have also been successful in Uganda's agricultural sector because of the presence of a ready export market, and the role of third parties. The role of third parties became important following the failure of both the government and the private sector to provide specialized support service for the various subsectors of the agricultural economy, resulting in the emergence of a third player. This third player fills the gap adequately on many occasions, by providing critical support to the industries. This support boosts the industries' performance in terms of revenues, export share, capacity and competitiveness. Two examples of such *third parties* are the Export Promotion of Organic Products from Africa (EPOPA) and the Uganda Flower Exporters' Association (UFEA). The former has been at the forefront in engaging with TNCs in the training of farmers on farming methods and organic certification procedures. The UFEA, on the other hand, plays a major role in addressing the policy issues faced by flower exporters. The association is an advocacy forum for flower exporters and has registered significant successes since it was formed in 1993.

7. Conclusions and recommendations

Foreign investment in commercial agriculture by individuals and TNCs, though growing since 2000, is still relatively low. Most of the companies engaged in commercial agriculture – about 70 percent of the total – are domestic-owned. This is also illustrated by the small number of planned projects in the sector that were registered by the UIA between 1992 and 2008. A total of 124 projects have been registered in the sector and they account for just 3.5 percent of all projects registered by the Authority. About half of the registered projects were in four subsectors: fish, general farming, flowers, and forestry.

This study has demonstrated that there was no notable difference in impacts of TNCs and domestic companies on employment; they were collectively high. Transnational corporations had positive impacts on: (i) increasing the production of rice and contributing to rice sufficiency in Uganda; (ii) increasing agricultural exports: for example TNCs accounted for 59 percent of coffee exports in 2008/2009; (iii) improving access to finance for smallholder farmers; (iv) introducing new hardy or high-yielding crop varieties (maize and rice); and (v) disseminating input technologies, providing farmer training to improve product quality, and enforcing production standards to ensure that the commodities comply with international export standards. Transnational corporations have also contributed to opening international markets to Uganda's export commodities and to creating linkages with local suppliers of raw materials and packaging materials. These corporations have created linkages – mostly in the flower industry – with local communities, and indeed, have supported community projects in health, education, recreation and infrastructure (roads and electricity).

Negative impacts of TNCs were noted in the following areas: (i) contributing to the depletion of fish stocks; (ii) creating dependence of farmers on the company for seed (in the case of the DEKLAB maize hybrid supplied by Monsanto (U) Ltd); and (iii) environmental degradation resulting from the conversion of a tropical forest into a palm oil plantation by Bidco (U) Ltd

With respect to the policy environment, it was noted that Uganda has policies in place to attract investment in commercial agriculture. However, the country does not have specific policies to benefit fully from investment in the sector through value capture. Companies are creating linkages along the value chain mostly through their own initiatives and through the necessity to ensure product quality and reliability of supply of raw materials.

References

Agricultural Review. Global Food Crisis. 2008. *Journal of the Agricultural Industry in Africa*, 13(2).

Asea, P. & Darlison, K. 2000. *Impact of the flower industry in Uganda*. ILO Working Paper, No. 148. Geneva: ILO.

Beule De, H. 2008. *Greenfields Nile perch and Tilapia: Sustainable Fish from Lake Kyoga, Kikalagania, Nakasongola, Uganda*. Project End Report Executive Summary. Kampala: Prepared for the Embassy of Sweden.

Business in Development Network. 2008. *Investing in Small and Medium sized Enterprises in Uganda*. Bid Country Guide series. (Available at: http://www.bidnetwork.org)

DANIDA (Danish International Development Agency). Agriculture Sector Programme Support (ASPS II) Annual Progress Report 2005/2006.

Dijkistra, T. 2001. *Export diversification in Uganda: Developments in Non-traditional Agricultural Exports*: ASC Working Paper 47 / 2001

Donohue, C. 2003. *Socio-Economic Impact Study of the Floriculture Industry in Uganda*. Prepared for the Agribusiness Development Component (ADC), Uganda's Investment in Developing Export Agriculture (IDEA) Project. Kampala.

Fedha, P.T. 2009. *Environmental and Health Challenges of the Floricultural Industry in Kenya*.

Kitakiro, R. K. 2000. *Best Practices in Civil Service Reform for Sustainable Development*, Uganda Electoral Commission Asian Review of Public Administration, Vol. XII, No. 1 (January-June 2000).

Lund T., Rahman M. H., Boye S. R., Johansen L. & Tveiten I. 2008. Evaluation of Africare Food Security Initiative in Nyabumba, Uganda. *Journal of Innovation and Development Strategy*. 2(1): 10–17

Obwona, M.B. 1996. Determinants of FDI and their impact on economic growth in Uganda.

Ministry of Agriculture, Animal Industry and Fisheries (MAAIF). 2006. *Development Strategy and Investment Plan (2005/06-2007/08)*. Kampala. (Available at: http://www.agriculture.go.ug)

Ministry of Agriculture, Animal Industry and Fisheries (MAAIF). 2008. *Policy Statement for the Financial Year 2008/09*.

Ministry of Agriculture, Animal Industry and Fisheries (MAAIF). 2004. *Department of Fisheries Resources. National Fisheries Policy*.

Ministry of Agriculture, Animal Industry and Fisheries (MAAIF). 1994. Agricultural Seeds and Plant Act.

Ministry of Finance, Planning and Economic Development (MFPED). 2007. *Background to the Budget 2007/08*. Kampala. (Available at: http://www.finance.go.ug)

Ministry of Finance, Planning and Economic Development (MFPED), 2007. *Poverty Eradication Action Plan 2004/5 - 2007/08*. Kampala: Ministry of Finance, Planning and Economic Development .

Ministry of Finance Planning and Economic Development (MFPED). 1998a. 1998 Statistical Abstract. Kampala: Ministry of Finance, Planning and Economic Development.

Mugoya, M. 2009. Case study: FICA Seeds Uganda Ltd. Paper prepared for Market Matters Inc. (Unpublished).

Nagawa, F. & Haike, R. 2006. ESCO Vanilla and Cocoa Organic Exports from Bundibugyo district. Project End Report, April 2002 - March 2006. Submitted to the Embassy of Sweden, Kampala.

Nyombi, K. & Bolwig, S. 2003: A Qualitative Evaluation of Alternative Development Strategies for Ugandan Fisheries

Sayer, G. 2002. *Coffee Futures: The Impact of Falling World Prices on Livelihoods in Uganda*. Prepared for Oxfam International.

The Daily Monitor. 25 March 2009. Kampala. (Available at: http://www.monitor.co.ug)

The Daily Monitor. 12 May 2009. Kampala. (Available at: http://www.monitor.co.ug)

The Monitor Business Directory. 2008. Kampala. (Available at: http://www.monitordirectory.co.ug/)

The New Vision. 11 September 2007. Kampala. (Available at: http://www.newvision.ug)

The New Vision. 21 September 2007. Kampala. (Available at: http://www.newvision.ug)

The New Vision. 10 November 2005. Kampala. (Available at: http://www.newvision.ug)

Tulip, A. 2005. Kawacom (U) Ltd *Coffee Organic Exports from Paidha, Sipi and Bushenyo Districts in Uganda*. April 2002–April 2005. Executive Summary of Project End Report. Submitted to SIDA, Sweden.

Uganda Bureau of Statitics (UBOS). 2007. Uganda National Household Survey 2005/2006. Report on the Agricultural Module. Kampala.

Uganda Bureau of Statitics. 2007. Report on the Uganda Business Register 2006/2007. Kampala

Uganda Bureau of Statistics. 2007. Kampala: UBOS 2008 Statistical Abstract.

Uganda Bureau of Statistics. 2005. Kampala: UBOS 2006 Statistical Abstract .

Uganda Coffee Development Authority (UCDA). 2008. The 16th Annual Report. 1 October 2006–30 September 2007). Kampala.

Uganda Coffee Development Authority. 2007. Volume 15. 1 October 2005–30 September 2006). Kampala.

Uganda Coffee Development Authority. 2006. Volume 14. 1 October 2004–30 September 2005). Kampala.

Uganda Coffee Development Authority. 2003. Volume 11. 1 October 2001–30 September 2002). Kampala.

Uganda Districts Handbook, 2005-2006 Fountain Publishers. 2005. Kampala (86-87)

Uganda Exports Promotion Board. 2007. Available at: http://www.ugandaexportsonline.com

Uganda Investment Authority (UIA).2007. UIA Guide to Investing. Kampala. (Available at: http://www.ugandainvest.com)

Uganda: **Uganda National Exports Strategy**, 2007

U**nited States Agency for International Development** (USAID). *Uganda, Moving from Subsistence to Commercial Farming in Uganda*. Agricultural Productivity Enhancement Program Final Report, 2008.

World Bank. 2004. Investment Climate Survey: Uganda Regional Programme for Enterprsise Development, Africa Private Sector Group, Washington: World Bank.

World Bank and the Ministry of Tourism, Trade and Industry, Uganda: 2006. Report of the Diagnostic Trade Integration Study of Uganda. Volume 1:14.

Websites

1. faostat.fao.org
2. http://bidco-oil.com
3. http://en.wikipedia.org/wiki/Agriculture_in_Uganda
4. http//en.wikipedia.org/wiki/Ugandan_Bush_War
5. http//en.wikipedia.org/wiki/Uganda-Tanzania_war
6. http://lcweb2.loc.gov/frd/ cs/ugtoc.html
7. http://ugandacoffee.org
8. www.volcafe.com
9. www.olamonline.com
10. www.kawacom.com
11. www.ugacof.com
12. www.nkg.net
13. www.fides.nl
14. www.wac-international.com
15. www.royalvanzanten.com
16. www.marineandagro.com
17. www.alphauganda.com
18. www.hwangsungbiz.com
19. www.warda.org
20. www.unctad.org
21. www.acdivoca.org

PART FOUR

BUSINESS MODELS FOR AGRICULTURAL INVESTMENT - IMPACTS ON LOCAL DEVELOPMENT

Cambodia:
Local impacts of selected foreign agricultural investments[1]

1. Introduction

Cambodia, situated on the Indochinese peninsular, is endowed with huge freshwater reserves and an immense area of arable land. The country is also the destination for investment of some food-importing countries, including China, Kuwait, Malaysia, Qatar, Republic of Korea and Viet Nam. State private land, in the form of economic land concessions (ELCs), is leased to concessionaires for agricultural exploitation for a maximum of 99 years (GTZ 2009). Currently, 85 companies, both domestic and foreign, have been contracted to exploit a total land area of 379 034 hectares (MAFF 2010).[2] Non-governmental organizations (NGO) and international organizations have expressed critical concerns as to the potential effects of ELC holders' activities on the poor local communities in the immediate vicinity. To date, there has been little research on the economic, environmental and social impacts of FDI inflow to agriculture, or on its benefits for Cambodia. However, global examples of the costs and benefits of such investments show that although large-scale agricultural land exploitation could restrict communities' access to land and water, it could also contribute to the country's economic development through investors' participation in developing local infrastructure needed for agricultural expansion.

1.1 Study objectives

This study aims to examine the validity of some of the concerns expressed in Cambodia, by shedding some light on the potential effects of FDI in agriculture on local communities and their environments. Initially, it investigates the extent and nature of FDI in agriculture and its subsectors, including crops, livestock, food-processing, forestry and fisheries. It then goes on to analyse the policy and regulatory environment and institutions governing and facilitating FDI as well as prevailing business models – in the acquisition of agricultural land. It concludes by providing some policy recommendations in response to the challenges facing the sector.

1.2 Methodology

Data on land acquisition, particularly data on contract arrangements and ex-post and ex-ante data on socioeconomic and environmental indicators in the selected project locations, are rather patchy. The study was based mainly on interviews with key informants and with communities in the concession areas; it applied a counterfactual approach, with the aim of providing policy-makers and other relevant players with a general overview of the likely impacts of certain FDI projects on local communities and their environment. Case studies of FDI in the subsectors were produced based on past research and on consultation with government officials and community representatives residing near concessions. Focus group discussions (FGDs) were held with local communities and village authorities to capture the main economic, social and environmental impacts. Economic indicators included income, employment, development of irrigation and roads; social indicators included health care, water and land access and land

---

1 This chapter is based on an original research report produced for FAO by Saing Chan Hang, Hem Socheth, Ouch Chandarany, Phann Dalis and Pon Dorina, Cambodia Development Resource Institute

2 For a detailed profile of each investment firm, see www.elc.maff.gov.kh/profiles.html

conflicts; environmental indicators included soil quality, water quality, use/overuse of pesticides and fertilizers and cutting down of trees (forest cover). The study also approached foreign investors to discuss the costs and benefits of their projects and the potential hurdles to their investment in Cambodia. The team also gathered secondary data from the Ministry of Agriculture, Forestry and Fisheries (MAFF), the Council for the Development of Cambodia (CDC), the National Institute of Statistics of the Ministry of Planning (MoP), the Ministry of the Environment (MoE), and international organizations.

1.3 Scope and limitations

The broad nature of its scope meant that the study did not set out to reveal critical details of FDI projects and investment hurdles in agricultural subsectors. Rather, the aim was to investigate selected projects and firms in those subsectors, based on consultation with government officials in charge of investment monitoring or facilitation, namely, officials from MAFF and CDC, and the expert judgement of the study team. More importantly, given the time constraints, the study strived to reveal the overall picture of FDI in those subsectors only, compiling the likely effects on local communities and their environment by applying a counterfactual approach.[3] Efforts were made to consult foreign investors, but this was difficult as they were hard to trace: only two were interviewed in the end.

[3] The pitfall of this approach is that measured impact could be either over or underestimated: asking respondents to compare their socioeconomic status before and after the project is highly subjective. However, the study aims mainly to provide only an overall picture of the likely effects of certain projects. In-depth impact analysis of specific projects can be investigated later, applying more sophisticated project evaluation techniques, such as propensity score matching, before and after, difference-in-difference and instrumental variables.

2. Role of agriculture in the national economy

Despite a significant reduction in the share of agriculture in the total national output during the past two decades – from around 46 percent in 1993 down to about 28 percent in 2009 (MEF 2010) – the sector remains one of the key growth-enhancing pillars as well as a poverty-reducing tool. This is because around 85 percent of the total population lives in rural areas, the majority of whom rely on agriculture (mainly paddy rice) as their primary income and source of livelihood. As outlined in the government's Rectangular Strategy Phase I and Phase II, the National Strategic Development Plan (NSDP) 2006–2010, and the NSDP Update 2009–2013 in pursuit of growth, employment, equity and efficiency, agriculture ranks high among the four broad strategic development priority angles. The other three are rehabilitation and construction of physical infrastructures; private sector development and employment generation; and capacity building and human resource development.

2.1 Contribution of agriculture to the national output

Prior to 2000, agricultural production accounted for almost half of Cambodia's national output, reflecting the country's agrarian nature. However, the sector's contribution has declined markedly over the past two decades. The latest data from MAFF show it contributed only 33.5 percent of the country's gross domestic product (GDP) in 2009, down from 45.3 percent in 1993. The sector's share of employment of the national workforce also shrank, from 67.4 percent in 2002 to 55.9 percent in 2007, although this remains substantial despite the slump. This significant change in the structure of the Cambodian economy is a result of a rapid expansion in manufacturing industry, namely, textiles and clothing and the services industry. Annual average growth (gross value added) in the sector was at about 5.6 percent from 2002 to 2009. Such slow growth can be attributed to weak rural-urban linkages; unsecured land ownership; sluggish investment, both public and private, particularly in irrigation, transport and agricultural research; and limited support infrastructure such as availability of and access to finance and affordable and reliable energy and telecommunication services (World Bank 2004a, 2004b, 2006). The sector is dominated by crop cultivation, mainly paddy rice: crops contribute around half the national agricultural output. Fisheries, including freshwater, aquaculture and marine, account for approximately 33 percent, livestock and poultry contribute about 16 percent, and forestry and logging around 8 percent of total agricultural output.

2.2 Production and harvested areas

Alongside rapid growth in the manufacturing industry in the past decade, an expansion of **paddy rice**, the staple food in Cambodia, has also been noticeable. The area under paddy rice increased from about 2.4 million hectares in 2004 to 2.7 million hectares in 2009, resulting primarily from the government's expansion plan, while production also surged significantly from 4.2 million tonnes in 2004 to 7.6 million tonnes in 2009. Substantial growth in paddy rice production has also produced a considerable paddy rice surplus. The subsector is estimated to employ around 2 940 000 people, which shows its significant potential to contribute to poverty alleviation in rural Cambodia (UNDP 2007: 5). There is also evidence of fast and stable growth in the production of **other main crops** such as cassava, maize and soybeans, and a slight increase in mung beans, between 2002 and 2009. This growth can be attributed to rising prices through increasing demand for these crops in neighbouring Thailand and Viet Nam, who are their traditional buyers. Cambodia also produces a wide range of specialized crops, including sweet potato, peanuts, sesame, sugar cane, tobacco, jute and vegetables.

A steep acceleration in **rubber** prices on international markets during the past decade has generated considerable interest from both domestic and foreign investors in the sector in Cambodia, making it the country's main industrial and strategic crop. There has also been considerable engagement by Vietnamese investors in recent years, but the exact magnitude of involvement is difficult to estimate. The latest data from MAFF show that the total area under rubber plantation (matured and immature) – including rubber estates, new investment in the form of ELCs and smallholders – was 130 921 hectares in 2009, up from 82 000 hectares in 2007.

Livestock has contributed around one-sixth of total agricultural production during the past decade, and the sector is estimated to have employed 400 000 workers in 2006 (UNDP 2007: 5). In terms of number of heads, poultry takes the largest share, despite a marked decline in 2004; the subsector later accelerated due to subsidies to counter slumps caused by avian flu and increased awareness among farm owners of prevention measures. Production of cows and buffalo also expanded during the period, with average annual growth rates of 2.9 percent and 2.5 percent, respectively. By contrast, there was a marked decline in pig production between 2006 and 2009, owing to rising fear of pandemic swine flu (AH1N1), substantial illegal imports of pigs from neighbouring countries, and high domestic production costs (MAFF 2010: 19).

Inland freshwater fish contribute the most to total fish production in Cambodia, due to the country's immense freshwater lake and its long stretch of the Mekong River. Total catch did not change significantly between 2002 and 2009. However, concern has been mounting as to the potential negative effects of the rising number of upstream hydropower projects, such as those in China and Lao People's Democratic Republic, on downstream catches, such as in Cambodia. On the marine and aquaculture sides, growth in production has been slow but stable. More investment in fisheries could help offset possible declines in fish catches in the future. This is especially critical as the sector's contribution to low-skilled income earners is substantial: it provides approximately 260 000 jobs (UNDP 2007: 5).

In the **forestry** subsector, there was large-scale illegal logging and a significant reduction in the country's forest cover in the 1990s, though it should be noted that there is no reliable source of data on forest cover in Cambodia. Given rampant illegal forest harvesting, the government imposed a moratorium on all logging activities and timber exports in the early 2000s, and cancelled about half of the total number of forest concessions. This resulted in a decline in forest production and exports but contributed to environmental conservation and wilderness protection. According to the MAFF, total forest cover in 2006, including evergreen, semi-evergreen, deciduous, wood shrub in dry land and several other types, was 10 864 186 hectares, that is, approximately 60 percent of the country (MAFF 2007: 94). Forestation efforts by the Forestry Administration and private tree planting companies have not made a significant contribution to the country's forest cover: the area under tree plantation in 2009 was 18 924 hectares, up from 11 250 hectares in 2005.

2.3 Foreign exchange earnings

Besides employment generation and production for domestic consumption, agriculture also generates a marked proportion of national exports. Wood, articles from wood and natural rubber played a leading role in the sector in generating foreign exchange earnings between 2002 and 2008, followed by edible fruits, vegetables and roots, cereals, fish and live animals. However, the average share of these products in total exports was only 4.48 percent, as Cambodia's national exports are concentrated largely in textiles and clothing. This latter sector has grown dramatically in recent years, except for in 2008 and 2009, when it was hit by the two crises, namely, the fuel price crisis and the global economic crisis.

2.4 Regional comparison: opportunities and challenges

Cambodia's paddy rice yield remains low compared with other countries in the region in the past decade. However, despite a low yield of 2.9 tonnes per hectare in 2009, there are signs of improvements between 2005 and 2009, which can be explained by better application of fertilizer and pesticide, and additional investment in irrigation (World Bank 2009: 8). Better application of inputs, use of better quality seeds, less reliance on traditional tools and equipment and reduced dependency on weather conditions through investment in irrigation (whether public, private or by farmers themselves), can help the country catch up with others in the region. As 80 percent of farmers grow rice, and as rural areas have high poverty incidence, government and private sector assistance in the form of Build-Operate-Transfer, such as irrigation facilities, and support from development partners and NGOs in terms of both hard and soft infrastructure are key to regional catch-up, and to help farmers move out of poverty. In August 2010, the government unveiled a policy to promote paddy rice production and milled rice export.

There is potential for growth in other crops too. Figure 1 shows that Cambodia was a champion in terms of its cassava yield in 2009 and was comparable with other countries in terms of its maize and soybean yields. Cambodia's cassava yield in 2009 was 22.3 tonnes per hectare, higher than the regional average (excluding China), of 15.1 tonnes per hectare. Maize and soybean yields were 4.3 tonnes per

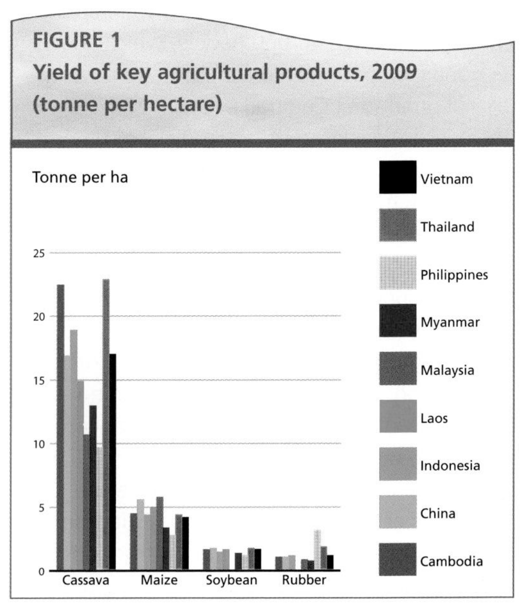

FIGURE 1
Yield of key agricultural products, 2009 (tonne per hectare)

Source: FAO, 2010.

hectare and 1.5 tonnes per hectare, respectively, slightly above regional averages of 4.1 tonnes per hectare and 1.4 tonnes per hectare. Cassava, maize and soybeans thus have possible production and export potential. Cambodia's natural rubber yield at 940 kg per hectare is lower than the regional average of 1.3 tonnes per hectare, but higher than in Malaysia (693 kg per hectare) and Myanmar (616 kg per hectare). Nevertheless, despite the low yield, growth in rubber has been impressive in recent years given the rebound in world demand for natural rubber since the global economic recession. However, several hurdles and bottlenecks to expansion exist, some of which are more binding and protracted than others and require immediate interventions from the government (Table 1).

3. Extent and nature of foreign investment in agriculture in cambodia

After its transition to a free market economic system in the early 1990s, Cambodia took steps to promote investment, both private domestic and foreign, through the privatization of state-owned resources and the promulgation of the Law on Investment in 1995, which provided tax and administrative incentives – and protection – to domestic and foreign investors. Investment started to flourish, gaining more momentum when Cambodia achieved genuine peace and stability in the late 1990s after the Khmer Rouge collapsed and rebel fighters were incorporated into the government defence forces.

Data from the Cambodia Investment Board (CIB) of CDC for the period from 2000 to June 2010, shows upward trends of investment. The critical turnaround in total investment occurred in 2005, which stemmed primarily from considerable engagement from China, Thailand and Republic of Korea, then rising substantially and continuing to expand into 2008. However, the pace of expansion slowed in 2009, as a result of the impacts of the global economic crisis on the Cambodian economy: total investment in that year fell sharply, to twice as low as that in 2008. Rostow (1960), among others cited in Todaro and Smith (2003: 115), indicates that countries which are able to save or generate investment of around 15 to 20 percent of GDP can grow at a much faster rate than those which save less, and that this growth will then be self-sustaining. It is difficult to apply this to the Cambodian case, though, given the absence of data on actual implementation of approved investment projects in the country.

TABLE 1
Likely constraints to agricultural development in Cambodia

Internal/domestic challenges	External constraints/factors
• High informal export cost • Lack of irrigation infrastructure and low level of technology in farming and processing • High input cost and low quality and capacity of milling/ processing • Inadequate storage and grain silos • Lack of low-interest credit • Lack of awareness of new and efficient planting techniques and lack of motivation to diversify production • Insufficient (or absence of) trademarks and geographical indications • High transportation costs due to infrastructure problems • Lack of marketing skills and market information system • No brand names • Deforestation due to the expansion of certain crops, like soybeans	• Many major consuming countries protect their markets (e.g. Japan, Korea Rep., some ASEAN countries) • Few countries offer preferential market access for Cambodia • Exports rely largely on demand and milling facilities in Thailand and Viet Nam • Importing countries often require a Special Purpose Ship Safety Certificate, which Cambodia lacks • Narrow export markets i.e. Thailand and Viet Nam, mostly for informal exports

Source: UNDP, 2007

TABLE 2
Investment approved by CIB, 2000–2010 (Fixed assets in million US$)

Country	2000	2001	2002	2003	2004	2005	2006	2007	2008	2009	2010	2000–June 10	Share
Cambodia	57	5	3	185	76	366	2 081	1 323	3 951	3 764	151	12 112	45.6
Japan	0	-	2	-	2	-	2	13	7	5	-	132	0.5
Korea Rep.	19	2	79	2	5	56	1 010	148	1 242	121	35	2 720	10.2
Taiwan	19	57	7	1	14	10	48	35	19	27	39	276	1.0
Hong Kong	5	1	2	5	-	1	4	26	-	7	17	69	0.3
China	28	5	24	34	83	454	757	180	4 370	891	60	6 887	25.9
Singapore	8	-	1	4	5	25	12	2	52	272	6	388	1.5
Malaysia	2	51	1	5	33	26	28	241	3	7	110	507	1.9
Thailand	26	15		7	1	81	100	108	26	178	2	544	2.0
Viet Nam	-	-	24	-	-	-	31	156	21	210	83	525	2.0
USA	12	6	-	-	2	4	62	3	672	1	2	764	2.9
France	5	-	-	6	3	8	-	35	38	50	-	145	0.5
UK	-	-	-	-	-	-	-	-	-	2	1	3	0.0
Others	35	3	5	1	5	18	320	298	415	330	83	1 514	5.7
Total	217	205	238	251	229	1 050	4 454	2 667	10 818	5 865	591	26 586	100
FDI	160	140	145	66	153	684	2 373	1 345	6 866	2 101	440	14 474	

Note: data for 2010 = January to June. Source: CDC, 2010

Private domestic engagement and foreign engagement were about equal between 2000 and 2010: the average share of FDI in total investment in the period was about 58 percent. FDI showed an upward trend but slowed in 2009 after the global economic crisis hit. Among foreign investors, China stood out, followed by the Republic of Korea, United States, Thailand, Viet Nam, Malaysia, Singapore and Taiwan, in that order. Traditional investors, like China, Republic of Korea, Malaysia and Thailand, have been injecting more funds into the Cambodian economy since 2000, which reflects their growing trust and confidence in the country's investment environment. Additionally, countries such as Viet Nam, Japan and United States have shown rising interest in the past couple of years. This is a promising sign for overall output growth in the medium and long term, despite the shocks of 2008 and 2009.

3.1 Approved Investment by sector

In terms of distribution by sector, tourism and industry have absorbed a great deal of investment during the past decade, followed by services. Tourism was champion, with a share in total investment averaging 35 percent between 2000 and June 2010, followed by industry (32 percent) and services (25 percent). Investment in agriculture was sluggish during the same period (Figure 2). In terms of accumulation of approved investment, tourism took the lead with 58 percent, followed by services (19 percent) and industry (17 percent); agriculture contributed the smallest share: around 6 percent. High capital inflows into tourism are attributable to the government's 1997 Open Sky Policy, the achievement of peace and political stability in 1998, and the gradual improvement of national infrastructures, particularly road connectivity and bridges.

Sluggish growth in agricultural investment between 2000 and 2010 can be explained by the fact that investment in agriculture is a long-term process with weak likelihood of return, and by the widespread nature of ill-defined property rights which prevent enterprises from using land and property as collateral to access finance (World Bank 2006: 74).

3.2 FDI in agriculture by nationality

Table 3 shows a significant contribution of FDI to agriculture in Cambodia between 2000 and 2010, with an average share of fixed assets of 78.4 percent, along with a slow but stable growth of private domestic investment, which averaged 21.6 percent in the same period. The significant rise in both private domestic and foreign investment has been more evident in recent years. This can be attributed to steep rises in global demand for natural rubber, particularly from China, India, Japan and the United States, as a result of a hike in the price of synthetic rubber after the petroleum price spike in late 2007 and 2008; for bio-fuel refined from common crops like palm oil and corn; and for food, such as rice, from food-importing countries in the region and the world.

Although the monetary value of foreign projects looks small compared with total FDI in the three major sectors of tourism, industry and services, the total size of secured land in agriculture may be substantial and thereby have negative implications for the environment and food security.

Looking at data for the period 2000 to 2010, illustrated in Table 3, China stands out as the second-largest investor in agriculture with a share of fixed assets of 17.6 percent, following Thailand, with the largest share of 21.7 percent. There is also evidence of rising interest in agriculture in Cambodia from such countries as Viet Nam (14.8 percent), Republic of Korea (6.5 percent), Singapore (4.8 percent), India (4.4 percent), United States (3.6 percent) and Japan (1.8 percent). All of these countries are mainly involved in crops and forestry, as discussed in the following section. Although this dramatic surge in foreign engagement in agriculture could be favourable at the macro level, negative trends could arise at the micro level if there is an absence of sound and prudent investment coordination mechanisms, environmental impact assessments and regular on-site investigations that are pro-poor, environmentally aware and consider food security as a priority issue.

3.3 Subsectoral breakdown of investment in agriculture

The CDC does not have a template giving a sectoral breakdown of agriculture data; categorization must be done instead, using different data sources. This study breaks agriculture down into crops, livestock, fisheries, forestry, fruit, food processing and others. This

FIGURE 2
Share of approved investment by sector, 2000–June 2010 (Million US$)

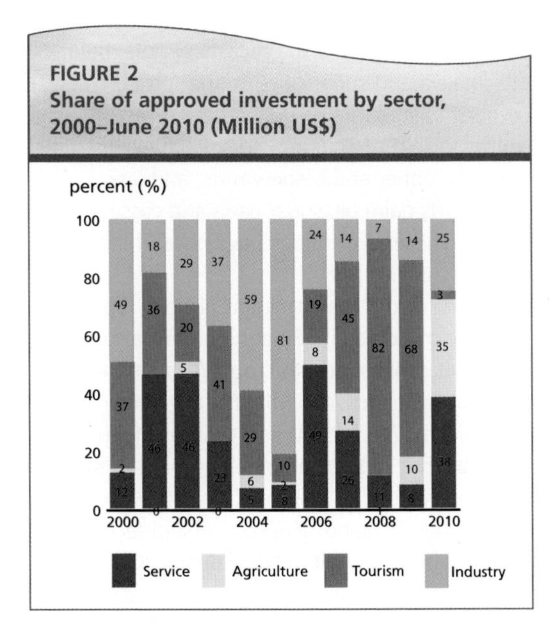

Source: CDC, 2010

FIGURE 3
Share of approved investment by sector, 2000–June 2010

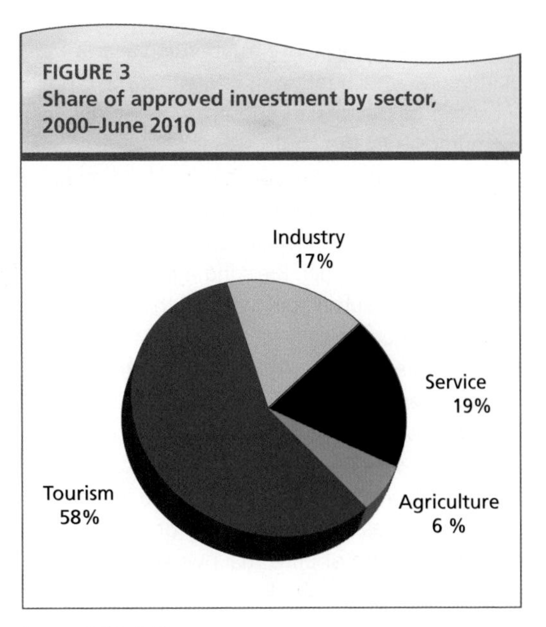

Source: CDC, 2010

TABLE 3

Agriculture investment approved by CIB, 2000–2010 (fixed Assets in million US$)

Country	2000	2001	2002	2003	2004	2005	2006	2007	2008	2009	2010	2000–June 10	Share
Cambodia	3.6	0.4	0.6	0.0	4.1	10.4	141	96.0	38.3	41.9	35.5	371.9	21.6
Japan	-	-	-	-	-	-	-	31.2	-	-	-	31.2	1.8
Korea Rep.	-	-	11.4	-	-	-	-	59.4	5.8	3.4	32.7	112.7	6.5
Taiwan	-	-	-	-	-	-	-	-	-	-	3.0	3.0	0.2
India	-	-	-	-	-	-	-	-	-	75.0		75.0	4.4
China	-	-	-	-	-	3.7	72.8	33.2	18.6	134	41.9	303.9	17.6
Singapore	-	-	-	-	-	-	-	-	-	82.1	-	82.1	4.8
Malaysia	-	-	-	-	8.2	-	-	-	-	-	-	8.2	0.5
Thailand	-	-	-	-	-	-	73.1	104	22.8	173	-	373.1	21.7
Viet Nam	-	-	-	-	-	-	27.4	43.1	20.9	104	59.8	254.8	14.8
USA	1.3	-	-	-	-	-	58.5	-	-	-	1.7	61.5	3.6
Israel	-	-	-	-	-	-	-	1.7	-	-	-	1.7	0.1
Canada	-	-	-	-	-	5.5	-	-	-	-	-	5.5	0.3
France	-	-	-	-	-	-	-	-	6.2	-	-	6.2	0.4
Uk	-	-	-	-	-	-	-	-	-	2.1	-	2.1	0.1
Denmark	-	-	-	-	-	-	-	-	-	-	29.5	29.5	1.7
Total	4.9	0.4	12.0	0.0	12.3	19.6	373	369	113	615	204	1 722.4	
FDI	1.3	0.0	11.4	0.0	8.2	9.1	232	273	74	573	169	1 350.5	
Share in FDI	25.5	0.0	95.0	0.0	66.6	46.6	62.2	74.0	66.0	93.2	82.6	78.4	

Source: CDC, 2010

is the first attempt, and the results may differ in future work given the constraints to subsector classification. For instance, rubber and acacia plantations, which might fit into either the crops subsector or the forestry subsector, is in crops for the purposes of this study.

As mentioned in Section 3.2, Thailand ranks first in agricultural investment in Cambodia, followed by China, Viet Nam and Republic of Korea, in that order. Breaking agriculture down into five subsectors, we find that foreign investors engage mainly in crops, forestry and other sectors. In crops, fixed asset investment is dominated by Thailand, Viet Nam and China. Thai investors tend to have a strong interest in sugar cane, which is estimated to have created approximately 13 500 jobs, and less so in rubber, palm oil and cassava. The majority of the firms come in the form of pure Thai ownership or partnership with Cambodians. In contrast, Vietnamese investors appear to have been focusing on rubber plantation in recent years, which is estimated to have generated around 11 000 jobs, and less so on cashew nuts, palm oil, cassava and sugar cane, in that order. The ownership structure of Vietnamese investors is similar to that of Thai investors. China is geared towards rubber and cashew nuts, and less so towards palm oil, sugar cane and cassava. Most Chinese investments have pure Chinese ownership.

Apart from these dominant players, Japan, Republic of Korea and Singapore have also made a marked contribution to crops. A Japanese investor has partnered with Thai (50 percent), Cambodian (20 percent) and Chinese (15 percent) investors in a sugar cane plantation in the Koh Kong province in southwest Cambodia, on the border with Thailand. This partnership received approval from the CIB in November 2007. Korean investors tend to be drawn to rubber, cassava and cashew nuts, in projects which take the

TABLE 4
Approved agriculture subsector investment accumulated over 2000–June 2010 (Fixed Assets in million US$)

Country	Crops	Livestock	Fisheries	Forestry	Others	Total	Share
Cambodia	174.0	3.6	0.4	72.2	121.8	371.9	21.6
Canada	-	-	-	5.5	-	5.5	0.3
China	194.6	-	-	77.3	32.1	303.9	17.6
France	6.2	-	-	-	-	6.2	0.4
Israel	1.7	-	-	-	-	1.7	0.1
Japan	31.2	-	-	-	-	31.2	1.8
Korea Rep.	90.1	-	-	-	22.6	112.7	6.5
Malaysia	2.6	-	-	-	5.6	8.2	0.5
Singapore	82.1	-	-	-	-	82.1	4.8
Taiwan	-	-	-	-	3.0	3.0	0.2
Thailand	327.0	-	-	46.0	-	373.1	21.7
USA	1.3	-	-	18.3	41.9	61.5	3.6
Viet Nam	236.6	-	-	18.2	-	254.8	14.8
India	-	-	-	-	75.0	75.0	4.4
UK	-	-	-	-	2.1	2.1	0.1
Denmark	-	-	-	29.5	-	29.5	1.7
Total	1 147.4	3.6	0.4	266.9	304.1	1 722.4	100.0
FDI	973.4	0.0	0.0	194.8	182.3	1 350.5	
Share in FDI	84.8	0.0	0.0	73.0	60.0	78.4	

Source: CDC, 2010

form of pure Korean ownership or partnership with Cambodians. Singapore has two projects, each of which is in both rubber and corn, with pure Singaporean ownership; the CIB approved both of these in July 2009. In addition, there is involvement from France, Malaysia, Israel and the United States. France engaged in rice plantation and milling in November 2008 as a joint venture. Malaysia started a joint venture with Cambodia (12.5 percent), to grow palm oil, rubber and cashew nuts in July 2004. Israel, as a joint venture firm, started cassava plantation in March 2007, and the United States (100 percent USA fixed asset), became involved in cashew and cassava plantation in February 2000.

Chinese and Thai investors are dominant in the forestry sector. Chinese investors focus on the production of pistachio and acacia trees, by investing their own fixed assets in the form of ELCs. In the case of Thai investment, it is not clear what tree groups are involved. There is also involvement from neighbouring Viet Nam, as well as from Denmark and Canada. Vietnamese investment comes in the form of a joint venture with Cambodia (30 percent), in acacia plantation, which was approved by the CIB in March 2010. Canada engages in teak plantation in partnership with China (40 percent), under the company name GG World Group Development;[4] this venture obtained approval from the CIB in December 2005. Denmark is also involved in teak plantation and uses 100 percent own fixed assets; this project was approved by the CIB in April 2010. In addition to foreign engagement in crops and forestry, there are a number of other foreign investments in other subsectors, namely, agro-industry in general, animal meal, corn-drying, crepe rubber processing, palm oil refinery and development, herbal tea production, hollyhock plantation and

[4] www.elc.maff.gov.kh/comprofiles/stgg.html

processing and integrated sugar cane plantation and sugar refining. Investors from China and the United States are involved in general agro-industry, whereas the Republic of Korea invests in corn plantation and drying, hollyhock plantation and processing and animal meal. Malaysia had a project in palm oil development and refinery approved by the CIB in June 2004 as a joint venture with Cambodia (40 percent), and Taiwan invested its own fixed assets in a herbal tea factory in March 2010. The United States has one project in hollyhock plantation and processing, and India has one in integrated sugar plantation and sugar, ethanol and power generation, with a licensed company named Kamadhenu Venture[5] growing sugar cane and producing ethanol in the Kratie province.

3.4 FDI in food processing

In the past decade there has been no significant involvement by foreign investors in food processing, whose share of fixed assets in total sector fixed assets is 28.7 percent. Cambodia thus takes the lead in the subsector, but in partnership with foreign investors like Viet Nam, Singapore and Thailand. Australia set up a soya milk manufacturing plant with own assets in 2003 and has joined with Cambodia (which holds 51 percent), in beer manufacturing. Canada started a project in beer manufacturing in 2010 with 100 percent fixed assets, and Singapore has partnered with Cambodian (10 percent), to produce pure drinking water. Thailand partners Cambodia (40 percent), in producing instant noodles and instant food; Viet Nam does so (Cambodia taking 30 percent), in producing beer, soft drinks and drinking water. China invested 100 percent of own capital in sea food processing in 2004. On the whole, only countries in the region appear to have an interest in food processing, except for Australia and Canada.

Overall, at approximately 6 percent, the share of approved fixed investment in agriculture has been minimal but stable during the past decade. Low appropriability of return on investment coupled with ill-defined property rights could hinder private and foreign investment in the subsector.

3.5. Engagement of foreign investors in ELCs

Given the absence of data on land/plantation size of foreign investments from the CDC and some

[5] MAFF has a different name for this: www.elc.maff.gov.kh/comprofiles/krtcarmad.html

TABLE 5
Approved fixed assets in food processing, 2000– June 2010 (Million US$)

Country	2000	2001	2003	2004	2007	2008	2009	2010	2000–June 10	Share
Australia	-		0.7		9.0				9.7	13.0
Cambodia	-	0.8			10.1	3.7	8.5	30.2	53.3	71.3
Canada	-							4.3	4.3	5.7
China	-			0.7					0.7	0.9
Singapore	-							3.8	3.8	5.1
Thailand	-	1.2							1.2	1.6
Viet Nam	-				1.7				1.7	2.3
Total	-	2.0	0.7	0.7	20.9	3.7	8.5	38.3	74.7	
FDI	-	1.2	0.7	0.7	10.8	0.0	0.0	8.1	21.5	
Share in FDI		60.0	100.0	100.0	51.5	0.0	0.0	21.2	28.7	

Source: CDC, 2010

of the discrepancies between CDC data and the MAFF website,[6] this study probed data recorded by the MAFF further in order to understand the scale of FDI in agriculture, particularly in the form of ELCs, which could shed some light on the potential effects of FDI on the socioeconomic situation, the environment and food security in the country. Public investment through the leasing of state private land in the form of ELCs to private domestic and foreign investors has been in evidence since 1995, prior even to the promulgation of the Land Law in 2001. According to the Sub-decree on Economic Land Concessions, dated 27 December 2005,[7] ELC refers to a mechanism to grant state private land through a specific contract to a concessionaire for use in agricultural and industrial agricultural exploitation, namely, the cultivation of food crops or industrial crops; raising animals and aquaculture; construction of plants, factories or facilities for processing domestic agriculture raw materials; or a combination of some or all of the above mentioned activities.

The principal aims of such initiatives are to develop intensive agricultural and industrial agricultural activities; to generate employment and diversify livelihood opportunities in rural areas; and to generate government revenue (US$1–10 per hectare per year based on four land categories). Table 8 indicates that private domestic as well as foreign investment in ELCs was not as considerable from the mid-1990s to the early 2000s as it has been from the early 2000s to the present. Cambodia's newfound peace and stability, along with a better investment environment and rising global demand for industrial crops like rubber and cassava, could be driving factors.

It should be noted, however, that the scale of concessions between 1995 and 2003 was massive. In 1999, 2000 and 2001, concessions were granted for 20 000 hectares, 315 028 hectares and 100 852 hectares, respectively. That such large-scale investment models were permitted before the adoption of the Land Law in 2001 and the Sub-decree on Economic Land Concessions in 2005, meant they were implemented in the absence of sound regulations and governance mechanisms, putting resources and local stakeholders at risk.

By the end of 2009, the MAFF had granted 86 ELC projects (excluding 12 which were cancelled); all but nine have profiles on MAFF's website. Article 59 of the Land Law in 2001 states that land concession areas shall not be more than 10 000 hectares. In response, the MAFF has been negotiating with companies that have a land area larger than 10 000 hectares, and are slow in implementing their business plans. Overall, 12 projects have been revoked, two of which have a land area above 10 000 hectares. The three mega projects mentioned earlier remain underway.

However, it is vital to note that, in order to secure larger tracts of state private land, some companies use two different names to obtain ELCs. For instance, Koh Kong Plantation Limited and Koh Kong Sugar Company Limited in Koh Kong province have secured land areas of 9 400 hectares and 9 700 hectares, respectively. This creates a total land size of close to 20 000 hectares. Even more importantly, the Council of Ministers can grant exemptions to the required reduction of concessions covering over 10 000 hectares, under conditions set out in Article 39 of the Sub-decree on Economic Land Concessions. As of late 2009, the total land size of reported ELC projects was 1 024 639 hectares. Ngo and Chan (2010: 6) indicate that approximately 500 000 hectares of ELC projects have been granted by and are under the administration of the Ministry of the Environment (MoE) but are not included in MAFF data. This pushes the total size of ELC projects up to around 1.5 million hectares, close to the size of the area under paddy rice of 2.7 million hectares in 2009 (MAFF 2010: 16). As Table 7 shows, ELC projects are mainly under private domestic ownership, according to data from the MAFF website. Foreign investors have acquired an ELC land size of 355 914 hectares, which is around 35 percent of the total and equates to 13 percent of the total paddy rice area in Cambodia in 2009. China is the dominant player among foreign investors, with 17 projects covering a total land size of 186 935 hectares

[6] Note that the CDC dataset has no record of several of the ELCs reported on the MAFF website.

[7] www.elc.maff.gov.kh/laws/subdecree.html

TABLE 6
Approved and cancelled ELCs, 1995–2009

Year	Permitted and ongoing projects				Permitted but later cancelled projects			
	Size (ha)	Min. size (ha)	Max. size (ha)	No. of projects	Size (ha)	Min. size (ha)	Max. size (ha)	No. of projects
1995	11 000	-	-	1	-	-	-	-
1996	2 400	-	-	1	-	-	-	-
1998	60 200	-	-	1	51 500	23 000	28 500	2
1999	33 400	3 000	20 000	4	4 100	-	-	1
2000	341 898	1 070	315 028	5	11 200	3 200	8 000	2
2001	128 275	5 000	100 852	4	-	-	-	-
2004	6 100	1 200	4 900	2	-	-	-	-
2005	67 043	3 000	10 000	8	10 000	-	-	1
2006	168 256	4 400	10 000	20	40 393	7 172	9 214	5
2007	29 001	6 436	8 100	4	8 692	-	-	1
2008	40 936	6 523	7 200	6	-	-	-	-
2009	136 130	807	9 820	21	-	-	-	-
Total	1 024 639	807	315 028	77	125 885	3 200	28 500	12
Projects with no reported date and land size				**9**				

Source: Authors' calculation based on data from MAFF website,www.elc.maff.gov.kh/profiles.html, October, 2010

(18 percent), including one mega project of 60 200 hectares secured in 1998 in Koh Kong province. It has a strong interest in rubber, acacia and pistachio plantation.

Viet Nam has seven projects with a focus on rubber plantation in Kratie, Ratanakkiri, Mondolkiri, Kompong Thom and Preah Vihear provinces, and Thailand has five projects concentrating on sugar cane in Oddar Meanchey and Koh Kong provinces. The United States has four projects growing teak trees in Kratie and Kompong Speu provinces, and the Republic of Korea has five projects investing mainly in rubber and cassava plantation in Kompong Thom, Ratanakkiri, Kratie and Kompong Speu provinces. India has interests in sugar cane, whereas Malaysia has interests in palm oil plantation.

As illustrated earlier, foreign investors tend to have a strong interest in such crops as rubber, cassava, sugar cane, teak, acacia and pistachio, which are specialities of Cambodia and well suited to the soil conditions in the country. By number of projects, their engagement is seen mainly in Cambodia's strategic crop provinces, such as Kratie (13 projects), Mondolkiri (5), Kompong Speu (5), Ratanakkiri (4), Stung Treng (4), Oddar Meanchey (4) and Kompong Thom (3).

Given the absence of data on actual implementation of foreign investment projects, it is difficult to generalize about the job creation and income generation benefits to Cambodia. If all the projects are fully realized, generation of income and jobs could be substantial. However, large-scale investment in agriculture in ELCs entails at least a degree of clearance of the forest, upon which many rural households depend as a major source of income. Concentration of projects in provinces like Mondolkiri and Ratanakkiri, where the majority of the population is made up of ethnic minorities and where most people depend on non-timber forest products as their major source of income, could endanger ethnic groups' livelihoods. Negative effects

TABLE 7
Distribution of ELCs by nationality, 1995–2009

Category	Size (ha)	% of total	Mean size (ha)	Min. size (ha)	Max. size (ha)	No. of projects	No. of projects >10 000 ha
Active projects							
Unreported	-	-	-	-	-	9	-
Cambodia	668 725	65	18 576	807	315 028	36	6
China	186 935	18	10 996	5 000	60 200	17	1
India	7 635	1	7 635	7 635	7 635	1	0
Korea Rep.	27 622	3	5 524	3 000	7 500	5	0
Malaysia	7 955	1	7 955	7 955	7 955	1	0
Taiwan	4 900	0	4 900	4 900	4 900	1	0
Thailand	37 436	4	7 487	6 523	9 700	5	0
USA	36 203	4	9 051	7 000	9 820	4	0
Viet Nam	47 228	5	6 747	2 361	9 380	7	0
Total	1 024 639	100	13 307	807	315 028	86	7
FDI	355 914	35				41*	1
Cancelled projects							
Cambodia	34 711	28	8 678	7 172	10 000	4	0
China	66 800	53	13 360	3 200	28 500	5	2
USA	9 214	7	9 214	9 214	9 214	1	0
Viet Nam	15 160	12	7 580	7 560	7 600	2	0
Total	125 885	100	10 490	3 200	28 500	12	2
FDI	91 174	72	-	-	-	8	2

Note: * excluding number of unreported projects. Source: Authors' calculation based on data from the MAFF website, www.elc.maff.gov.kh/profiles.html (retrieved in October 2010)

on community livelihoods can also be seen particularly in provinces like Kratie and Stung Treng, whose populations also depend on non-timber forest products plus fishing.

In addition, though Cambodia is now experiencing rice production surplus, it should be noted that long-term competition with regard to land use between paddy rice production and other crops of foreign and private domestic investors could lead to concerns related to food security.

Effects of investment and ELCs in particular should be netted using the above-mentioned indicators i.e. generated income and employment, loss of communities' major sources of income from forest products, the competing use of water and land between rice and other crops under foreign investment projects, and companies' contribution to developing local infrastructure such as roads and irrigation facilities. Section 6 offers a preliminary impact assessment of FDI on agriculture in Cambodia.

3.6 Barriers to FDI in agriculture

A survey of firms, to examine motivations for and barriers to investment in agriculture in Cambodia, was conducted in November 2010. A semi-structured questionnaire was used to capture information on the firms' contribution to infrastructural development and perceptions of mechanisms to mitigate investment barriers. CDC data for the period 2000 to June 2010 indicates that total of 59 firms were involved in agriculture and food processing. CDC provides no contact details for these firms, and the majority of the firms are not listed in the Yellow Pages, which means it was immensely difficult to approach them. Some of the firms listed by CDC can also be found on the MAFF website, along with their contact details. Using the Yellow Pages, as well as contact information from MAFF, the authors compiled a list of 31 firms with contact information. Because most of the addresses obtained from the MAFF website are now redundant, we were able to get in touch with only two firms, namely HLH Agriculture (corn plantation and drying), and Kogid (corn-drying and rice-milling). This meant that seeking answers to questions on barriers to investment in agriculture was a daunting task. Based on responses from the two companies,[8] five factors stood out as key hindrances to investment applications and operations in agriculture in Cambodia. First, land tenure and securing a lease agreement remain a challenge. One company claimed it took several transactions to complete the leasing arrangement. Land brokers with bad intentions can also cause disputes with local communities in terms of overlapping claims.

The second problem relates to unclear guidance as to which institution a firm should approach to obtain a business licence. It took one firm two months to have its ELC approved by the MAFF by using the right Cambodian broker, without whom it would have taken at least two years. Third, law enforcement remains weak. Fourth, administrative procedures are still long but somehow acceptable. Finally, dispute settlement and problem-solving mechanisms are seriously limited. For instance, there is no department in a specific ministry to help investors deal with problems arising during application for a business licence. Note, though, that firms do not see paying tips to public officials as a bad thing or a major problem, as it helps to speed up the application process, and firms accept that public servants are low-paid professionals.

In response to the above obstacles, firms recommended that the government consider adopting computerized investment or business licence application procedures and other related administrative procedures in order to reduce paperwork, time and cost and make it easier for firms to deal with problems during application. Firms also recommended setting up a feedback department in each responsible ministry, to help investors address problems arising during application or operations.

No inclusive business model or contract arrangement between firms and farmers has been introduced. The real practice is that firms buy corn from farmers at the market rate without entering into any future contract. Farmers either get their produce to the factory gate or sell it to brokers, who later sell it to factories or firms. Using brokers is more common, particularly when brokers are the farmers' creditors. Overall, despite a certain number of barriers, the two investors questioned appeared to be supportive of the business environment in Cambodia. They remarked, however, that Cambodia, as a developing country, has a great deal to do to improve its investment environment, and that there are steps that should be taken immediately.

4. Cambodia's investment policies and regulations by subsector

The Cambodian Government has set comprehensive strategies to promote investment in agriculture. Being an agrarian country, Cambodia has a huge potential in this sector, as there is plenty of land available. However, the sector has not been sufficiently developed, partly due to limited capital investment in irrigation networks, technology, energy and fertilizers. The global economic crisis in late 2008 hit Cambodia's major sectors: clothing, tourism and construction. Although MAFF figures show a rather stable trend in agricultural production, investment in the sector declined because of low capital investment and low appropriability of returns from agricultural investment projects. Since Cambodia's economic growth is narrowly based, promoting agriculture plays a crucial role in strengthening

[8] It is crucial not to make generalizations using the results of these discussions, as a sample of 2 out of 59 companies does not give a true picture across the subsectors. This evidence should therefore be viewed as anecdotal.

the country's economy. It is therefore prioritized in the government's Rectangular Strategy as a key pillar in achieving growth, employment, equity and efficiency. Under this strategy, agricultural policies are playing an increasingly important role.

The Rectangular Strategy emphasizes that to improve agricultural sector productivity, diversification and intensification, land management, fisheries and forestry reforms must go hand in hand with the development of rural infrastructure, energy, credit, markets and technology. To further promote agricultural investment and improve productivity, land reform is particularly crucial. The government is determined to implement the Land Law, the Law on Expropriation, the Law on Pre-emption and Land Development, the Law on Construction and Urbanization and the country's National Construction Standards. However, the implementation of the Land and Expropriation Laws has sparked some controversies, as certain land or farm owners are not entitled to compensation, because according to the 2001 Land Law, some roads and rights of way are treated as state public property. They are not even entitled to improvements made during their occupation of such property. The reconstruction of National Highway No. 1 is a good example[9]. To support the sector, the government has also waived taxes on imports of agriculture-related materials, and suspended taxes on agricultural land. Despite this support, Cambodian farmers still face production difficulties because their crops depend on rainfall and are vulnerable to natural disaster.

In accordance with Cambodia's investment law, promulgated in 1994 and amended in 2004, and Sub-decree No. 111 on the implementation of the amendment of investments dated 27 September 2005, the CIB of the CDC is responsible for overseeing the development of investment (domestic and foreign) activities. The CIB's registration mechanism is used for investment projects costing in excess of US$2 000 000, located in two provinces/municipalities, and special economic zones, while the sub-committee on investment is responsible for granting permits for projects costing up to US$2 000 000.

4.1 Fisheries

Fish is the second staple food source after rice in Cambodia. It plays a crucial role in people's livelihoods and is one of the main sources of income and food security in the country. There are two levels of fisheries management in Cambodia, namely, central and local governments. At the central level, the Department of Fisheries of the Ministry of Agriculture, Forestry and Fisheries is in charge of policy formulation and the conduct of research and inspection, while at the local level, the sector is under the control of the Provincial-Urban Fishery Authorities, which have the powers to ensure compliance with the law in the areas under their jurisdiction.

Several laws govern the sector, including the Fisheries Management and Administration Law (1987), Proclamation on Competent Authorities in issuing permission to fish in open water, aquaculture, fish processing and special permissions (1989), and the Sub-law on Transportation of Fishery Products (1988). It should be noted that Cambodia's master plan for fisheries 2001-2011 was developed in 2001, while the Fisheries Law promulgated in 2006 was to better manage the sector. Despite the introduction of necessary regulations, illegal fishing and habitat destruction along the Mekong River and in the Tonle Sap Lake continue more or less unabated, and conflicts over fishing rights between communities and politically and economically more powerful commercial fishing lot owners are common (FAO 2011: 6, 10). Law enforcement is observed to be *ad hoc* or case-by-case. For instance, the recent tough action taken by the prime minister on 1 July 2011 ordering the Minister for Agriculture, Forestry and Fisheries to remove fishery chiefs in five provinces around the Tonle Sap Lake, on suspicion of fishery infringements by irregularly selling fishing lots, is part of an effort towards fishery reform (Yang 2011)[10].

[9] Source: ADB (2007), "RETA 6091: Capacity Building for Resettlement Risk Management—Cambodia Country Report".

[10] http://news.xinhuanet.com/english2010/world/2011-07/01/c_13961037.htm

On the investment regulation of the sector, no specific rule is set out to promote investment in this subsector. However, this subsector's investment incentives i.e. exemption of taxes and duties, are set forth in Cambodia investment law. Investments include fish hatcheries of more than 2 hectares and shrimp farming and other aquaculture production greater than 10 hectares. The sector's investment procedures also fall under the procedure highlighted generally above. In order to develop the subsector, the government has distributed marine and freshwater fishing lots to the people for both consumption and commercial purposes, with the aim of facilitating fishing operations, sustaining catch sources and preserving natural resources. Further, given the increasing demand for fish from the rising population, the government has encouraged people to shift their focus from natural catches to aquaculture. To better manage the resource, the government is establishing an effective price mechanism by ensuring proper demarcation of fishing lots and making the process of fishing lot bidding more transparent. This will help increase state revenue from fisheries. Tough measures are being taken to prevent and crack down on illegal fishing activities and the encroachment of flooded forests. Fish farmers and communities are given technical assistance, credit and market facilitation to improve their capacity and increase their revenue. To increase competitiveness and market access, the government has encouraged large-scale fishery investments by improving infrastructure.

4.2 Forestry

Generally, the Forestry Administration (under MAFF) is in charge of the general governance of forests and forest resources in Cambodia, in accordance with the National Forestry Sector Policy and the Forestry Law (2002). The sector comes under two levels of management i.e. central and local government[11]. Prior to 2000, forest harvesting was rampant, rapid and widespread. The government's cancellation of 40 percent of all forest concessions in the early 2000s – equivalent to almost half of the original area under concession – as well as the its moratorium on logging in concession areas and log transportation in January 2002, has significantly reduced rampant logging (World Bank 2004a:19, 76). Another measure was introduced to clamp down on illegal natural forest products export through the introduction of a subdecree on forest and non-timber forest products allowed for export and import, dated 20 November 2006. This new rule allows the export of all kinds of processed and non-processed timber products derived from man-made forests, which offers room for private/foreign investors to engage in the form of economic land concession (ELC), which is under the governance of the MAFF (see discussion on ELC in section 4.5). It should be noted that there is no specific rule to promote investment in the forest sector.

In addition, the government has put efforts into forestry reform by establishing forestry policies, including the law on Concessions, subdecrees on Economic Land Concessions and Forestry Concession Management, Forestry Community formation and other regulations related to environment preservation, such as the law on Environmental Protection and Natural Resource Management and the subdecree on Environmental Impact Assessment. These aim to ensure the livelihoods of local communities improve, as a large number of people in rural communities depend on forests. With help from the international community, people are educated on how to make proper use of and also protect the forests. In this way, sustainable development can be achieved and biodiversity protection guaranteed. Once forests are well protected, local communities are the ones who benefit both economically and non-economically[12] from community forestry, as they have secured access to land and legal rights

[11] For the detailed function and structure of Forest Administration see: http://www.forestry.gov.kh/AboutFA/MandateEng.html

[12] Non-economic benefits include spiritual/customary values (for ethnic minorities) and training, social capital and networking.

on forest use[13]. Across the country, the forestry communities have reported significant reduction in illegal activities within forest areas under their management. Protecting forests has a direct impact on wildlife and biodiversity, which in turn is conducive to sustainable development and poverty reduction. Therefore, the government has shifted more focus onto raising environmental awareness by educating people and students on the conservation, protection and sustainable management of natural resources. In addition, the serious punishment meted out to those who are involved in illegal logging has been in the media headlines; the offenders have been jailed. In spite of such enforcement, illegal logging has been reported subsequently and the law enforcement personnel themselves are alleged to be behind these acts, indicating the limitations of law enforcement in the sector.

4.3 Livestock

Livestock farming i.e. pigs, cattle and poultry, needs to grow as demand for food increases, and the government is now focusing on better food quality and safety. To attain this, the department of Production and Animal Health of MAFF, which also has branches in the provinces, has set out the tasks related to the policies on animal health and production subsector, improving services for animal health and production and credit support for livestock farmers. It is also responsible for enacting laws and regulations on the quality of animal products, controlling the import and export of livestock at border gates and inspecting sanitary and phytosanitary standards at slaughterhouses. With support from USAID and FAO, draft law on animal health and production which covers aspects of animal and public health issues was initiated in early 2009. Cambodia is the transit point for cattle from Thailand to Viet Nam, and itself exports and imports cattle and pigs from the two neighbouring countries; therefore, it is vital to have such a law. Large-scale cattle and pig trading companies appear to be highly influential in the long-distance movement of livestock both within Cambodia and between neighbouring countries. They are well protected through high level connections to relevant ministries, through which they are able to influence decision-making among not only livestock traders, but also the police, military police and veterinary officials (FAO, ADB, OIE SEAFMD 2009:21-22).

No specific rule is observed in promoting investment in the industry; however, as set out in Cambodian Investment law, investors qualify for custom duty exemption if they produce more than 1 000 head of livestock; manage a dairy herd larger than 100 head; or produce more than 10 000 poultry and eggs. Despite the absence of a specific investment promotion policy, the overall investment process is in line with the general investment procedure stated above. Issues arising in the sector are often brought up at the government private sector forum, which is a trouble-shooting mechanism, by the Technical Working Group on Agriculture and Agro-industries. For instance, the issue of rampant pig smuggling – which harms domestic producers – was addressed through instruction no.001 dated 13 August 2007 on the prohibition of meat and live pig imports.

4.4 Water resources and technology

Water resource management is crucial to agricultural development. To date, Cambodia's irrigation system is still not sufficiently developed; the vast majority of farming is rainfed, making it highly vulnerable to variable climatic conditions. The government has formed water-user communities across the country by expanding reservoir capacity to meet demand. In addition, the MAFF and the Ministry of Water Resources and Meteorology (MoWRAM) have, as required by the NSDP, prepared a Strategy for Agriculture and Water through the newly established Technical Working Group on Agriculture and Water. The government, in order to scale-up productivity in agriculture for own consumption and for sale, has intervened further, such as by providing high-quality seeds yielding high-quality produce;

[13] Ministry of Foreign Affairs of Denmark, *Community-based Natural Resources Management: Lessons Learned from Cambodia*, Technical Advisory Services, December 2010.

facilitating the adoption of better technology by farmers; reducing harvest and post-harvest losses; and promoting innovative agricultural practices, including integrated crop management. Reducing the price of electricity will also help lower the cost of production.

4.5 ELCs and commercial production

As Cambodia's potential to export agricultural products to the world market grows, the government is placing higher priority on commercial agricultural production, especially rice and other agribusiness crops such as rubber and cassava. Unlike subsistence agriculture, commercial agriculture can expand the revenues of rural people and thus lift them out of poverty. It will also provide new opportunities for children in rural areas to go to school, which will in turn better rural people's livelihoods. To this end, government interventions relate to market information; new product opportunities to fit customer requirements; value-added processing facilities; quality assurance and food safety; profitable business promotion; and infrastructural development.

The Cambodian Government proposed the agriculture sector intervention in 2003 by targeting structural reforms – with support from ADB through its Agriculture Sector Programme Loan – under which the government is in charge of disseminating wider information related to agricultural marketing and technology, liberalising fertilizer pricing and marketing, formulating rural credit policy, divesting the rubber subsector, establishing local rural development committees and improving property rights. Further, the government has continued to improve access to productive land under secure title for the rural poor and commercial development in urban areas with support from several aid agencies such as AusAID on mine clearance and agricultural extension services, and the World Bank on land titling.

As seen above, ELCs are a mechanism for providing state private land[14] to concessionaires for industrial agricultural exploitation, including tree plantation, animal raising, aquaculture and the building of factories for agricultural raw material processing. Alongside the granting of large tracts of unused and/or unauthorized lands as ELCs to both foreign and Cambodian investors, rural infrastructures such as roads, bridges, markets and hospitals are being developed by both government and investors. Income from jobs created by these companies has to some extent allowed local communities to improve their living conditions.

The subdecree on Economic Land Concessions was introduced on 27 December 2005 in order to tap the opportunity of developing intensive agricultural and agri-industrial activities, increasing employment in rural areas, and generating state or provincial revenues. Article 29 of the subdecree states that the MAFF, as chair of the technical secretariat and an inter-ministerial body, is authorized – and responsible for – granting economic land concessions with total land area greater than 1 000 hectares, and the provincial/ municipal governor is authorized and responsible for granting economic land concessions with total area less than 1 000 hectares. However, the MAFF always seeks approval from the Office of the Council of Ministers before granting approval to a company.

By law, feasibility studies and environmental and social impact assessments are required prior to contract approval. Interview with one ELC company indicates that to obtain quick ELC investment approval a company has to engage local consultants – who are closely connected to officials at the MAFF – to conduct feasibility study and environmental impact assessment. The process could be protracted if a company were to engage only international consultants. Despite the carrying out of feasibility studies and project border demarcation prior to approval, overlapping claims between villagers and ELC companies prevail. Provincial governors and the commune council often act as mediators once conflicts arise; however, villagers are often disappointed with the solutions offered by the company, a sentiment echoed by local authorities (see section V. for further analysis).

[14] By subdecree, the ELC can also be used as a legal instrument to convert state public land into state private land (Articles 14 and 15, Land Law 2001).

The National Strategic Development Plan Update 2009–2013 sets forth other policy measures to promote agriculture, including a focus on land reform and clearance of land mines. These measures include strengthening land management, distribution and use; securing land ownership; curbing illegal landholding; and preventing concentration of unused land in few hands. Moreover, small farmers have been provided with social land concessions (SLC)[15] in order to foster their production and ability to diversify. Clearance of land mines and unexploded ordnance is a top priority for the government to help enable better access for farmers and investors to larger land and more remote areas.

4.6 Rice

Cambodia has tried to grasp new opportunities in the rice subsector, following food price increases in 2007 and 2008. Rice is not only a main source of food for the Cambodian people but also a potential source of income for the country's economy. As such, the government has encouraged investment in rice, by waiving tariffs on seeds, fertilizers and pesticides to promote paddy rice production. In addition, Prime Minister Hun Sen called for further investment by both local and foreign investors in rice mills[16]. The promotion of rice processing, especially milling and packaging is aimed at upgrading rice quality, reducing cost of production, increasing value-added and marketing for export. Paddy rice production is expected to reach around 7 million tonnes in 2010/11, with domestic consumption of around 3 million tonnes in the same period. This increased production is a result mainly of investment in irrigation, expansion and intensification of cultivated land, the use of better farming techniques and improved seeds.

Recognizing the important role of paddy rice in enhancing growth and reducing poverty, the government came up with the policy on "Paddy Rice Production and Promotion of Milled Rice Export" in mid-2010, aiming to achieve a paddy rice production surplus of 4 million tonnes and milled rice export of at least 1 million tonnes by 2015 by continuing to invest in irrigation facilities; encouraging private sector investment in paddy rice processing and export; and improving procedures for export and transport facilitation. However, the rice subsector is challenged by weak governance and institutional support. Ear (2009), examining dynamic governance and the growth of Cambodia's rice industry, finds that the sector's export is markedly imperfect as two entities, the state-owned Green Trade Company (GTC) and the National Cambodian Rice Millers Association, headed by the director of GTC, are allowed to export milled rice above 100 tonnes without an export licence. Another dominant exporter, Angkor Kasekam Roongroeung, acquired an export licence through its governmental channels.

4.7 Rubber

In rebuilding the rubber subsector, the government started by investing in production for domestic consumption. Later, Cambodian rubber began to penetrate foreign markets, especially those in Viet Nam and China. The government then introduced important institutional and policy reforms for market-based agricultural growth with the assistance of external funding agencies, especially from the Asian Development Bank (ADB). The government began to withdraw from direct intervention in the production and marketing of agricultural and agrobased products, though state-owned rubber estates continued to constitute about 80 percent of the total rubber-exploited area of the country until 2004, when the world price of rubber increased (ADB 2003).

Continued state ownership was impeding the growth of the whole subsector by restricting the development of smallholder and family-scale rubber production and constraining smallholder processing and marketing (ADB 2003). Quality was also an issue, leading to Cambodian rubber being 15–20 percent lower in price on the

[15] Criteria for SLC have no time restriction with a maximum 1.250 m2 for residential use and 2 hectares for agricultural use, and it can be transformed into private property.

[16] Radio Free Asia, March 22, 2011.

global market than that of other countries. To address these problems, and in response to ADB conditionalities, the government initiated an overall review of the subsector and examined the rules and regulations for marketing rubber products. From 2000 to 2006, several policy and regulatory reforms were undertaken to promote smallholder rubber plantations and private sector processing factories and collection points. These liberalized private rubber production and marketing (Circular 2826 SCN.KSK on the "Announcement on Trading and Buying Stations for Rubber from Family Plantations" on 13 June 2005), and provided mechanisms for the standardization of rubber grades to enable Cambodian rubber to fetch higher prices on the global market (Prakas 086 RBK.KSK on the "Use of Regulation on Grades of Rubber in Cambodia" on 17 March 2004).

5. Preliminary impact assessment of FDI in agriculture

The past years have witnessed a growing interest in investment in land and agriculture in particular, in Cambodia. Concerns over potential risks of such investment have been echoed by various interest groups. A study by German Technical Cooperation (GTZ) in 2009 identifies both opportunities and risks from foreign investment in land in Cambodia, which are highlighted below.

Socioeconomic aspects: at the macro level, there has been evidence of job creation in the production of biofuel providing average monthly wages of US$100; improvement in local roads, but also degradation of community roads by heavily loaded trucks of the investment projects; generation of foreign exchange earnings through export of wood and wood products, rubber, cotton, essential oils, fish and live animal; and a contribution to government budget through land concession rental payment of between US$0–10 per hectare/year. At the microlevel certain risks arise. For instance, there is evidence of negative effects on indigenous people's access to land; loss of community opportunity to collect non-timber forest products; absence or lack of transfer of technical skills from foreign firms to farmers, such as breeding, use of seeds, improving soil conditions or using fertilizers. Opportunities do also emerge, which range from double income for unskilled construction and agricultural labourers to the development of schools and healthcare centres.

Food security: the status of community food security could be affected through the loss of community access to non-timber forest products. Hansen *et al.* (2006) indicate that non-timber forest products contribute about 42 percent of poor household income and 30 percent of medium household income in rural communities in Cambodia. However, this study found that foreign investment in land is unlikely to have a negative effect on rural community food security in the short term.

Environment: foreign investment in land creates both environmental gain and risk. A host country could gain from the import of new technology and environmentally friendly agricultural production methods, and reduction of soil erosion through agricultural production on formerly abandoned land. However, benefits also bring risks in their wake. Environmental concerns over large-scale foreign investment include climate change and soil erosion, water security and quality, biodiversity and local ecology.

Given the concerns highlighted above, this study with a specific focus on foreign investment in agriculture, but not in land in a broad sense, is intended to shed more light on likely impacts of foreign investment in agriculture on local community livelihoods, the environment, food security, and land and water use. Taking stock of CDC data from 2000 to June 2010, the study found that foreign investors in agriculture engage mainly in crops and forestry in the form of ELCs, whereas in food processing they engage mainly in the production of drinking water, soya milk and instant noodles. This section focuses on crops (sugar cane, rubber, corn) and forestry in order to examine the likely impacts of such foreign investments.

This assessment is based on results from focus group discussions (FGDs) conducted by the study team in December 2010 in the Kompong Thom province for the case study on rubber, consultation with corn plantation and drying company HLH Agriculture, previous FGDs conducted by the Cambodian Development Resource Institute (CDRI) on the impact of Chinese investment in natural resources on communities in the Kratie and Stung Treng provinces (April 2010), case studies on rubber in Mondolkiri province, and sugar cane in Koh Kong province conducted by the Cambodia Economic Association (CEA) in May and June 2010 (Table 8).

Generally, each FDI project has both negative and positive effects on the environment and livelihoods of local communities, and their scale varies from one subsector to another. Rather than adopting a full subsector-wide impact assessment, this study develops case studies to generate ideas and raise awareness among various stakeholders, particularly policy-makers. As such, it is important to refrain from making sectoral generalizations using these results.

6. Projects

6.1 Crops

The case studies on crops covered one sugar cane project, two rubber projects and one corn project. The projects were not selected randomly using factors such as land size, geographical location, crop type or company nationality, but through considerations of their importance and a study of their budget limitations.

Sugarcane plantation in Koh Kong Province

Koh Kong Plantation Ltd and Koh Kong Sugar Company Ltd

According to the CEA (Cambodia Economic Association) (2010), two ELC companies under the same representative's name—that of Ly Yong Phat—have been granted a licence for sugar cane

TABLE 8
Summary of fieldwork by the CDRI and CEA in 2010

No.	Company	Size (ha)	Location	Subsector	Source
1	Koh Kong Plantation Ltd (Thailand, Japan, China, Cambodia)	9 400	Botomsakor district, Koh Kong province	Sugar cane plantation	CEA June 2010
	Koh Kong Sugar Company Ltd (Thai)	9 700	Chi-Khor Leu commune, Sre Ambel district, Koh Kong province	Sugar cane processing	CEA June 2010
2	Socfin KCD (Belgium-Cambodia)	10 000	Bousra commune, Mondolkiri province	Rubber plantation	CEA May 2010
	DAK LAK (Viet Nam)	4 000	Busra commune, Mondolkiri province	Rubber plantation	
3	Tong Ming Group Engineering (China)	7 465	Kbal Damrei commune, Kratie province	Forestry (acacia)	CDRI April 2010
4	Tan Bien-Kompong Thom Rubber Development (Viet Nam)	8 100	Kraya commune, Kompong Thom province	Rubber plantation	CDRI December 2010
5	HLH Agriculture (Singapore)	9O ral) 4500 Amleang)	Oral district and Amleang commune, Kampong Speu province	Corn plantation and drying	CDRI November 2010

Source: Ngo & Chan 2010: Nos. 1, 2; Hem &Tong 2010: Nos. 3, 4

plantation from the Cambodian Government. The land covers about 20 000 hectares, of which 9 400 hectares is under Koh Kong Plantation Ltd in Botomsakor district and 9 700 hectares under Koh Kong Sugar Company Ltd in Sre Ambel district. The sugar cane goes through the processing factory of Koh Kong Sugar Company Ltd, which has a capacity of 6 000 tonnes a day, and is packed for export to European Union markets. CEA carried out its survey in Trapaing Kandal and Chi-Khor Leu villages, interviewing 143 households in order to assess impacts on employment, livelihood transformation and land transactions.

Socioeconomic impacts: a number of skilled, semi-skilled and unskilled jobs have been created for the local people and for those living in the surrounding areas and other provinces. These jobs include preparing land, planting, applying fertilizer, controlling pests, weeding, harvesting, collecting and transporting, which are dependent on the season. The planting and harvesting period represents the high season for employment, stretching from November to May. During this high season in 2009, approximately 3 400 workers were employed, around 30 percent of whom were from Koh Kong. Only around 1 300 daily workers were employed between June and October 2009. These labourers receive a daily wage of around US$2.50. In addition, accommodation and transportation are provided to migrant workers from such provinces as Banteay Meanchey, Kampot and Kompong Thom. The two companies also employed a total of 511 Cambodian and Thai office staff, earning an average of US$6 per day, but it should be noted that a slim proportion of the local people employed were working as office staff.

Prior to the arrival of the company, the local people earned a living by farming wet season rice and cash crops and raising cattle. After the company arrived, some of the land once used for cash crops was annexed by the sugar cane plantation. This has triggered serous disputes between the local people and the company, as a source of revenue for the former has been taken away. Further, local people's cattle can no longer roam freely on farm and forest land that has become part of the company's investment zone. If the company's guards see the community's livestock on the company's plantation, they will catch, detain and sometimes shoot the animals. It should be noticed that the company, after being awarded its ELC licence in 2006, started clearing the land without giving prior notice to the local community. This provoked strong protests by the local people: 449 households filed a petition letter on the loss of their farmland and other property. Even though the company has offered compensation to the affected households, there have been many complaints that this is insufficient and unfair: villagers who have relatives working for the company tend to get better compensation, with overall compensation ranging from US$25 to US$350. One critical reason villagers cannot make a strong argument for their land is that most of them do not possess a land title for the land they have been cultivating.

The central challenges behind the disputes include the absence of clear guidelines or procedures to resolve the disputes, uneven compensatory choices offered by the company to affected households, replacement of common dispute settlement mechanism by non-transparent and unfair case-by-case, household-by-household solutions, and limited or no assistance from the local authorities. A community representative claimed that local authorities are not helpful, and thus the community has little support in finding a way to deal with the company.

Environmental impacts: the sugar cane plantation and processing factory reportedly generate two types of pollution. First, water pollution seriously affects the daily life of the local community, as disposed chemical substances contaminate the water upstream, which then passes downstream to the villagers. This water is vital to the villagers, so this pollution harms both humans and animals. Second, factory emissions lead to air pollution, which makes it hard for the people living nearby to breathe.

Infrastructural impacts: the CEA report did not capture information on the development of local infrastructure by the investing company. However, it is likely that roads have been built in order to transport materials and workers in and out.

Rubber Plantation in Mondolkiri and Kompong Thom Provinces

Socfin KCD and DAK LAK (Mondolkiri)

The CEA (Ngo and Chan, 2010) conducted a survey in May 2010 on rubber plantation in Mondolkiri in two communes—Bousra and Krang in Peach Chinda district. Eight ELC companies are licensed for rubber plantation, each with a land area of 3 000–5 000 hectares. Most of them have cleared land and planted rubber trees. All of the ELCs are reported to be active, except for Sarmala Company. CEA examined the potential effects mainly of Socfin KCD and DAK LAK.

Socioeconomic impacts: at the start of its operations in 2008, Socfin KCD employed around 10 000 people—around 2 000 between May and August and around 8 000 people for the rest of the year. About 20–25 percent of the total workforce is made up of local people; 60 percent female. The company will employ at least 1 500 workers when the plantation is fully planted. Unskilled workers can earn around US$5 a day; their jobs include weeding and applying fertilizers; they work 10 to 15 days per month. Skilled workers can earn a bit more per day – around US$6.50–8. All the skilled employees are migrant workers, particularly from Kompong Cham province, with work experience in the rubber industry. They are satisfied with their jobs even though they have to live far away from their families.

Overlapping land claims between villagers and companies are prevalent, as villagers traditionally move their farmland from one place to another every few years and do not have certificates of land ownership. All villagers in the Bousra commune have been affected by Socfin KCD and DAK LAK investment projects. Villagers were shocked when Socfin KCD cleared their lands without giving them any notice. Disputes erupted which resulted in villagers burning the company's tractors. The Land Conflict Resolution Committee, headed by the provincial deputy governor and comprising members from district and commune authorities, was set up to tackle the problem. Around 172 affected households were on the list endorsed by the government. However, it was claimed by a local NGO (CLEC: Community Legal Education Centre) that 362 households in Bousra were initially affected in 2008. Ultimately, the company came up with compensation schemes, such as cash payment of around US$200–250 per hectare, land exchange and land exchange and development. Among the options, villagers tended to opt for cash payment, as the land offered in exchange was too far from the village and not fertile enough for cultivation. Some initially selected the third option, but later switched to cash payment given the delays in the process. Overall, cash payment was the best option.

On the DAK LAK project, around 40–50 percent of the concession belongs to villagers, with whom the company negotiated the deal in advance. Half of the villagers' section of the concession is near the village, which is convenient for raising cattle. The company has also provided loans to villagers with a grace repayment period of 10 years to allow them to plant rubber on the land allocated to them, and has agreed to buy latex from villagers at 80 percent of the international market price. The company even allows landless farmers to cultivate crops between rubber trees on its concession land before rubber trees are mature enough for tapping. Villagers were satisfied with the company's offer.

Before the companies arrived, people in Bousra made a living by farming lowland and highland rice and cash crops, collecting forest by-products, raising livestock, fishing, hunting, gold mining and small businesses. However, the ELCs have affected highland cultivation badly, and some have also lost their land to the companies. This has brought hardship, as they can no longer generate an income from collecting non-timber forest products, which used to generate US$10–15 per week. They now have to go further from home to hunt and to find forest by-products in order to survive. Moreover, some have lost an income source from livestock rearing given the loss of grassland to ELCs.

Environmental impacts: the study captured no critical aspect of the local environment. It might be assumed, however, that the investment has caused a natural imbalance, given the drastic and large-scale influx of rubber trees. Rubber trees store carbon, but they require a larger volume of water, which could lead to water use that competes with that of local ethnic communities. It could also result in loss of natural habitat, as forest land has been turned into private estates.

Infrastructural impacts: the companies have rehabilitated roads and contributed school buildings. For instance, DAK LAK donated a school building to the community in Koh Nhek district. However, there have been complaints about the destruction of community-funded roads since the arrival of the companies.

Tan Bien-Kompong Thom (Kompong Thom)

The Cambodian Development Resource Institute conducted fieldwork in Kraya commune in December 2010, consulting a commune councillor on the overall impacts of ELCs on the commune and holding a Focus Group Discussion (FGD) in Thmor Samleang village in order to examine the effects of the investment project. It should be noted that the number of private investments, both foreign and domestic, in the form of ELCs in Kraya commune has been rising. There have been several accounts of disputes between Vietnamese investments i.e. Tan Bien Investment and Phuek Fa Investment (Phuek Fa was formerly known as Mean Rithy Investment, owned by a Cambodian investor), and the local communities.

The major occupations of the people in Kraya commune are growing paddy rice and cassava and collecting forest by-products. Villagers claim that an income solely from growing paddy rice is not enough: farmers generate extra revenue by collecting wood resin and vines or selling labour on rice fields or plantations. Some catch porcupines, as they are in high demand for medicinal purposes. Some use their income from forest by-products to buy fertilizer in order to increase their paddy yield (the average yield per hectare is between 2 and 4 tonnes), as soil quality is poor. Given the rising demand for cassava in recent years, about half of the total population of the commune has decided to clear state forest land to grow this crop. Aware of the hardships facing the villagers, the commune councils have turned a blind eye to their clearance of state forest land. On average, a tonne of cassava fetches around KHR300 000 (about US$73). Some clear state forest land to grow cashew nuts or jack fruit.

Socioeconomic impacts: at the onset of Tan Bien's business operations there was evidence of significant job creation as the company employed a large number of people in the commune (no data confirmed by the commune council). Tasks included mainly weeding, digging, spraying pesticide and watering, at an average daily income of around KHR12 000–13 000 (approximately US$3.0-3.5). However, the majority of employed villagers left their jobs, complaining of unbearable hardship. Problems ranged from insufficient skills to grow rubber saplings and apply pesticide, to fraud in wage disbursements. In response, the company gathered a workforce from neighbouring communes and other provinces, mainly Kompong Cham, as farmers from there are more familiar with producing rubber sapling and maintaining rubber trees. Another impact has been the loss of traditional sources of income, particularly resin and vine collection, as firms have bulldozed the commune's forest. In addition, several households have lost their cassava plantations to ELCs, as these had been acquired through illegal clearance of state forest land. Note that this clearance originally occurred with the tacit acceptance of the commune council, as noted above. As such, the project has been a double blow for households: loss of income sources but no employment and skills acquisition opportunities. In compensation for the loss of the land, the company provided households with US$200 per hectare for cleared land with crops and US$100 per hectare for newly cleared land without crops/ vegetation. However, for the most part, one striking characteristic of dispute settlements between the company and villagers is that the company representative agreed to follow requests made by the commune facilitator but in practice the company broke this promise. For instance, the company agreed to keep streams in place, both small and big, for animals to drink from, and for other agricultural purposes, but later filled them in and made no serious effort to keep its promise. The commune council has little power over the company, which is obliged to get permission from provincial level to implement the project. Meanwhile, if villagers let their livestock roam onto the plantation, the animals are detained and can be retrieved only after paying a fine of KHR100 000 (approximately US$25).

Environmental impacts: there is no evidence of competing water use between the company and Thmor Samleang villagers. Villagers' crops of paddy rice, cassava and other crops are rainfed, and the company has excavated its own reservoir for watering rubber tree saplings. In addition, there are no complaints from villagers as to the overuse of pesticides that could degrade soil quality and pollute groundwater, or the costs of harvesting the forest around the village and particularly the commune. However, this could be due to the limited knowledge of villagers regarding the environmental impacts of the development.

Infrastructural impacts: the company has built roads connecting its land to the commune centre. Although it seems these roads have been built for the sake of the company's business operations alone, Thmor Samleang, which is near the company's land, benefits from their construction. There is no other evidence of the company's contribution to local infrastructural development, such as hospitals, schools, irrigation and the like.

Business model: no formal and genuine business model has been applied, but the company has cleared forests and expropriated households plantations; in return it has employed villagers on the rubber plantation on an irregular basis only.

Corn plantation in Kampong Speu Province

HLH Agriculture

The HLH Group Limited, which is 100 percent Singapore-owned and was listed on 21 June 2000 on the SGX Mainboard, operates in agriculture, property and investments, construction and agriresearch and development. In Cambodia, the company received investment approval in 2007 and started maize plantation in 2009 on 9 800 hectares in Oral district and 4 500 hectares in Amleang district, Kompong Speu province, 48 km from Phnom Penh. The HLH Group produces animal feed on a 70-year concession with a capital investment of about US$45 million. It has one processing factory with production capacity of around 600 tonnes per day, and four plantation farms, each covering between 450 and 2 000 hectares. The company has imported high-tech machinery from Singapore. All farms are overseen by farm supervisors from the Philippines, Taiwan, Singapore, China and Myanmar, who use different management styles. All the assistant supervisors are Cambodians, who are supposed to take over the management jobs in the future, once they have gained enough skills and experience. Thirty of the company's circa 450 full-time staff are office staff, with only two from Singapore. On the plantations, there are about 200 workers in low season and around 1 000 in high season; the majority are Cambodian.

Socioeconomic impacts: when the project began in 2007, a significant number of jobs were created, providing an average daily income of between KHR10 000 and 12 000 (around US$2.5–3.0), with villagers having to spend around KHR1 000 (around US$0.33) on transport to and from the plantation. Recently, the number of workers employed has declined considerably, as the company has replaced them with imported machines, for example for sowing and harvesting. This has limited villagers' opportunity to access this new income source. The company claims that it has not been able to apply a business model because farmers cannot afford to buy seeds and machines and lack appropriate technology for cultivating maize. Meanwhile, the project has severely affected local people's main and traditional sources of income, namely, rice growing and charcoal production: HLH Agriculture has claimed around 40 percent of villagers' paddy rice fields and also cleared forest land. Another large private investment (Kompong Speu Sugar), belonging to Cambodian tycoon Ly Yong Phat (of the Koh Kong sugar cane plantations detailed above), has affected around 90 percent of villagers' paddy fields.

Resolving the disputes has been difficult as villagers do not have certificates to prove they own the land. The company claims that there have been no big problems with the local community as it has worked in a consultative manner, inviting the local people, authorities and other relevant bodies to come together to find a solution to any issues arising. In terms of compensation, the company promised new plots of land but, as these are very far from the village, only some people have agreed to this. In terms of cash compensation, according to villagers, the company pledged to provide each family US$1 000–2 000 per hectare, but this money has not yet materialized. Villagers can grow paddy rice only in the rainy season, given the lack of irrigation. Therefore, the majority of villagers also produce charcoal by going to the forest and chopping down small trees. Each household has on average two to four kilns; a small kiln can process about 5 m^3 wood in 15 days and generate circa KHR350 000 (approximately US$85), while a big one processing around 10 m^3 wood and taking roughly 20–25 days can make about KHR700 000 (circa US$170) However, right now, villagers can no longer produce charcoal because the forest near the village has been cleared and the company does not allow people to enter its demarcated land. Villagers can get wood now only by going further into the forest, which most of them do not want to do. They feel pessimistic about their future livelihoods since the arrival of the company, and some have left their company plantation jobs. Meanwhile, the company claims the forest was already cleared when it arrived.

Environmental impacts: as more land has been cleared, forest coverage has declined. This could have adverse impacts on the ecological system and the biodiversity of the area. People in the village are very worried about this environmental degradation.

Infrastructural impacts: according to group discussions, so far the two companies mentioned (HLH Agriculture and Kompong Speu Sugar) have built no significant physical infrastructure. A few roads and bridges have been built for company use only. Most of the roads are government built.

6.2 Forestry in Kratie Province

Tong Ming Group Engineering and Eight Other ELCs

There are eight ELCs in the Kbal Damrei commune according to an interview with the second deputy chief of the commune in April 2011. Distribution of the project by nationality is shown in Table 9. It is important to note that this account of the preliminary impact assessment has been compiled from the investigation of one investment project owned by Tong Ming Group Engineering in early 2010, and does not reflect the full-scale or commune-wide impact of the total number of investment projects listed in the table below.

Socioeconomic impacts: communication between villagers, companies and authorities is weak: in some cases, villagers were shocked to see the companies turn up to clear land close to their backyards. There are disputes between the local community and the companies over both land-grabbing and restrictions on farmers entering the forest. Villagers can no longer go to collect vines, wood resin, rattan and bamboo or to hunt, which has cut off a main source of income. The forest plantation is protected by armed guards, and if villagers are found entering it, they are arrested and fined and their belongings seized. Meanwhile, villagers are no longer in favour of working at the company owned by Tong Ming Group Engineering, given its strict working conditions (e.g. monthly salary is reduced if workers take a day off and workers are not allowed to go home if overtime work is necessary), and its land-grabbing activities. Workers are mainly from Prey Veng and Svay Rieng provinces.

Environmental impacts: the CDRI team observed that the companies in the commune were using heavy machines to clear the forest, often violating the regulation that requires companies to keep intact 200 m of forest on each side of a large stream (one that is 20–30 m wide) and 100 m of forest on each side of a smaller stream to ensure the sustainable use of water and protection of the environment. The real practice is that companies leave around 10 m of forest on each side of the stream. Villagers also mentioned the frequent transportation of timber at night to Viet Nam.

Infrastructural impacts: there is evidence of road construction, but no school buildings or pagodas have been built. New roads have been built mainly for the company's use and not for the public, who are not allowed to intrude on the company's property.

6.3 Overall assessment

Overall, based on the authors' preliminary examination, the costs of FDI projects seem to outweigh the benefits to an extent that is hard to estimate. Foreign direct investment projects have both positive and adverse effects. On the positive side, some projects have created significant employment for local communities and for communities in nearby provinces. However, others have not. Land conflict is quite common across ELC projects, given the lack of a sound system of land tenure and limited consultation with local communities prior to the granting of ELCs. In addition, some ELC projects involved in clearing the forest eliminate major sources of community income, such as collection of non-timber forest products and hunting. The filling in of streams was also evident in some projects, which could lead to water shortages.

Notably, Cambodia has not experienced a food deficit during the past decade. Paddy rice production and the total area under cultivation have grown quite favourably, with average growth rates of 9.4 percent and 2.7 percent, respectively, between 2004 and 2009. In 2009, the total rice cultivated area reached just over 2.7 million hectares, generating total production of 7.6 million tonnes and a surplus of 3.5 million tonnes (Table 10). Therefore, at first glance, and from a short- to medium-term perspective, Cambodia does not seem to have grave concerns related to food security. It is more important to note that the drastic rise in the size and total number of ELC projects does not seem to threaten national

TABLE 9
Profiles of ELCs in Kbal Damrey Commune, Sambour District, Kratie Province

No.	Company	Year of approval	Duration of contract	Size (ha)	Purpose of investment	Nationality
1	Great Island Agricultural Development Co., Ltd	2006	70 years	9 583	Teak plantation and processing factory	American
2	Global Agricultural Development Co., Ltd	2006	70 years	9 800	Tectona/teak plantation and processing factory	American
3	Asia World Agricultural Development Co., Ltd	2006	70 years	10 000	Teak plantation and processing factory	Chinese
4	Great Asset Agricultural Development Co., Ltd	2006	70 years	8 985	Pistacia Chinasis Bunge and other tree plantation	Chinese
5	Great Wonder Agricultural Development	2006	70 years	8 231	Pistacia Chinasis Bunge and other tree plantation	Chinese
6	Tong Ming Group Engineering	2007	70 years	7 465	Rubber, acacia, jatropha plantation and processing factory	Chinese
7	Agri-industrial Crops Development	2008	70 years	7 000	Rubber and acacia plantation	Chinese
8	Carmadeno Venture Limited	2009	70 years	7 635	Sugar cane plantation	Indian

Source: Interview with deputy chief of Kbal Damrey Commune April 2011; MAFF ELC profiles 2011 (http://www.elc.maff.gov.kh/profiles.html)

TABLE 10
Evolution of paddy rice production area and ELC land size, 2004–2009

Category	2004	2005	2006	2007	2008	2009
Rice cultivated area (ha)	2 374 175	2 443 530	2 541 433	2 585 905	2 615 741	2 719 080
Rice area expansion (%)	2.6	2.9	4.0	1.7	1.2	4.0
Paddy rice production (tonnes)	4 170 284	5 986 179	6 264 123	6 727 127	7 175 473	7 585 870
Growth of paddy rice production (%)	-11.5	43.5	4.6	7.4	6.7	5.7
Paddy rice surplus (tonnes)	650 184	2 061 830	2 240 438	2 577 562	3 164 114	3 507 185
Milled rice surplus (tonnes)	416 118	1 319 571	1 433 880	1 649 640	2 025 033	2 244 598
Accumulated ELCs (ha) (total)	583 273	650 316	818 572	847 573	888 509	1 024 639
ELC expansion (%)	1.1	11.5	25.9	3.5	4.8	15.3
Accumulated ELCs ha (FDI)	73 100	97 480	188 499	217 500	258 436	355 914

Source: MAFF, 2010

food security in the medium term because the current stable surge in paddy rice cultivated areas will continue to produce surplus for domestic consumption in the short and medium term. In addition, in the short and medium term, ELC expansion does not appear to compete for water use over paddy cultivation. However, in the longer term competing use for water and land could be detrimental to national food security if size of ELC projects keeps growing at the current pace.

However, it is important to note that rising world demand for food (rice) and industrial crops like rubber, cassava and sugar cane for the production of biofuel energy could place Cambodia in a perilous situation if its trade and investment policies do not take food security seriously into account. In the long term, land use conflicts between villagers and ELCs producing rice and industrial crops for export, fierce competition for water currently used for paddy rice production, and loss of traditional sources of income for local communities could contribute to a decline in household food consumption and thereby a reduction in nutrition.

7. Conclusions and recommendations

Agriculture has been a constant contributor to the national economy, employing a significant proportion of the rural workforce and generating substantial foreign exchange earnings. In its fourth legislature, the Cambodian Government has focused even more strongly on promoting the sector, by relaxing taxes related to agricultural products and developing rural infrastructures such as roads and irrigation. New measures have been taken to help people in local communities, including removing big fishing lots that once were under private ownership. The government, with the support of development partners, has provided technical assistance to rural people in rice farming, fisheries (aquaculture) and livestock production. It has also built up the irrigation system so the farmers can become less dependent on rain, particularly in rice production, and more resilient to climate change.

The government has also undertaken forestry reform in order to facilitate investment in forestry and crops through the establishment of legislation on concessions, forestry community formation and environmental protection. The subdecree on Economic Land Concessions, adopted in 2005, helps in the granting of land concessions to foreign and local investors to exploit unused and/or infertile land. Laws relating to sanitary and phytosanitary issues and animal health have been enacted, to control livestock production, prevent animal losses and contain animal diseases. Recently, the government, through the

MAFF, prohibited the import of livestock from neighbouring countries, to prevent swine flu. This measure not only reduces the risk of animal disease pandemics but also helps local producers compete with imported products.

On the trade side, a market mechanism has been set up to channel agricultural products to local and international markets, as the sector is one of the main drivers of economic growth in Cambodia. To help the sector become more competitive, there have been improvements to soft infrastructure, related to rules and regulations, red tape and costs of doing business. Attracting investment in the energy sector is also deemed important, as the price of electricity in Cambodia is still high compared with in other countries in the region.

The share of total agricultural investment to total investment is small, at around 6 percent between 2000 and June 2010, although interest rose during this period from investors from countries such as Thailand, China, Viet Nam, Republic of Korea, Singapore, Japan, Malaysia, Canada, America, India, France, United Kingdom, United States and Denmark. Investors engage mainly in crops, namely, rubber, cassava, corn, sugar cane and cashew nuts, and forestry, such as teak and acacia. The dramatic rise in interest in recent years has sparked concern from various stakeholders as to the potential effects of foreign ELC projects on community livelihoods, local environment quality and national food security.

Preliminary examination using data from both the CDC and the MAFF shows both positive and adverse effects from FDI projects. Some projects have created significant employment for local communities; others, however, have not. In addition, land conflict has been common, resulting from a weak land tenure system and limited consultation with local communities prior to the granting of ELC projects. Moreover, some projects involved in forest clearance have eliminated traditional community income generation through collection of non-timber forest products, such as vines, wood resin, bamboo and rattan, and hunting. The filling in of streams by some projects could lead to water shortages. Overall, then, it seems the costs of FDI projects tend to outweigh the benefits.

With a growing population and land becoming more limited, food security is becoming a concern in Cambodia. According to an examination of CDC investment data (2000–2009) and MAFF/ELC investment data, and preliminary fieldwork on a number of FDI projects, the authors find that, in the short and medium term, Cambodia will not suffer from food insecurity, despite the existence of a traditionally widespread informal paddy rice export to Thailand and Viet Nam. Nevertheless, in the long term, conflicts over land use, water shortages and loss of other sources of income, all contributed to by a dramatic expansion in investment in ELCs in recent years, could lead to a decline in household food consumption and thereby a reduction in nutrition. Particularly at risk are subsistence farming households and those that cannot earn enough from growing rice, such as in Kraya commune in Kompong Thom province.

7.1 Policy recommendations

In order to ensure that opportunities for foreign investment in agriculture in Cambodia are sustainable and beneficial to all stakeholders involved, the government and concerned stakeholders should consider taking the steps outlined below.

7.2 Central and local government

- Environmental impact assessment should be conducted with wide participation from concerned stakeholders, particularly from communities residing close to project sites, prior to granting ELC approval. As the impact assessment becomes more reliable and transparent, the number and scope of land conflicts will be reduced.
- In order to avoid land disputes and overlapping claims, the MAFF and related institutions should demarcate ELC borders in consultation with communities adjacent to each project.
- The government should monitor ELC operations more closely to prevent sub-standard forest clearance activities, such as filling in of upstream water sources and excessive logging activities.

- Authorities at both central and provincial levels should hold frequent consultations with communities and companies so as to be able to pre-empt problems.
- The MAFF should update progress on the operations of various ELCs on its website – and through other public media – on a regular basis, to ensure more transparency and generate more credibility.
- The government should take food security into account seriously in the provision of future ELCs in order to avoid problems of competing land use between industrial crop production for export and paddy rice production for domestic consumption.
- Also on food security, the government needs to review overall policy on investment in agriculture, particularly in relation to ELCs, given their drastic expansion, and to ensure agricultural trade policy is not geared solely towards foreign exchange earnings.
- The government must tackle the large-scale informal export of paddy rice and other crops with neighbouring countries.
- To reduce land conflicts and ensure benefits are derived from large-scale agricultural investment, whether private domestic or foreign, the government must take swift and prudent action to provide land titles to all rural and remote communities, at little or no cost, to be implemented on a sporadic basis by prioritizing those affected by projects.
- Future rules and regulations should put increased stress on the protection of rural communities through implementing social impact assessment prior to project approval, and they should be in line with regulations on investor protection.
- It is vital to eliminate unofficial fees and set up a computerized investment and business licensing application process, and related administrative procedures to improve the investment climate; this should be done through a step-by-step or ministry-by-ministry approach.
- Dispute settlement mechanisms should be reviewed in order to build confidence among private sector firms.
- Relevant ministries should set up a department for feedback in order to be able to provide clear guidance and assistance to investors when they need it.

7.3 Private companies

- Existing and future ELC holders should be more transparent and accountable in their operations to nearby communities and the public by contributing more to the development and maintenance of local infrastructures.
- Companies should maintain good, frequent and direct communication with communities through various community social gatherings and the like.
- Resolution of conflicts, such as land disputes, should be based on a win-win approach, according to the model used on the DAK LAK rubber plantation project in Mondolkiri.
- Companies, ELCs and processing plants in particular should be more responsible for the quality of the environment and the ecological system of the project area.
- Chemical substances used in factories should comply with environmental regulations so as not to harm people and animals and contaminate water in the surrounding area.

7.4 Affected communities

- Communities should maintain good, frequent and direct communication with companies through various community social gatherings and the like.
- Communities should be more involved with education programmes provided by the authorities and NGOs regarding their rights to property and how to resolve land conflicts.
- Communities should report any irregular operations of the ELC companies to the commune or provincial authority immediately, for example the filling in of streams or excessive logging.

7.5 NGOs/Civil society

- Local NGOs should actively engage in raising community awareness regarding civil rights and how to exercise those rights.
- Community NGOs should closely monitor potential conflicts between local communities and ELC companies and compile accounts to inform the public.
- Civil society groups should advocate for better recognition of community rights by ELC companies and local authorities.

References

ADB (Asian Development Bank). 2009. *Cambodia: Agriculture Sector Development Program*. Progress Report on Tranche Release. Manila.

ADB. 2007. *RETA 6091: Capacity Building for Resettlement Risk Management - Cambodia Country Report*. Manila.

ADB. 2003. *Report and Recommendation of the President to the Board of Directors on Proposed Loans and Technical Assistance Grant to the Kingdom of Cambodia for the Agriculture Sector Development Program*. Manila.

Amos, A. 1986. *Prospects and Problems for Foreign Investment in Nigerian Agriculture.* London, UK, Commonwealth Secretariat.

Braun, J. & Meinzen-Dick, R. 2009 *Land-grabbing by investors in developing countries: risks and opportunities*. Policy Brief 13. Washington, DC, International Food Policy Research Institute.

CDC (Council for the Development of Cambodia). 2010. *Cambodian Investment Board: Projects by Sector Approved.* Phnom Penh.

Clay, J. 2003. *World Agriculture and the Environment: A Commodity-by-Commodity Guide to Impact and Practices.* Washington, DC, Island Press.

Ear, S. 2009. *Sowing and Sewing Growth: The Political Economy of Rice and Garments in Cambodia.* (unpublished)

FAO. 2011. *National Fishery Sector Overview: Cambodia*. Fishery and Aquaculture Country Profiles. March 2011. Rome.

FAO. 2010. Food and Agriculture Organization's Online Database (available at http://faostat.fao.org/site/567/default.aspx#ancor).

FAO, ADB & OIE SEAFMD. 2009. *FAO, ADB, OIE SEAFMD Study on Cross-Border Movement and Market Chains of Large Ruminants and Pigs in the Greater Mekong Sub-Region*.

IMF (International Monetary Fund). 2009. *Cambodia: 2009 Article IV Consultation—Staff Report; Staff Supplement; and Public Information Notice on the Executive Board Discussion*. IMF Country Report No. 09/325. Washington, DC.

IMF. 2007. Cambodia: 2007 Article IV Consultation—Staff Report; Staff Supplement; and Public Information Notice on the Executive Board Discussion. IMF Country Report No. 07/290. Washington, DC.

GTZ (German Technical Cooperation). 2009. *Foreign Direction Investment in Land in Developing Countries.* Eschborn.

Hansen, K. & Top, N. 2006. *Natural Forest Benefits and Economic Analysis of Natural Forest Conversion in Cambodia*. Phnom Penh, Cambodian Development Resource Institute.

Hem Socheth & Tong Kimsun. 2010. *Maximising Chinese Investment Opportunity in Cambodia's Natural Resources*. Phnom Penh, Cambodian Development Resource Institute.

MAFF (Ministry of Agriculture, Forestry and Fisheries). 2005. *Agricultural Sector Strategic Development Plan, 2006-2010*. Phnom Penh.

MAFF. 2007. *Agriculture Statistics 2006–2007*. Phnom Penh.

MAFF. 2008. *Report on the Achievement of the Ministry of Agriculture, Forestry and Fisheries in Implementing the Government's Rectangular Strategy 2004-2007.* Report for National Conference on the Progress of the People over various Achievements of Implementation of the Government's Rectangular Strategy 2004-2007 on 29-31 May – 02 June 2008. Phnom Penh.

MAFF. 2009. *Annual Report for Agriculture, Forestry and Fisheries 2008–2009*. Phnom Penh.

MAFF. 2010. *Annual Report for Agriculture, Forestry and Fisheries 2009–2010 and Goals 2010–2011*. Phnom Penh.

Mihalache-O'Keef, A. & Quan Li. 2010. *Modernization vs. Dependency Revisited: Effects of Foreign Direct Investment on Food Security in Less Developing Countries*. Paper for American Political Science Association Meeting 2010. Washington, DC.

MoP (Ministry of Planning). 2010. *National Strategic Development Plan Update 2009-2013*. Phnom Penh.

MoP. **2006**. *National Strategic Development Plan 2006-2010*. Phnom Penh.

MRB. **2010**. Malaysian Rubber Exchange (available at http://www3.lgm.gov.my/mre/MonthlyPrices.aspx#).

Ngo, S. & Chan, S. 2010. *Does Large Scale Agricultural Investment Benefit the Poor?* Phnom Penh, Cambodia Economic Association.

Rostow, W.W. 1960. *The Stage of Economic Growth: A Non-Communist Manifesto* Cambridge, UK, Cambridge University Press.

SOFRECO & CEDAC. 2005. *Study on the Evolution of the Cambodian Rubber Sector*. Draft report. Clichy-Cedex France and Phnom Penh.

Smaller, C. & Mann, H. 2009. *A Thirst for Distant Lands: Foreign Investment in Agricultural Land and Water. Manitoba,* International Institute for Sustainable Development.

Stockbridge, M. 2006. *Competitive Commercial Agriculture in Africa: Environmental Impacts*. Rome, FAO.

Tasker, P. 2003. *A Review of the Current Status of the State Owned Rubber Estates in Cambodia*. Agriculture Sector Development Programme Appraisal. Phnom Penh, ADB.

Todaro, M.P. & Smith, S.C. 2003. *Economic Development.* Edinburgh, Pearson Education Limited.

UNDP. 2007. *Cambodia's 2007 Trade Integration Strategy: Executive Summary and Action Matrix.* Phnom Penh.

Vermeulen, S. & Cotula, L. 2010. *Making the Most of Agricultural Investment: A Survey of Business Models that Provide Opportunities for Smallholders*. London, UK, IIED.

World Bank. 2004a. *Cambodia at the Crossroads: Strengthening Accountability to Reduce Poverty*. Phnom Penh.

World Bank. 2004b. *Seizing Global Opportunity: Investment Climate Assessment and Reform Strategy for Cambodia*. Phnom Penh.

World Bank. 2006. *Halving Poverty by 2015? Poverty Assessment 2006*. Phnom Penh.

World Bank. 2009. *Sustaining Rapid Growth in a Challenging Environment*. Cambodia Country Economic Memorandum. Phnom Penh.

Yang, L. (ed.) 2011. *Cambodian PM sacks fishery chiefs in 5 provinces*. Xinhua News Agency (available at http://news.xinhuanet.com/english2010/world/2011-07/01/c_13961037.htm).

Ghana:

Private investment flows and business models in Ghanaian agriculture[1]

1. Introduction

This chapter is divided into two sections. The first section presents an analysis of FDI flows bringing out key issues and impacts on the Ghanaian economy. In the second section, an assessment is made of two business models used by private investors in Ghana drawing on case studies from: i) the Integrated Tamale Fruit Company (ITFC) to illustrate the nature and issues in a model involving collaboration between a private company and local farmers through a nucleus estate and out-grower scheme for the production of organic mangoes; and ii) the Solar Harvest Ltd (formerly Biofuel Africa Ltd) which provides an example of a production model centred on large-scale plantations.

2. Analysis and impacts of private investment flows to Ghanaian agriculture

Several studies[2] have provided some excellent reviews of trends in FDI inflows to the Ghanaian economy from 1970 to 2003. However, in the period since 2003 which has witnessed increased attention by private firms to invest in large tracts of land for agricultural related activities and a huge turnaround for the Ghanaian economy, very little reliable information is available.

In Ghana, the period of 2003-2008 witnessed a significant increase in FDI inflows compared to the previous decade, with FDI growing from the level of US$167 million in 2003 to US$2.1 billion in 2008 within a span of about 5 years (Figure 1). With this trend, Ghana has outperformed several competitive West African FDI destinations including Senegal and Côte d'Ivoire, the two leading economies of the West Africa CFA zone. This remains true despite the onset of the 2007/8 global economic downturn which negatively affected global FDI trends.

Furthermore, in a survey conducted for Ghana in 2008 (see Aryeetey et al. 2009), 35 percent of the Multinational Enterprises companies covered indicated that macroeconomic and political stability were the most important factors influencing their decision to locate in Ghana. Other reasons given were market size, potential for growth and extent of natural and physical resources in the country, etc.

Tables 1, 2 and 3 provide a summary of the FDI related projects registered by the Ghana

TABLE 1
Values and number of FDI projects registered with GIPC in Ghana

	Number of projects	(%)	Values, millions $	(%)
Agriculture	78	5	110.7	1.0
Building	129	8.2	2 221.8	19.8
Export Trade	64	4.1	21.8	0.2
General Trade	318	20.2	987.7	8.8
Liaison	79	5	10.5	0.1
Manufacturing	401	25.5	7 211.4	64.3
Service	370	23.5	595.9	5.3
Tourism	136	8.6	52.8	0.5
TOTAL	**1 575**	**100**	**11 213**	**100**

Source: Ghana Investment Promotion Centre (GIPC)

[1] This chapter is based on original research reports produced for FAO by John Bugri and Adama Ekberg Coulibaly.

[2] Foreign direct investment flows to Ghana, Yaw Asante. p.102. In foreign direct investment in sub-Saharan Africa: *origins, targets, impact and potential*. Edited by s. Ibi Ajayi. African Economic Research consortium. Nairobi. Kenya. 2004.

TABLE 2
Ghana annual registered FDI projects by sector, 2003–2009, in US$

Sector	No. of projects	Total FDI	2003	2004	2005	2006	2007	2008	Jan–Jun 2009
Agriculture	78	110 739 277	7 935 771	5 531 725	2 282 900	6 100 832	30 262 277	55 612 844	3 012 928
Building/Construction	129	2 221 887 925	1 807 617	7 829 047	57 136 311	47 582 520	27 901 785	2 075 645 145	3 985 501
Export Trade	64	21 836 218	1 071 150	266 417	3 277 917	9 283 240	2 018 955	5 828 539	90 000
General trade	318	987 732 417	11 420 732	16 065 349	34 271 283	32 758 167	58 823 951	819 877 682	14 515 254
Liaison	79	10 535 285	0	0	0	0	361 200	9 524 085	650 000
Manufacturing	401	7 211 482 509	17 802 282	23 512 748	33 053 721	2 170 553 339	4 754 832 226	191 305 841	20 422 352
Services	370	595 994 195	44 715 593	74 681 649	33 863 215	42 218 080	48 594 748	286 377 708	65 543 201
Tourism	136	52 821 869	3 891 502	25 846 797	3 274 357	8 970 833	7 245 554	2 657 043	935 782
	1 575	11 213 029 696	88 644 646	153 733 732	167 159 704	2 317 467 010	4 930 040 696	3 446 828 888	109 155 019

Source: GIPC

TABLE 3
FDI agricultural projects as registered in the first six months of 2009 with GIPC

Name of company	Year business start	Nature investment (c: creation; e: extension)	Location	Region	Origin foreign equity	Ownership (f: 100 % foreign; jv: joint venture)	Sector (plantation, fisheries, forestry, livestock)	Activity/target products	Target markets (l: domestic; e: export)
Ahimsach Ghana Limited	27/04/2009	Creation	Opp. Kaadja	Eastern region (ER)	Lebanon	F	Fisheries	Fisheries (production of fingerlings, tilapia, catfish). production of table size fish	Local
Bionic Palm Limited	12/03/2009	Creation	Mim	Brong Ahafo Region (BR)	British Virgin Islands	F	Plantation	To produce & process cashew nuts& agricultural products of high quality for export.	Export
Ghana Sumatra Limited.	18/06/2009	Creation	Ashaiman	GR	Lebanon	F	Fisheries	Fish farming	Local
Haeshim Fisheries Limited	05/01/2009	Creation	Kusi	Eastern Region (ER)	Indonesia	JV	Fisheries	To produce & market oil palm seed to meet the demand of the Ghanaian oil palm in	Local
Inga Farms Limited	27/02/2009	Creation	Agogo & Kwamang	Ashanti Region (AR)	India	JV	Plantation	Large scale commercial farm and plantation.	Export
Maleka Farms Limited	23/02/2009	Creation	Tema	GR	Korea Rep.	JV	Fisheries	Tuna fishing	Local/export
Mim Cashew & Agricultural Products Ltd.	22/05/2009	Creation	SEGE	GR	Singapore	F	Plantation	General farming, jathropha with suitable crops like maize, sunflower, soya, etc.	Export
New Agricultural Services(Agricserv) Ltd.	15/06/2009	Creation	LATERBIORKORSHIE	GR	India	JV	Plantation	Cultivation of sunflower	Export
Suraj Agro Limited.	08/06/2009	Creation	VOLTA	Volta Region (VR)	India	JV	Plantation	General farming and exporters	Export
Triton Aquaculture Africa Limited	15/06/2009	Creation	Tema	Greater Accra Region (Gar)	Uk	F	Fisheries	Aquaculture Farming	Local

Source: GIPC. Note: ER: Eastern Region; VR: Volta Region; GAR: Greater Accra Region.

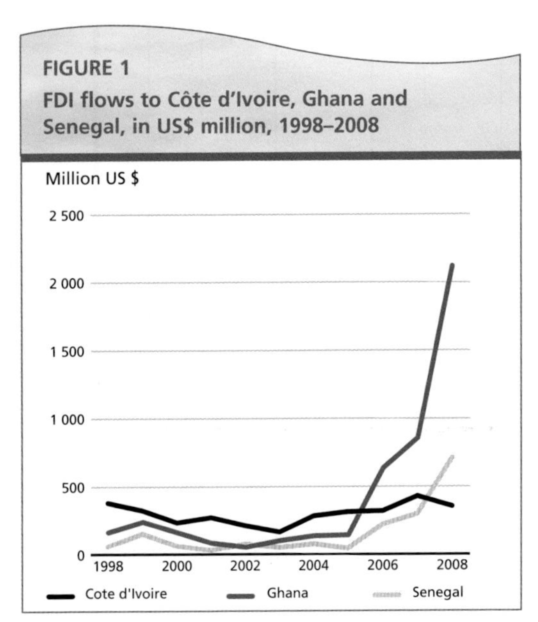

FIGURE 1
FDI flows to Côte d'Ivoire, Ghana and Senegal, in US$ million, 1998–2008

Source: Unctad 2009

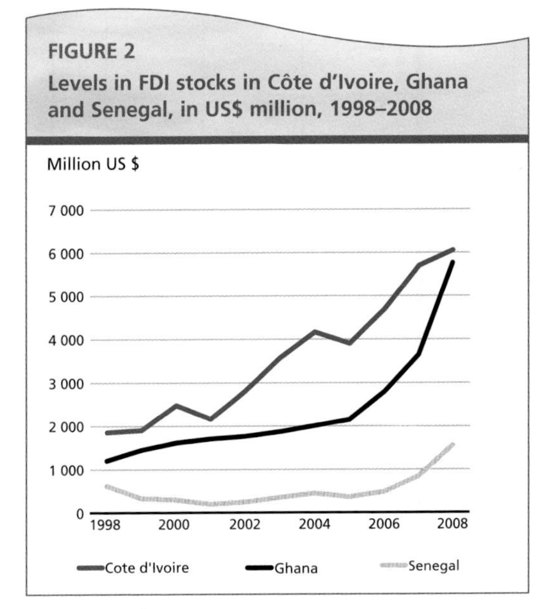

FIGURE 2
Levels in FDI stocks in Côte d'Ivoire, Ghana and Senegal, in US$ million, 1998–2008

Source: Unctad 2009

Investment Promotion Centre (GIPC) over the period 2003–2009, by economic activity. From Table 1, the cumulative value of investments registered for the period amounted to US$11.2 billion which was used to finance 1 575 projects. The greatest share of the investments went to the following subsectors: manufacturing, building and construction, general trade and services. Agriculture accounted for only about 5 percent although this does not account for projects in the food and beverages, fishing, and others related to agriculture.

From Table 3, joint-venture projects appear the most dominant modes of investment types into the Ghana economy with the values of these activities amounting to US$144 million, while the 100 percent foreign-owned enterprises were valued at US$71.5 million.

2.1 Sources of FDI to Ghanaian agriculture

FDI in Ghana comes from several countries in almost all parts of the world. Drawing from FDI data available for 1994-2008 from GIPC, FDI inflows to Ghana amounted to about US$13.5 billion, of which US$322.7 million (2.4 percent) went into agriculture (Table 4). Of this total amount, France invested about US$57 million followed by India (US$47.3 million), Switzerland (US$46.4 million), United States (US$43.3 million), Denmark (US$23.6 million), Belgium (US$19.8 million), Republic of Korea (US$14.6 million), Netherlands (US$9.2 million), China (US$5.1 million), United Kingdom (US$4.1 million) and Germany (US$4.1 million). The largest African investor country in Ghana is Nigeria and although it invested only US$3 million in agriculture, Nigeria's invested US$795 million in general trade category – the highest for that category. Nigeria also invested significant amounts of US$10 million in building and construction category and a similar amount in the Ghanaian manufacturing sector. Investments from Asia accounted for 37 percent out of the 184 agriculture related projects registered over the 1994–2008 period. In 2008, for example, China topped the list of investors with the highest number of projects[3] registered with the GIPC followed by India, Lebanon and Nigeria.

[3] All sectors considered.

TABLE 4

Cumulative value of registered projects classified by country, economic activity, September, 1994–December, 2008, in US$ million

Country	Total value of projects	Agriculture	Building/ construction	Export trade	General trade	Liaison	Manufactur-ing	Services	Tourism
UK	4883.5	4.1	40.4	3.6	12.9	0	4747.7	67	7
USA	2507.5	43.3	47.7	0.6	0.9	0	2240.2	150.9	23.95
United Arab Emirates	2077.3	0	2077.3	0	0	0	0		0
Nigeria	1120.8	0.03	10.8	0.1	795.2	0.01	10.5	0	0.3
Malaysia	546.3	0	0.3	0	0	0	7.4	538.6	0
China	237.5	5.1	5.8	1	49.5	0	157.22	13.9	4.9
India	156.5	47.3	8.7	8.5	44.1	0	33	14	1
Lebanon	118.8	0.6	12	0.3	50.8	0	45.7	3.4	6
Switzerland	110.4	46.4	2.1	4.2	2.8	0.1	27.6	23.1	4.2
Italy	104.7	0.2	40.7	1.2	1.5	0	58	0.2	2.9
Ireland	91.5	0	0	0	0	0	0	91	0.2
France	83.9	57	1	1	1.5	6.9	8.2	7.6	1.3
Netherlands	77.1	9.2	9.4	4.2	4.3	0	8.7	37.2	4.1
Republic of Korea	59.0	14.3	8.2	0.03	3	0	9.7	22.7	1.1
South Africa	58.4	0	2.8	0	9.1	0	3.1	41.9	1.5
Cayman Island	53	0	52	0	0	0	1	0	0
Norway	47.1	1.4	0.3	0	0	0	45	0.3	0.1
Mauritius	46.6	0	0	0	12	0	3.6	31	0
Denmark	41.6	23.6	1.1	0.1	0.5	0	8.8	7.3	0.2
Britain/India	40.6	0	0	0	0	0	40.6	0	0
British Virgin Islands	40.5	2.4	0.6	0	5.8	0	6.8	25	0.1
Canada	38.8	0.4	13	0.01	1.5	0	18.5	5.3	0.2
Germany	37.6	41	11	0.4	1.2	0.4	12.5	6.4	1.6
Africa	37.3	0	0	0	0	0	0	0	37.3
Belgium	34.1	19.8	0.3	0.2	2.2	0	9.6	1.3	0.8
Others	605	43.6	91.5	13.4	104	2.5	171	166.1	13.8
Total	13 255	323	2 437	38	1 103	10	7674	1255	113
% of total	**100**	**2.4**	**18.4**	**0.3**	**8.3**	**0.1**	**57.9**	**11.8**	**0.9**

Source: Ghana Investment Promotion Centre (GIPC)

The investments have been unevenly distributed among Ghana's regions and economic subsectors. All forms of investments in the various sectors, even in the agricultural sector, have concentrated in the Greater Accra Region (GAR), which recorded 1 504 projects. The next major recipient region of investments is the Ashanti Region (AR) which recorded 86 projects followed by the Central Region (CR) with 45 projects and the Western Region with 41 projects. Some regions have had little (Upper East Region (1 project); Upper West Region (4 projects).

2.2 Institutional and regulatory framework for investments in Ghana

In terms of economic policy objectives, Ghana aims at achieving a middle-income status by 2015, and become a leading agro-industrial country, thereby substantially reducing its poverty and hunger levels. Sustainable economic growth is to be private-sector driven, and achieved a conducive investment environment, macroeconomic stability, and pro-market reforms. In line with this developmental strategy, Ghana has sought to promote the private sector as the engine of economic growth by creating an enabling environment for private investors, both domestic and foreign. To meet these goals, Ghana has taken several FDI related measures, discussed below.

While there is no specific legislation on domestic investment, provisions on foreign investment are laid down in the Ghana Investment Promotion Centre Act of 1994 and in various sector specific laws and regulations.

- Both domestic and foreign investors are required to register their investments in accordance with the companies' code of 1963 or the partnership Act of 1962. Any enterprise with foreign participation must also register with the GIPC, indicating its activity, the amount of capital to be invested and where, the origin of the funds and plan for financing. Registration usually takes five working days. The registration fees are zero for wholly Ghanaian owned companies, US$100 for joint-ventures, and US$2 500 for foreign owned companies[4]. Registration must be renewed every two years for a fee of US$1 500. Additional fees apply for work permits for foreigners. In addition, foreign investors have to prove transfer of the required capital, and submit information on the proposed investment project, including equity structure, major activities, employment, and environmental impact.
- However, it is worth noting that the GIPC Act excludes foreign investors from participating in four economic sectors. These comprise the retail trade, operation of taxi and car services with fleets of less than ten vehicles; lotteries; and the operation of beauty and barber salons. Outside these areas, FDI is subject to a minimum capital of US$10 000 for foreign investors in joint-ventures, and US$50 000 for projects wholly owned by foreigners. Trading companies, whether partly or fully foreign-owned, require a minimum foreign equity of US$300 000 and must employ at least ten Ghanaians.
- Specific rules apply in the minerals, fishing, maritime transport, and postal services sectors, as well as to companies listed on the Ghana Stock Exchange (GSE). There is compulsory local participation in mineral and oil projects, whereby the government acquires 10 percent equity in ventures at no cost. Non-Ghanaians are not allowed to engage in small scale mining. As established by the 2002 Fisheries Act, ownership of fishing operations is restricted to Ghanaian citizens, but foreigners may own up to 50 percent of vessels engaged in tuna fishing. Only Ghanaian companies can engage in domestic maritime activities. Furthermore, the share held by any single external resident in a company listed on the GSE is limited to 10 percent, and the maximum

[4] Higher fees apply to foreign-owned trading companies (US$5 000) and establishments in the tourism subsector (between US$1 000 and US$10 000, depending on investment volume).

level of foreign ownership for each firm is 75 percent. Foreign insurance subsectors were abolished with the entry into force of a new Insurance Act in 2006.

- There are other relevant areas of investment incentives available to foreigners and Ghanaians alike. Certain machinery imported for investment purposes are eligible for reduced import tariffs and VAT rates. Tax rebates are available for investments in specific regions. There are also medium- and long-term credit facilities made available to investors through designated financial institutions under the Ghana Investment Fund (GIF) scheme.
- The GIPC Act guarantees foreign investors "unconditional transferability" of dividends or net profits, and remittance of proceeds on sale or liquidation of their enterprises.
- Ghana has also ratified investment promotion and protection agreements with a number of countries[5]. Ghana has signed and ratified double taxation agreements (DTAs) to rationalize tax obligations of investors, with the view of preventing double taxation. Double taxation agreements have been signed and ratified with a number of countries actively involved, such as France and the United Kingdom. They have been signed also with Germany and concluded with Belgium, Italy and countries of the Former Yugoslavia[6].These agreements are seen as complementing the investment legislation in force to help attract foreign investments. The Ghana Arbitration Centre, a private initiative established in 1996, provides a forum for the resolution of disputes with a view to bolstering investors' confidence.
- Ghana is also a signatory to the World Bank's Multilateral Investment Guarantee Agency (MIGA), a convention which provides insurance against non-commercial risks. The country is also a member of the International Centre for the Settlement of investments Disputes (ICSID).

The Ghana Free Zone schemes related FDI drive

Ghana established a free zone board (Ghana Free Zone Board – (GFZB)) along with the country free zone schemes in September 1995, following the promulgation of the Ghana Free Zone Act, in order to accelerate the exploitation of the country's general export potential. This Act has offered an extensive package of incentives including the following to companies operating in the zone:

- 100 percent exemption from payment of direct and indirect duties on all import for production and export from free zones;
- Exemption of free zones developers from income or profit tax for 10 years;
- Income tax after ten-year tax holidays not to exceed a maximum of 8 percent;
- Exemption from withholding taxes on dividends emanating from free zone investments;
- Relief from double taxation for foreign investors and employees[7];
- Freedom from a foreign investor to hold a 10 percent shares in any free zone enterprise; and
- Various guarantees in respect of repatriation of profits and against unreasonable nationalization of assets. In this connection, there are no conditions or restrictions on repatriation of dividends or net profit; payments for foreign loan servicing; payments of fees and charges for technology transfer agreements; and remittance of proceeds from sale of any interest in a free zone investment.

GHANA

[5] Agreements have been signed and ratified with : China, Denmark, Germany, Malaysia, Netherlands, Switzerland and United Kingdom. Agreements have been signed, but are not yet ratified with Benin, Burkina Faso, Côte d'Ivoire, Cuba, Egypt, France, Guinea, India, Mauritania, South Africa, United States, the Former Yugoslavia and Zambia.

[6] See Sector Profile - Agriculture & Agroprocessing. Source: GIPC. 2009.

[7] Ghana has currently ratified double taxation agreement with France and Netherlands.

An investor who wishes to establish an enterprise in the Ghana free zones will, however, require various licenses and permits to enable him to operate. Detailed descriptions of these arrangements including the general requirement for foreign direct investor applying for the free zones can be found on the GFZB website.

The GIPC and the Free Zones have the same mandate to attract foreign investors, encourage, promote and facilitate investments in all sectors of the economy[8] except mining and petroleum. But unlike the GIPC which comes under the office of the President and licenses companies with foreign investors, the GFZB falls under the Ghana Ministry of Trade and Industry (GMTI) with the added ability to license local companies who can meet the 70 per cent export requirement to which foreign investors are bound to adhere. As the results show, the GFZB has been effective and should be supported in the assisting and monitoring of the activities of the export processing zones (EPZs) whose establishment is being carried out across the country.

The free zones enclaves and enterprises can be located anywhere in the country upon the approval of the GFZB. In these locations, the EPZs provide buildings and services for manufacturing including the processing of local as well as imported raw and intermediate materials into finished products prepared primarily for exports. A certain proportion of these outputs may end up in domestic markets subject to meeting normal duty. The EPZ is thus a specialized industrial estate located physically and/or administratively outside the custom barrier, oriented primarily to meet export demands. The EPZ facilities have served as a showcase to attract investors and as a convenience for their being established. The Ghana Free Zones Board also requires a free zone enterprise to show evidence of possession or lease of real property or intent to acquire such property before issuing a free zone enterprise license[9]. This is not without consequences for land prices, considering the difficulties of accessing land in the country, especially for the poor.

The Land Commission and FDI related land concerns

In Ghana, it remains fundamental to address the problem of land tenure insecurity to help meet the effective quest for attracting and providing a safer environment for foreign investors. In the opinion of many observers, land acquisition, ownership and management of land relations remain among the biggest challenges for Ghana to meet its development agenda. This has arisen from the fact that the Ghanaian land tenure system straddles two vastly different pillars of law and practice. It has been estimated that about 80 percent of the land in Ghana is held by customary authorities who provide land for residential or other economic activities[10]. To access these land resources, investors must proceed with the terms of a customary regime which dates as far back as the days of the old Ghana Empire. Indeed, there is a long-standing complex customary regime governing the assignment of tribal lands generally referred to as "stool land" in the South, and "skin land" in the Northern part of the country. This regime stems from centuries of oral tradition and practice and varies by location, chiefdoms, and systems of lineal succession.

Alongside this customary system, there is a statutory regime of the Government of Ghana (GoG), codified in the constitution of 1992 and supported by an extensive body of law and regulations dealing with the ownership and use of land. Under this law, Article 266 of the constitution establishes that foreigners may not own land in Ghana. However, they may lease residential, commercial, industrial or agricultural land for renewable periods of up to 50 years. The dominance of this communal system, the imperfect fit between its regulations and those of the national government and the insufficiency of close coordination between these two autonomous land administration regimes have resulted in rampant

[8] See www.gipcghana.com and www.gfzb.com.gh

[9] See Procedures for establishment of free zone development projects and enterprises. Source: Ghana free zone board. 2009.

[10] Antwi, Yaw Adarwah, strengthening customary land administration: A DFID/World Bank sponsored project in Ghana, paper presented at the Fifth FIG Regional Conference, Accra, 8–11 March 2006.

land disputes, slowing and complicating the process of securing land for agribusiness purposes. Seeing multiple claims on the same piece of property are the rule rather than the exception. In July 2004 for example, there were about 66 000 land dispute cases before Ghana courts, resulting mainly from the inability of traditional or customary authorities to identify the extent of land boundaries among potential investors, thus suggesting the need for better understanding of the way landed property is registered, especially as it concerns foreign investors.

TABLE 5
Land tenure associated with crop type in Ghana

Products	Lease period
Tree crops (citrus, cocoa, pineapple	50 years
Staples (cassava, rice, cocoa, yam)	10 years
Cattle	25 years
Small ruminants	17 years

Source: USAID Ghana, November 2009, P. 64

Land acquisitions in Ghana by statutory means

The constitution makes specific reference to "public lands, vested in the President on behalf of, and in trust for, the people of Ghana"[11] and stool lands, which "shall vest in the appropriate stool on behalf of, and in trust for, the subjects of the stool in accordance with customary law and usage"[12]. Thus, in line with the Constitution, the management of the land related responsibilities appears shared between the Lands Commission and the Office of the Administrator of Stool Lands as follows.

The Land Commission is primarily responsible for the management of public lands, coordination between state and customary authorities on the establishment of development policies, and development and execution of a comprehensive land title registration system throughout Ghana. The Office of the Administrator of Stool Lands is tasked for the establishment of a stool land account for each stool, to serve as a repository for all incomes generated by the land of that stool, to effect collections and disbursements of such incomes, and to assure that stool land dispositions are in conformity with national and regional planning authority programs. Income derived from stool lands is distributed, once 10 per cent of the total has been withheld to cover the office's administrative expenses, according to the following schedule: 25 per cent to the stool through the traditional authority for the maintenance of the stool in keeping with its status; 20 percent to the traditional authority; and 55 per cent to the district assembly, within the area of authority of which the stool lands are situated. Under these provisions, the customary holders of lands are ultimately granted only 40.5 per cent of total proceeds from land rents, while the state, through its office of the Administrator of Stool Lands and the District Assemblies, receives 59.5 percent. These arrangements best explain the development of a parallel and undocumented system administered by customary authorities through which they are able to capture a far larger participation in the transaction values involved in the extension of leases to third parties.

Land acquisitions in Ghana under the customary regime

There are three types of customary land rights recognized in Ghana: i) *allodial* or freehold title, held by the community as a whole; ii) *usufruct*, held by individuals or groups of individuals who form part of the community; and iii) mixed types of tenancy, including leaseholds, that are allocated either to community members or to foreigners.

Unlike the straightforward provisions regarding land transactions and land tenure contained in the constitution of 1992 and laws involved in the establishment of a formal registry for land titles, the rules by which kings and paramount chiefs allocate the land they hold in trust are based on customary practice, handed down by oral tradition, and vary according to tribal group and location within the country. While this approach lends itself to certain variances in interpretation

[11] Art. 257 (1), Constitution of the Republic of Ghana, 1992, Chapter 21 (Lands and natural resources).

[12] Id. (art 267 (1))

over time, and thus contains a high propensity for dispute, there appears to be a considerable richness in the range of situations these rules cover, and a general understanding among those who are subject to them.

The government has responded to this wide spectrum of situations by enacting in September 2008 the Lands Commission Act, thereby revising and consolidating into one piece of legislation the existing laws on the various public institutions that manage and administer land in the country. This Act intended to create a "one stop-shop" for land management while improving the delivery of services of a long list of implementing institutions including the following involved with property registration.

Despite the plethora of institutions currently acting to address the land related issues, Ghana has not yet succeeded in simplifying the process of acquiring land in a safe and transparent manner. Most land transactions, especially in the agribusiness, have involved leases of stool or skin lands of varying lengths, involving multiple agreements struck with multiple landlords of sometimes questionable reputation. The multiplicity of agreements derives also from the fact that most agricultural holdings in Ghana are of three acres or less and that the availability of land under the control of any particular paramount chief is limited. In the case of the Golden Exotics project, for example, which involves 875 acres of pineapples and 2 050 acres of bananas, the land acquisition process left the company with some "hundreds" of lease agreements, suggesting the need to continue support for programmes/projects such as the 2003 Ghana Land Administration Project (GLAP), the Ghana Land Bank (GLB) project which all aim at strengthening the country's land administration while streamlining the current tiring steps involved in the land title registration process.

Gauging the overall business climate in Ghana

A number of recent studies have extensively discussed[13] the business framework presented above and provided good insights about the way business has been carried out in the Ghana economy in general and its agriculture sector in particular. The following shows in tabulated form the summary data provided for Ghana[14] relative to other countries. As can be noted, for 2010, Ghana is performing quite well on the global scale when it comes to registering properties, protecting investors and enforcing contracts in the country. Leaving aside these areas, the country has a long way to go in many other areas crucial in the FDI attractiveness as the global ranking provided below suggests. It is also worth mentioning that in this same year and for the first time, a sub-Saharan African country – Rwanda – was the world's top reformer, based on the number and impact of reforms implemented between June 2008 and May 2009. There is no reason why West African countries with enormous resources such as Ghana, Côte d'Ivoire and Senegal cannot perform effectively in a way to rank among the top 10 or 20 world reformers. For example[15], it now takes a Rwandan entrepreneur just two procedures and three days to start a business. The import and exports systems in this country is now seen as one of the most efficient in the world with the transfer of properties taking less time thanks to a reorganized registry and statutory time limits. In this country, investors have now more protection, insolvency reorganization has been streamlined, and a wide range of assets can be used as collateral to access credit.

[13] See Doing business in Ghana 2010 report. The doing business project of the World Bank provides objective measures of business regulations and their enforcement across 183 economies and selected cities at the subnational and regional level.

[14] This table lists the overall "Ease of Doing Business" in Ghana (out of 183 economies) including the rankings by each topic. The detailed summary tables discussing the key indicators for each of the topic listed under review along with the benchmark information gauging Ghana against regional and high-income economy (OECD) averages are also available.

[15] Doing business 2010 report – Rwanda; website: www.worldbank.org or www.ifc.org

TABLE 6
Ease of doing business in Ghana and selected competing FDI destinations - 2010

	Senegal	Côte d'ivoire	Nigeria	Ghana
Doing business overall (183 economies surveyed)	157	168	125	92
Starting a business	102	172	108	135
Dealing with licenses	124	167	162	153
Employing workers	172	129	37	133
Registering property	166	145	178	33
Getting credit	150	150	87	113
Protecting investors	165	154	57	41
Paying taxes	172	152	132	79
Trading across borders	57	160	146	83
Enforcing contracts	151	127	94	47
Closing a business	80	71	94	106

Source: www.worldbank.org, doing business 2010 report
Note: Doing Business 2009 rankings have been recalculated to reflect changes to the methodology and the addition of two new countries.

3. Impacts of FDI in Ghana agriculture

Ghana has run a structural trade deficit of US$2.1 billion on average over the 2000–2008 periods underscoring the importance of the contribution the country can gain from export earnings mobilized from any other areas of the economy. Although traditionally Ghana has relied heavily on the exports earnings derived from a limited set of export products such as cocoa, minerals, timber[16] to improve its trade balance, the country has made significant strides in the past ten years towards attracting FDI in support of the production and exports of non-traditional products – one of the economic areas where FDI have been most felt in recent years. Figure 3 and Table 7 show the exports earnings performance of non-traditional agricultural commodities such as fruits[17], vegetable, fish, sea foods and others.

[16] Which brought about 26.3 percent, 43.3 percent, 6 percent respectively in the country total 2008 exports revenues

[17] Major non-traditional agricultural export commodities includes roots/tubers/plantain, cereal crops, fruits, vegetables, fish/sea foods, others.

Another key area where FDI has also made a significant impact on exports earnings in Ghana is the Ghana Free Zone Programme (GFZP). As can be noted (Table 8), exports earnings from the free zone companies increased 36 percent annually on average from 1998–2008 to reach US$1 305 million. The same programme has also helped the

FIGURE 3
Trend of export earnings of Ghana Non-Traditional- Export (NTE) Products, 2000–2008, in US$ millions

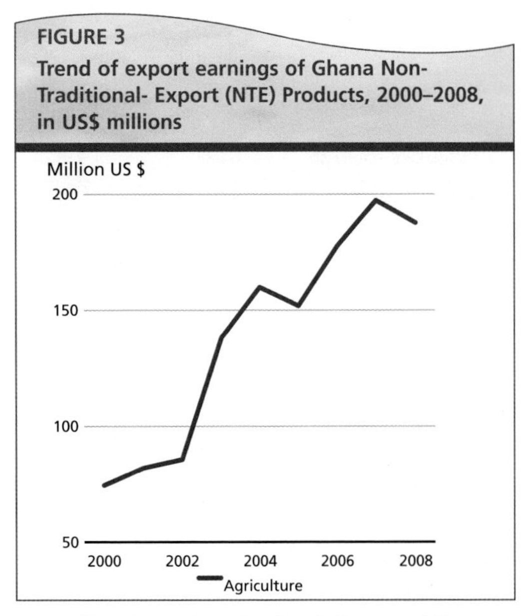

Source: Ghana Export Promotion Council. The state of Ghana Economy, P126, ISSER, Ghana

GHANA

TABLE 7
Volumes and values of major non-traditional agricultural export commodities, 2007 and 2008

	Volume (tonnes)			Value ($ 000)		
	2007	2008	% change	2007	2008	% change
Roots/tubers/plantain						
Yam	19 716	20 842	5.71	14 551	14 889	2.31
Cocoyam	234	273	16.53	114	212	8.47
Plantain	230	312	36.28	103	202	97
Cereals						
Maize Seed & maize	367	1 097	199.81	106	102	-3.7
Rice	6 109	744	-87.83	1 256	256	-79.6
Millet	3	0	-88.37	1	0	-70.87
Sorghum	1	1	42.25	0	1	105
Fruits						
Pineapple	40 456	35 134	-13.15	13 475	11 842	-12.11
Banana	52 069	69 773	34	9 965	12 717	27.61
Mangoes	824	858	4.1	998	522	-47.72
Pawpaw	1 194	968	-18.93	1 020	334	-67.21
Oranges	3 674	10 991	199.18	333	1 647	394.5
Limes/Lemon	0	18	5 903.3	2	53	2 530.2
Vegetables						
Dried Pepper	1 578	1 533	-3	539	627	16.3
Spinach	66	89	36.26	50	65	30.67
Aubergine/Green Eggs	92	249	169.18	34	128	276.37
Onions/Shallots	6 636	1 918	-71.09	112	227	102.58
Fish/Sea Foods						
Tuna	45 119	26 816	-40	539	627	16.3
Frozen Fish	66	89	36.26	50	65	30.67
Dried/Smoked/Salted Fish	6 636	1 918	-71.09	1 144	410	-64.2
Others						
Cotton Seed	5 337	3 711	-30.47	3 010	1 624	-46
Shea Nuts	57 166	55 489	-2.93	27 009	24 940	-7.7
Coffee Robusta	1 464	2 023	38.22	1 800	3 070	69.77
Kola Nuts	6 753	4 966	-26.47	12 957	976	-24.69
Cashew Nuts	23 616	81 190	243.79	10 779	20 426	89.47
Palm Nuts & Kernels	1 684	3 563	111.62	297	1 044	251.52

Source: Ghana Export Promotion Council; The State of Ghana Economy, P126, ISSER, Ghana

TABLE 8
Selected indicators of the impacts of the Ghana free zone programme, 1996–2008

Schedule/year	1996/97	1998	1999	2000	2001	2002	2003	2004	2005	2006	2007	2008
Companies Registered (number)	14	15	12	10	11	25	37	23	26	30	30	26
Total Existing Companies		*20*	*28*	*36*	*42*	*54*	*77*	*94*	*116*	*145*	*176*	*202*
Investment Capital (000 000 US$)	-	117	81	44	164	42	161	131	186	152	166	1 632
Cumulative Investment Capital (000 000 US$)	-	117	198	242	405	447	607	739	924	1 076	1 242	2 874
TOTAL EMPLOYMENT (people)	3 483	3 968	5 523	6 895	7 745	9 459	14 071	23 928	28 334	25 820	28 875	28 559
- National	-	*3 672*	*5 405*	*6 797*	*7 350*	*9 329*	*13 855*	*23 663*	*27 973*	*25 403*	*28 246*	*27 671*
- Expatriates	-	*296*	*118*	*98*	*95*	*130*	*216*	*265*	*361*	*370*	*634*	*924*
Yearly figures	3 483	485	1 555	1 372	850	1 714	4 612	9 857	4 406	(2 514)	3 055	(316)
Production (000 000 US$)	1	158	167	134	104	222	287	349	603	507	2 128	1 065
Cumulative (000 000 US$)	*1*	*160*	*327*	*461*	*565*	*788*	*1 074*	*1 423*	*2 026*	*2 532*	*4 660*	*5 725*
Exports (000 000 US$)	-	145	184	165	291	174	414	627	510	523	1 206	1 305
Cumulative Exports (000 000 US$)	-	*145*	*329*	*495*	*786*	*960*	*1 374*	*2 001*	*2 511*	*3 034*	*4 240*	*5 545*
Taxes/Duties on Local Sales (000 000 US$)	-	-	1	2	2	3	3	3	4	4	54	4
Cumulative Taxes/Duties(000 000 US$)	-	-	*1*	*2*	*4*	*7*	*10*	*13*	*17*	*21*	*75*	*79*
Salaries/Wages (000 000 US$)	11	11	8	7	10	15	17	29	49	52	71	86
Cumulative Salaries/Wages (000 000 US$)	*11*	*22*	*30*	*37*	*47*	*61*	*78*	*107*	*155*	*207*	*278*	*365*

Source: Ghana Free Zone Board (GFZB)

TABLE 9
Expected employment creation by FDI projects as registered with GIPC, (in thousands)

Subsectors	Cumulative (Jan.2001–Dec 2008)			2007		2008	
	Ghanaian	Foreign	Total	Ghanaian	Foreign	Ghanaian	Foreign
Manufacturing	26.6	2.2	28.8	8.2	594	3.1	399
Service	19.3	1.8	21	1.9	233	2.9	642
Building and construction	16.1	1.6	17.6	1.6	532	7.7	430
Agriculture	184.3	435	184.7	772	67	178.9	132
Tourism	4.9	503	5.4	869	84	390	91
General trade	10.6	1.2	11.8	1.7	205	2.6	313
Export trade	2.1	245	2.4	421	30	329	56
Total	263.8	7.9	271.7	15.5	1.7	195.9	2.1
% of total	97.1	2.9	100	89.9	10.1	99	1.04

Source: Ghana Investment Promotion Centre (GIPC)

government to reduce, among others, its fiscal deficits in the range of US$7.8 million annually.

FDI impacts have also been felt across other subsectors of the Ghanaian economy especially in agriculture and food security. According to the GIPC, the expectation were for 263 798 jobs and 7 889 jobs to be created for Ghanaians and foreigners respectively, between January 2001 and December 2008. Although agriculture received a very small amount of investments relative to other sectors of the economy, it is the sector where most of the job creation was set to take place. Thus, agriculture alone was expected to create 184 296 jobs (Table 9) underscoring again the crucial role FDI can play in reducing poverty in rural areas given the relatively high labour- intensive nature of agricultural related activities therein.

The surge in FDI into in Ghana in recent times has led to an increase in the levels of agricultural outputs, the quality of a wide range of readily available food as well as cash crop products and the purchasing power of certain classes of people, who are now in much better position to generate demand for the agricultural products produced in the country. All this is to suggest an improvement in the state of food insecurity in the country. This is evidenced by the fact that Ghana is one of the few countries in Africa slated to meet its MDG goal of eradicating poverty and hunger by 2015.

The next section presents an assessment of two business models used by private investors in Ghana drawing on case studies from: i) the Integrated Tamale Fruit Company (ITFC) to illustrate the nature and issues in a model involving collaboration between a private company and local farmers through a nucleus estate and out-grower scheme for the production of organic mangoes; and ii) the Solar Harvest Ltd (formerly Biofuel Africa Ltd) which provides an example of a production model centred on large-scale plantations.

4. Agricultural investment business model and impacts – evidence from case studies

This section examines key issues and impacts from two forms of business models used by private investors in Ghana drawing on case studies from: i) the Integrated Tamale Fruit Company (ITFC) to illustrate the nature and issues in a model involving collaboration between a private company and local farmers through a nucleus estate and out-grower scheme for the production of organic mangoes; and ii) the Solar Harvest Ltd (formerly Biofuel Africa Ltd) which provides an example of a production model centred on large-scale plantations.

4.1 The Integrated Tamale Fruit Company

Overview

The Integrated Tamale Fruit Company is a Ghanaian company with the head office in Gushie, 45 kilometres north of Tamale on the Bolgatanga trunk road in the Northern Region. Incorporated in 1999, the main activity of ITFC is the cultivation and trading of certified organic mango in fresh form. Activities are centred on a nursery for seedlings, a 155-hectare nucleus estate farm, an outgrower scheme and processing facilities that include a drying unit. The company's main target is the export market (8090 percent) and local market (10–20 percent). The main export destination countries are the United Kingdom, Netherlands and France.

The venture involves fully private companies' capital from the Netherlands and from Ghana. The largest shareholder of ITFC, with 50 percent of shares, is Wienco Ghana Limited, a leading Ghanaian-Dutch fertilizer and agrochemicals manufacturer. Wienco (Ghana) Limited, in turn, was established in 1979. Among other things, it delivers organic inputs to ITFC.[18] The second largest ITFC shareholder, with 30 percent shareholding, is Comma, a Dutch company. The remaining shareholders are Tamale Investments (a collection of local Tamale-area investors) at 5 percent, African Tiger Mutual Fund (a Ghanaian investment company) at 5 percent and Alhaji (the Nanton chief) at 10 percent.

The Company enjoys strong local support in its operational area and a positive reputation throughout the country. Since the beginning, ITFC has worked with United Kingdom-based Soil Association and has obtained organic certification for its mangoes. The Soil Association is the United Kingdom's leading certification body and campaign organization for organic food and farming. The Integrated Tamale Fruit Company is also GLOBALGAP (Global Good Agricultural Practices) certified; GLOBALGAP aims to develop widely accepted standards and procedures for the global certification of good agricultural practices.

The nursery

The nursery is located in the village of Dipale, in Savelugu/Nanton District. It has the capacity to produce 347 648 seedlings per year, which can withstand the harsh environmental conditions in subSaharan Africa. The nursery has three black nylon shade nets that protect the seedlings from 60 percent of the sun's rays. There is a total of 16 blocks under the shade nets and each block contains 21 728 seedlings. The nursery uses drip irrigation to supply the plants with water that is sourced from the White Volta River (with the permission of the Water Resources Commission), as well as from the company's private borehole.

The nursery is currently focusing on the production of the Kent and Keitt varieties but has the capacity to work with other popular varieties as needed. The Integrated Tamale Fruit Company uses a portion of its nursery space for experimental planting techniques for future improvement of current methods. Currently, experiments are taking place with planting medium and planting density, while the propagation of indigenous tree species is also being pursued.

Headed by a manager and an assistant, the nursery staff have been trained by international nursery specialists and pride themselves in their state-of-the-art techniques for grafting mangoes.

The nucleus farm

Like the nursery, the nucleus farm is also located in the village of Dipale, in Savelugu/Nanton District. The climate of the area is ideal for mango. The 155 hectares (about 383 acres) of land for the nucleus farm was acquired after protracted negotiations between chiefs and affected land users or owners. The land was officially handed over to ITFC in 2000. However, the author's search for the land transfer documents at the Lands Commission in Tamale revealed that they could not be traced. Besides the 155 hectares used for mango, an additional

[18] www.wienco.com

20 hectares of land is used for jathropa. The mango farm is divided into 16 equal blocks which are planted with four varieties, namely, Kent, Keitt, Amelie and Zille.

In order to meet organic standards, the farm uses a system of integrated pest management and organic disease control scheme. An internal control system has also been put in place to monitor all activities on the farm and ensure the best quality of organic mangoes and maintain their organic certification status. A micro-irrigation system is operated. This means that there is one sprinkler per plant. The system allows every plant to receive the required amount of water, which is pumped from the White Volta River with the permission of the Ghana Water Resources Commission. Water passes through a filtration system before reaching the sprinklers. The company has employed qualified personnel with requisite experience to manage the plantation, as well as sourcing farm workers from the surrounding communities. The staff strength of the nucleus farm is about 85, mostly labourers.

The out-grower scheme

The Integrated Tamale Fruit Company has been working with farmers in the communities surrounding the nucleus farm since 2000. The establishment and expansion of the outgrower scheme has received considerable development assistance. In the initial stages, PSOM, an NGO based in the Netherlands, assisted ITFC with partial financing in the form of a loan for the planting and irrigation of the 155-hectare nucleus farm and 50 hectares for outgrowers. Since the initial set up, ITFC has continued to expand its operations at a steady pace. It has also welcomed the addition of 400 outgrowers with the assistance of Dutch NGO Cordaid in 2004, and has begun to provide consultancy services to other development projects in the area that are dealing with agroforestry. In 2005, the United Nations Development Programme (UNDP) sponsored 100 additional outgrowers and the African Development Foundation sponsored 200 more. These sponsorships entailed payment for seedlings and cost of publication of educational materials. The Ministry of Food and Agriculture, through support from the World Bank, also assisted the outgrower scheme with a grant for partial funding of the seedlings and for an office complex. In total, the company currently has some 1 200 outgrowers who are organized in an Organic Mango Outgrowers Association (OMOA). All registered outgrowers become members of OMOA. The assistant general manager of ITFC indicated that "*recently some 155 outgrowers' names have been struck off the list of association membership because they have – on repeated occasions – allowed their farms to be destroyed by bush fires*". The 1 200 outgrowers are 60 percent of a 2007 target of 2 000 outgrowers achieved so far. Given the challenges faced by the company in increasing the number of outgrowers, the assistant general manager was unable to say when in the future the 2 000 outgrowers target will be fully achieved. Each outgrower cultivates an acre of mangoes, for a total of about 1 200 acres (486 hectares), with a carrying capacity of 100 mangoes per acre.

The aim of OMOA as provided in its constitution is: "to control and manage the affairs of … members with a view to improving the general earning capacity and well-being of the members". The specific objectives of OMOA are to:

i. foster friendship and solidarity among members;
ii. maintain a lasting harmonious link between its members and ITFC;
i. ensure the most beneficial marketing system for the produce of its members which OMOA will negotiate and maintain competitive prices with ITFC;
ii. promote and protect the welfare and interest of its members;
iii. have full power to do all things necessary or expedient for the attainment of any or all of its objectives in keeping with the constitution of the company;
iv. follow organic standards as prescribed by the Soil Association; and
v. expand and develop the association to its utmost potential.

The association is governed by an Executive Committee (EC) comprising an elected chairman, vice-chairman, secretary, assistant secretary and a treasurer. Other members of the EC are two representatives appointed by ITFC and one representative each from Diare East, Diare West, Pong-Tamale, Savelugu, Kumbungu, Karaga, Janga and Gushie zones. The functions of the EC as stipulated under the OMOA constitution are:

a. Keep an up-to-date account of the income and expenditure of the association and to submit the same for auditing after every financial year.
b. Present an annual report of the activities of the association at an annual general meeting.
c. Convene all meetings of the association as and when it is deemed necessary.
d. Secure and protect the well-being advancement of the interests of the members of the association.
e. Put in place a quality control monitoring system during the harvesting and packaging of the product.
f. Ensure strict compliance of GLOBALGAP rules and regulations by all members.
g. Ensure strict compliance of regulations of the Soil Association of the United Kingdom on organic standards.

Unfortunately, the constitution of OMOA neither provides for how a member can bring grievances to the attention of the association nor how the association can address its collective grievances with management of ITFC. This to a large extent could be a limiting factor for the voice of OMOA in the discharge of its functions.

According to the company, the outgrower option was motivated by recognition of the challenges and costs involved in acquiring land on this scale. In this area, land is held under customary tenure. Acquiring some 2 000 acres of land would have meant dealing with many individuals and family heads, and possibly multiple chiefs. The outgrower scheme was seen as a way to secure greater productive capacity without needing to acquire additional land. It should, however, be noted that another motivation for the outgrower scheme could have been the support received from development agencies and NGOs in building the capacity of outgrowers in organic mango production.

Outgrowers are provided with a long-term, no-interest loan in the form of inputs, to be used only for farming one acre of organic mango. It is the responsibility of the farmers to provide labour for their farms, including for digging, fencing, weeding and watering. The ITFC provides support on technical issues relating to farming organic mangoes, such as disease and pest control, pruning or shaping of the trees and water provision. Once the trees have reached maturity and they begin to bear fruit (three or four years after planting), ITFC provides the farmers with the technical assistance for harvesting and transports the fruits to the pack house for processing.

To render this arrangement operational, ITFC has a contractual agreement with the farmers. These are smallholder farmers who have agreed to put their customary lands to organic mango cultivation as outgrowers. The costs of the inputs are debited to individual accounts and will be paid back annually from the fifth year after planting. Each farmer must provide one bag of maize as registration fee and a sign of commitment, and pay 30 percent of his organic mango sales income towards the total debt repayment. The farmers are organized into groups and ITFC provides one assistant manager for every 400 farmers and one field assistant for every 40 farmers. The assistant managers and field assistants provide the farmers with the necessary technical support to farm organic mangoes. The total number of workers overseeing the outgrowers is about 70. During the rainy season, ITFC encourages the farmers to intercrop with groundnuts – not only as a cover crop to promote farm hygiene and for nitrogen fixing, but also as an intermediate income benefit to the farmer. It is also important to note that intercropping mango with groundnuts reduces the threat to food security posed by tree crop cultivation.

The nature of the contractual arrangement between ITFC and the individual outgrowers is as follows:

- A commitment fee of one bag of maize is required from the outgrower (value of about US$15) to begin the process of working with ITFC.

- ITFC gives the outgrower farmer an interest free loan. This is not a cash loan – it comes in the form of farm inputs, such as fertilizer, water tanks for watering the farms, seedlings and technical assistance.
- ITFC assists farmers in obtaining licences and certifications, which are a requirement for the organic export markets. As discussed, one of the key certifying organizations is the Soil Association, based in the United Kingdom.
- The outgrower has a four-year grace period to begin repaying the value of the loaned inputs. This means the outgrower only starts repaying the loan in the fifth year (mangoes take approximately three to four years to mature and bear fruit).
- From the fifth year, the outgrower pays 30 percent of their sales to ITFC until the debt is repaid. The outgrower is expected to pay the Ghanaian cedi equivalent of the US dollar amount of the loan.
- Until the outgrower finishes repaying the loan, all mangoes must be sold through ITFC. After the outgrower finishes paying the loan, they are free to sell to ITFC or any other buyer they choose.
- In case of any conflict arising from the contract, both parties shall resolve such conflict through the traditional arbitration set up in the community. Should this fail, the OMOA will appoint a three-member committee of elders, to issue a written advice in fourteen days. This advice is binding; if this is still not accepted, both parties may go to court. However, the assistant general manager confirmed that there has not yet arisen any conflict between OMOA and management to require the use of any of the above mechanisms of conflict resolution.

The outgrower incurs a start-up cost of approximately US$2 236 (comprising farm inputs such as cutlasses, fertilizer, water tanks for watering the farms and seedlings). This initial cost outlay covers the gestation period of the investment until returns are made. The mango tree takes about three to four years to bear fruit. The annual operating cost of about US$944 is mainly in the form of technical assistance. These costs, which exclude labour costs, are financed by ITFC for the first five years. The mango sales start at around US$150 in the third year and rise to about US$3 000 by the tenth year (Table 10). Note that the cash flow plan is in US dollars to serve as a hedge against inflationary tendencies over time. The plan is based on the conservative assumption that only 50 percent of the expected yields are attained for the various years. This also assumes that 40 percent of the outgrowers' produce

TABLE 10
The ITFC outgrower cash flow plan in US$

Year	0	1	2	3	4	5	6	7	8	9	10	...	20
Expenditures	2 236	944	944	944	944	944	944	944	944	944	944		944
Direct exports	-	-	-	85	283	453	623	907	1 190	1 417	1 700		1 700
Sales to local processor	-	-	-	55	182	291	401	583	765	911	1 093		1 093
Sales to local market	-	-	-	12	40	65	89	130	170	202	243		243
Total sales	-	-	-	152	506	810	1 113	1 619	2 126	2 530	3 036		3 036
Total debt	2 236	3 165	4 110	5 054	5 998	6 791	6 548	6 214	5 728	5 090	4 331		
Servicing of debt	15				152	243	334	486	638	759	911		
Cash flow	(15)	-	-	152	354	567	(165)	189	544	827	1 181		2 092

Source: Adapted from ITFC and Osei, 2008

are exported, 40 percent are sold to a local processor, and 20 percent are sold on the local market. The total loan of about US$6 956 (i.e. start-up cost plus operating costs for five years) owed to ITFC is expected to be paid off by each individual farmer at the end of the fourteenth year. After this period, individual farmers are expected to earn annual profits of approximately US$2 000 each. This revenue stream from the mango farm after debt repayment represents a substantial increase in income over that gained from subsistence farming. The average farm income in the Tamale area is about US$300 per year for most smallholder farmers who are not outgrowers. This suggests that, after year 14, the outgrowers' expected income of US$2 000 per annum would most likely put them in a better position than other farmers or labourers. But this reward is acquired in full only after 14 years in the scheme – a considerably long wait, with all the associated risks of production and general economic uncertainties.

The assistant general manager disclosed that outgrowers generally meet their debt obligations to ITFC and that some outgrowers are even willing to pay more than 30 percent of their mango sales to ITFC with the view to redeeming their indebtedness earlier than scheduled. This is hardly surprising, since only 50 percent of expected yield was used in the estimation of cash flows. However, the contractual arrangement of 30 percent of sales as annual repayment of loan was said to be strictly enforced.

A 66-year old outgrower made the following comments, which reflect the general mood expressed by the outgrowers interviewed for this study:

> "IFTC came into the community to help us in the cultivation of mangoes. Their aim is to improve and better our yields as well as provide a ready market for our produce. The company's operations are very beneficial to us. Teachers' bungalows have been built and health education is provided. We work in groups and use pipes to water the mango seedlings. We have also been promised electricity although we have yet to receive that."

However, concerns were also expressed by outgrowers in terms of delays in payments for the mangoes sold to the company. One farmer said that "produce sold to the company can take three to four weeks before payment is effected and this does not help with emergency situations where cash is needed by farmers".

The main risk that threatens the sustainability of the outgrower scheme is produce diversion to other market sources (side selling), especially given complaints of delay in payments. There is, in other words, the possibility that an exporter could come and offer ready cash for the outgrowers' mangoes. This could potentially threaten ITFC's export volumes and, accordingly, its command over the market. The diversion of produce to other buyers after the loan repayment of an outgrower can also affect the sustainability of the business model. The company believes this will be a real challenge, because they pay the farmers 20 percent when they collect the fruit from them, and it is only after ITFC has sold the produce that they pay the remaining 80 percent.

Another threat to sustainability is low production capacity among outgrower farmers. The study found that IFTC had to buy mangoes from Burkina Faso to augment local supplies. Even though the agreement between ITFC and the outgrower allows ITFC to deduct 30 percent of the proceeds from sales, outgrowers have to produce mangoes before these can be sold and the debt repaid. Therefore, low productivity would negatively affect the repayment of the loans. For example, a 39-year old male outgrower complained that production has suffered *because the company has not yet honoured all its responsibilities under the OMOA contract*, and another 43-year old female outgrower lamented that: *our group has not been provided with water under the terms of the contract and as a result the growth of our mango trees is negatively affected*. These concerns are indicative of low outgrower production locally as a result of ITFC not honouring all its contractual obligations to outgrowers, and hence the inability of some outgrowers to pay off their indebtedness, at least on schedule.

TABLE 11
Key features of IFTC's business model

Ownership	ITFC is a partnership between local and foreign companies. While ITFC nucleus farm documents could not be traced at the Regional Lands Commission, the nucleus farm is in the hands of the company. The outgrowers operate on their own lands, which are held under customary tenure.
Voice	The establishment of Organic Mango Outgrowers Association (OMOA) has given collective voice to the outgrowers. It is unclear what formal procedures are available for an outgrower to bring grievances to the company's attention.
Risk	Both the company and the outgrowers bear the risks associated with weather and the lack of sustainable water supply, with complaints from OMOA about high water charges. ITFC bears the risk of loss of market share in case of side selling. Prices are largely determined by the company, which implies that outgrowers suffer the risk of lower prices.
Reward	Based on the cash flow pattern of the outgrower as shown in Table 2, the outgrowers scheme of ITFC provides major financial rewards to outgrowers, especially when compared with annual incomes of other smallholder farmers in the area (US$2 000 as against US$300). But this reward is acquired in full only after 14 years on the scheme, which is a long-term timeframe and subject to economic uncertainties. Yet outgrowers were generally of the view that they are better off even when they are still indebted to ITFC than when they were ordinary farmers or labourers. Most of the OMOA members interviewed also indicated that they periodically receive training on how to improve their farming practices. ITFC has a total of 330 employees, comprising 70 outgrower staff, 175 for the pack house and 85 for the nucleus farm.

The pack house

The ITFC has built a pack house facility in Gushie, approximately 9 km from the nucleus farm. Like the nucleus farm and the outgrower scheme, the facility is also GLOBALGAP and Soil Association certified. With a staff strength of 175, the facility is the first of its kind in the Northern Region of Ghana, equipped with a refrigeration unit, a dump bath, a brush washer, a hot water bath, a brush dryer, a sorting table and a sizing machine. Solidaridad, a Dutch organization dedicated to a fair economy for all, in conjunction with AgroFair Assistance and Development for CTM Bolzano of Italy, provided funding for the packaging equipment.

The facility has the capacity to pack 5 tonnes of produce per hour or two 40-feet containers per day. The facility employs approximately 90 workers during the peak of the packaging season. These workers are mostly recruited from the local communities and Tamale. The water used in the facility is pumped from an underground borehole. The company has expanded the facility to include a drying unit whereby nonexportable fruit can be processed and also sold for bulk export.

Corporate social responsibility (CSR) activities

In addition to the support that ITFC has given the outgrowers, the company has also teamed up with Mondiaal Platvorm Venlo (MPV) Solidaridad to start the Children to School Project (CTSP). The project's objectives are to improve the infrastructure of primary schools in the project area. The CTSP has since become an NGO, with donor support from ITFC and from Roemond of Holland (funding), from Nordox Norway (teachers' quarters and kitchens) and from Mang-go Project of Holland (text books). Mang-go project also participates in an exchange programme in which university students spend time in the project area volunteering their services. A food programme is in place to ensure that all the students get one nutritionally balanced meal per day at school. The meals consist mainly of local dishes, and one chocolate drink. Clean drinking water is also provided to each school.

As an incentive for teachers to move to the rural areas to teach, housing facilities have been built for them. At the same time, the school environment was renovated to create friendlier educational surroundings for the students.

The ITFC also has a programme to support its staff in fighting the HIV/AIDS pandemic. In addition, the company has a biodiversity programme that provides participating communities in the project area with education on the protection and propagation of indigenous tree species and responsible medicinal herb harvesting. The company is a strong advocate of bush fire prevention and trains its staff and the outgrowers in methods of protecting their lands.

4.2 The Solar Harvest Limited Company

Overview

Solar Harvest Ltd, originally Biofuel Africa Ltd. (Ghana), is a private company with Norwegian capital, which was incorporated in Ghana in 2007. In 2009, the Ghana operation went into liquidation following the global economic downturn (Tsikata and Yaro, 2011) and was directly acquired by two of the original founders of the Norwegian company. It was at that stage that the company was renamed Solar Harvest Ltd.

The company has acquired large tracts of land in the Northern Region for the cultivation of jatropha. The acquisitions include 4 844.20 acres of land at Kpachaa, 13 156 acres at Jimle and 8 803.10 acres at Kpalikori, all in the Yendi District of the Northern Region, for a total of 26 803.2 acres (10 847 hectares). The Regional Lands Officer described the transaction as:

> a lease agreement drawn and executed between the lessor – the chief of Tijo (Tijo-Na) – and the lessee Biofuel Africa Ltd. The lease is for 25 years with option to renew for another 25 years. The rent is 2 Ghanaian Cedi [US$1.20] per acre of land and a two-year rent advance is to be paid after execution of a statutory declaration. Rent review is every seven years and any upward review of rent shall not exceed 2 percent of current rent.

Thus, Solar Harvest Ltd has leased the land for a term of 25 years, renewable for additional 25 years. Documentation from the Regional Lands Commission in Tamale, which was reviewed by this study, suggests that at least part of the lease agreements have not been formalized due to disputes between competing customary authorities on the one hand, and between customary authorities and community members on the other (see below). The duration of the leases is consistent with the restrictions placed on land access by foreigners under Ghana's 1992 Constitution. According to article 266 of the Constitution, foreigners cannot acquire interests in, or rights over, land for a term of more than 50 years.

The company's jatropha cultivation started in the last three years and there is yet to be a harvest. While jatropha remains central to the business plan for these plantations, since 2010, following the global economic downturn, the company has been increasingly diversifying into food crop cultivation, with an emphasis on maize production. It is understood that 400 hectares of land are under jatropha cultivation and 220 hectares of land has been used to cultivate maize at Kpachaa, and a further 25 hectares of land have been acquired in the Dipale area for vegetable cultivation (Tsikata and Yaro, 2011). The target markets for the produce is both local and international but it is unclear how processing is going to take place since there is no evidence of processing plants being built and the plans of management in this direction have not been disclosed. The jatropha plantation at Kpachaa was recently ravaged by bush fire, adding to the uncertainty over when there will be a harvest and the future of Solar Harvest Ltd operations.

Bush fires are common in the area during the dry season and are mostly caused by locals burning bushes in search of game. Referring to the burned plantation, the negotiator for the Solar land acquisition and assistant manager of the company deeply regretted the lack of funds which prevented the company from creating a fire belt around the plantation. However, although the burning of the jatropha plantation could have been caused through the spread of a common bush fire and therefore accidental, sabotage cannot be ruled out given the recent lack of cordial relations between the community members and Solar Harvest Ltd as reported in the Focus Group discussions.

In addition to the plantations, the company has also developed plans to expand and upgrade irrigation facilities for the cultivation of food crops on land belonging to local farmers. These plans are embodied in a Memorandum of Understanding (MoU) that the company signed in September 2011 with Ghana's Millennium Development Authority (MiDA). The Authority is the national entity responsible for implementing activities funded by the US Millennium Challenge Corporation and included in the Millennium Challenge Compact between the United States and the Government of Ghana. Under the MoU, Solar Harvest Ltd will become a partner on the Bontanga and Golinga irrigation sites in the Tolon-Kumbungu District of the Northern Region. These farmers are smallholders in control and use of their own customary land rights but in need of assistance to improve their production capacities. Solar Harvest Ltd will assist farmers with inputs like seeds, fertilizers, tractors and harvester services, on a credit basis, on some 1 321 acres (535 hectares) of existing irrigated land and water ponds for the cultivation of cotton, sugarcane, vegetables, fish and pasture for local and export markets. Should this proposal materialize, it would entail the company adding to its plantation portfolio activities that involve collaborating with local farmers through a contract farming arrangement. Under the same MoU, in addition to the land already under irrigation, 1 161 acres (470 hectares) of irrigable land are currently planned for expansion works to be led by Solar Harvest Ltd. An additional land area of 12 844 acres (5 200 hectares) is planned for future expansion in the hope that water would be pumped from the White Volta river. The company will manage irrigation infrastructure, provide the above support services to farmers and buy the produce of farmers and in turn sell to both local and external markets.

With the first phase of the above developments involving Solar Harvest Ltd and MiDA, the company is to invest approximately US$2 000 000 and this will be scaled up to more than US$30 000 000 when fully developed. According to company sources,[19] these investments are expected to bring about an increase in cereal production from some 18 000 metric tonnes per annum to 110 000 metric tonnes per annum at full potential and thereby increase cereal production 8-12 times the yields of traditional rain-dependent farming practices. These measures, if successful, will improve on the state of food insecurity in Northern Ghana. However, the focus in this study is on the jatropha operations of the company.

Land acquisition, stakeholders' roles and responses

Having taken place within the institutional context discussed on pp 199-200 of this chapter, the company's land acquisition remains contentious. Tsikata and Yaro (2011) explain the land acquisition process following the investor's placement of an advertisement in the media for land to cultivate jatropha as follows. First, visits were made by the investor's negotiator to the local chiefs to ask for the land. However, as the sub-chiefs only act as caretakers of the land on behalf of the divisional chief, Tijo-Na, these requests were then referred to the Tijo-Na. The company's negotiator and assistant manager, who also happens to be the grandson of Tijo-Na, organized a *durbar* (community meeting) of stakeholders, including chiefs, land users, NGOs, the District Assembly and other interested ordinary citizens at the Tamale Cultural Centre. At this *durbar* the benefits of Solar Harvest's operations were explained to participants. Two public hearings involving stakeholders were also organized in Yendi where the paramount chief and overlord of the Dagbon traditional area the Ya-Na resides, and Tijo where the divisional chief overseeing the communities – where the lands to be acquired – are located. The Tijo-Na is answerable to the Ya-Na and therefore also required the latter's permission to release the lands. The regent of Dagbon granted that permission in the absence of a substantive Ya-Na (the Ya-Na was murdered in a communal conflict in 2001). At the public hearings the environmental and social implications of the undertaking were explained by consultants to the stakeholders.

The Tijo-Na then hired the services of the lawyer of the regent of Dagbon to prepare contractual documents after the land was surveyed. An initial payment of 13 800 Ghanaian Cedi was paid by the investor, 40 percent of which went to the regent of Dagbon and the remainder (between 500 to 1 000 Ghanaian

[19] http://www.comuniq.com/news_viewer.php?news=20110926

Cedi) was shared among sub-chiefs and divisional chiefs, irrespective of whose land was part of the transaction (Tsikata and Yaro, 2011).

During this study, however, responses from local community members pointed to the lack of a consultative process that preceded the acquisition of the land from the local chiefs in 2008, and the displacement of a number of families from their source of livelihood, the land. Examples of these responses from three community chiefs are given in box 1.

The above local accounts suggest that land acquisition by Solar Harvest Ltd lacked wide stakeholder consultation. This has generated tension in the communities, and a potential for conflict. In the case of Jimle, for example, community members have returned to occupy and use company lands that are currently uncultivated and this corroborates the assessment by Tsikata and Yaro (2011) that only 400 hectares of the total of 10 847 hectares acquired by the company are currently used for jatropha cultivation. There arises, therefore, the question of security of tenure as tensions were reported between community members and the company over ownership and use rights over the land. CICOL (2008) also noted tenure insecurity in the case of the Solar Harvest Ltd land acquisition. The likely consequence of this is rising tensions and ultimately conflict in the area. Perhaps the reverence for the chiefs in the area, and particularly for the Tijo-Na, is what has prevented the tensions escalating into conflict between community members and the company. The inability of the company to keep to its promise of job creation has further strained relations between the company and the locals. Under these circumstances, some community members claim that the lack of compensation to farmers who lost their land had negatively affected their livelihoods. This analysis points to the need for Solar Harvest Ltd to improve on implementing the principles of responsible agricultural investment, especially in respect of land acquisition. The current guidelines being developed by the Lands Commission are therefore critical.

For example, a 45-year old widow of Kpachaa told her story as follows:

> "They [Solar Harvest] came to grow jatropha for fuel and as result I have lost my three acres of farmland which I depended on for a living. There was no proper acquisition of my farmland. It was just taken away from me under the instructions of the chief. In the beginning, the company employed me as a casual worker, but I have now been laid off and am suffering because I cannot get alternative land to farm and I was not given compensation. Even eating is a problem now but I am powerless to fight the company."

The above highlights the implications of Solar Harvest Ltd's operations for secondary land rights holders such as women. The situation worsens when the environmental resource base – on which they rely for alternative sources of income, e.g. fruits from trees – is cleared for large-scale plantation farming.

Available evidence collected during the study suggests loss of land (of around 400 acres) for both resident farmers in Kpachaa and for famers who reside in a different community and commute to Kpachaa to farm ("nonresident farmers").

From the evidence gathered, it was determined that the average size of farmers' land lost was 9.5 acres and 4.1 acres for non-resident farmers and resident farmers respectively. The non-resident farmers were mostly from Tamale, while resident farmers lived in the communities where the land was acquired by Solar Harvest Ltd. The difference in land size can be explained by the fact that most non-resident farmers from towns have commercial motives and better resources to acquire and cultivate larger sizes of land than resident farmers in the villages, who lack these resources and cultivate smallholdings on a subsistence basis. So the non-resident farmers lost more land because they had more to lose.

It is also important to note that no women were determined to have lost land. This conforms to the customary practice in Dagbon traditional area, and northern Ghana as a whole, that land inheritance is patrilineal, but women generally have access to use of land and in some cases, widows can lay ownership claims to land as the example of the 45 year-old widow of Kpachaa above illustrates.

BOX 1
Local perceptions on land acquisition

"Biofuel [Africa Ltd] approached me for land through one of my elders. The amount of land asked for was huge. Later, I was taken to a hotel in Tamale called Picorna Hotel. At the hotel were other chiefs and some educated people. We were made to thumb print some documents to show that we had agreed to the deal for further action by the Tijo-Na. Those whose lands were taken were neither consulted nor paid compensation and naturally this led to disputes. I can count up to ten families that have left this village because their lands were taken by Biofuel. We, the sub-chiefs, were asked to speak to these people and calm them down, drawing their attention to promises of jobs, schools, water, corn mills and other forms of development to be provided. I was encouraged because as a white man making promises their delivery was assured. However, as I speak, only two small dams have been provided and a corn mill. The corn mill is even a Biofuel business because we pay for its services."

[Interview with the chief of Kpachaa]

"The Tijo-Na gave out the land to Biofuel. As caretaker chief under his jurisdiction, I was only told about his action at a gathering in which we were informed that a development project was to take place on our land and we should embrace it for our own good. Jobs, dams, schools and other benefits were promised us. Of these, a small dam has been provided in Jimle but I am told those who had jobs have lost them. The loss of jobs and land on which one once farmed has brought misery to us in Jimle. I am unaware of any compensation that was paid to any farmer who lost his land, though monies were shared among the chiefs. Ironically, when some farmers who lost their land to Biofuel returned to these lands to cultivate them because they were not used by the company, the manager reported the matter to Tijo-Na and imposed four conditions on such farmers to select one for compliance. The first condition was for such farmers to discontinue the use of the land and be employed by the company; second was to continue to use such land but be enveloped by company farmlands such that they cannot expand their farms; the third condition was to be given some money as compensation and when the land is vacated; and finally, resettlement on alternative land elsewhere that is not acquired by Biofuel. People are cultivating their lands without complying with these conditions and there are regrets for having the company here but we lack the power to do anything about it since the Tijo-Na is regarded as our father and he brought the company to us."

[Interview with the chief of Jimle]

"It was one morning that we woke up to find earth-moving vehicles on our lands. When the operators of the vehicles were contacted, the community members were told by a white man that it was Tijo-Na who gave them permission to work the land. A large amount of our land was taken and this led to disputes between our people and Biofuel. Subsequently, when the Tijo-Na was contacted he confirmed that as true and apologized to me as the sub-chief of Jashee for not notifying me earlier, and further explained that it was a development project which promised benefits of schools, electricity, roads, corn mills and water and jobs for communities that embraced its activities. Of all these, however, only a small dam has been constructed and ten (10) members from this community were initially employed. The Tijo-Na is revered and therefore no collective action can be taken to argue the injustice done to local farmers whose lands were taken."

[Interview with the chief of Jashee]

The Assemblyman of Jimle, a locally elected position for community development purposes, spoke of Solar Harvest Ltd operations as follows:

> "There was lack of consultation in the process of acquiring the land for Biofuel. The Tijo-Na controlled the process and suppressed all forms of opposition. I was personally warned by the Tijo-Na for opposing the lack of consultation. Yes, I was warned by him to desist or face the consequences."

The position of a chief in the Ghanaian society has been described by Brobbey (2008) in the following words:

> "The people of this country cherish chieftaincy as an institution of such significance that it is inconceivable to think of a situation where the subjects of a chief will refuse his order. Ghanaians have such great respect, in some cases bordering on reverence, for chiefs that what the chief tells his people is in many cases instinctively obeyed."

The above observation underpins article 270 of the 1992 Constitution of Ghana which guarantees chieftaincy and traditional institutions which were in existence before the promulgation of the Constitution. This heightens the need for a regulatory framework for land acquisitions that will respect and protect the land rights of community members and provide also for economic, social and environmental outcomes that are equitably distributed to all stakeholders.

In the focus group discussions with community members at Kpachaa, Jimle, Jashee and Cheegu communities, the findings of the consultation process of the land acquisition were consistently described in phrases such as: '*the chiefs brought the company without consultation with land users*'. On payment of compensation to farmers who lost their lands the typical response was: '*there was no compensation to farmers for the loss of their land*'. Regarding promises made by the company to the communities it was mostly observed that, initially, jobs were created especially for casual labour for land clearance and general land husbandry activities. A former company employee, who recorded the names of the recruited workers, produced a notebook in which 64 employees of the company comprising 37 men and 27 women were documented as casual labourers. They earned 75 Ghanaian Cedi each (about US$50) per month as of 2010. By February 2012, when these focus group discussions were conducted, it was estimated that the only remaining company employees were three watchmen, the managing director and his assistant; totalling five. The relations between the company and locals were no longer cordial and a typical opinion expressed was that: '*Biofuel Ltd should leave our lands and allow us to do our farming as its presence has made us worse off*'.

GHANA

Socio-economic outcomes

The company's operations have been negatively affected by the global economic downturn. This has led to difficulty in accessing international sources of finance and to heightened environmental concerns of jatropha cultivation. While at some point the company employed around 400 workers, as of February 2012 the company's workforce has been reduced to five people, namely the managing director, the assistant manager and three watchmen. At the peak of operations when some 400 workers were employed, these were mainly casual labourers from local communities and Tamale on comparable wages (75 Ghanaian Cedi each per month) to other staff of similar grades in the localities. The company's manager explained that he had to invest about one million US dollars of his own resources to assist in operations and to diversify into food crop production. This raises real questions as to where the resources necessary to scale up the investment will be coming from, especially given limited local sources of finance for jatropha cultivation (see Technoserve, 2007). According to the assistant manager, the manager of Solar Harvest Ltd is currently in Norway, trying to mobilize an amount of US$500 000 for the company to revitalise its activities. Given the scale of the financial difficulties faced by the company and the fact that such a large-scale agricultural investment business has huge inherent risks, one wonders if the business plans of Solar Harvest Ltd were carefully evaluated before the start of operations.

As part of its biofuel operations, Solar Harvest Ltd plans to produce jatropha through plantations – there is currently no outgrower scheme. While jatropha cultivation on a plantation basis has food security risks, the fact that currently only a small amount of the land acquired by Solar Harvest Ltd has been put to cultivation and some community members are returning to unused land for farming, the risk to food security is minimal in the short term. However, in the long term, when more land is put to jatropha cultivation, this might negatively affect food security in the area if no proper remedial measures are taken to address this risk.

The company has developed a CSR (corporate social responsibility) programme. This includes providing two small dams and a corn mill to the Kpachaa community; and a small dam each at Jimle and Jashee. At the time of this study in February 2012, however, the corn mill at Kpachaa was out of order and community members had to travel several kilometres to the nearest towns to access the service of a corn mill. It was also indicated that the services of the corn mill are paid for by community members. The four dams were inspected by the author of this study and seen to serve as important sources of drinking water and other domestic purposes including the watering of animals. It was impossible to access contractual documentation to assess the extent to which these benefits are part of clear and enforceable commitments for the company, or to assess these benefits in the broader context of the economic deal embodied in the contract (for example, comparing these benefits to benefits promised to customary authorities or other groups).

The company liaises with the community through a central committee made up of two members selected by each community to represent it and this serves as a link with the management of the company. The son of Tijo-Na is the chairman of the central committee. In the attempts of Action Aid to present the negative consequences of the company's operations, the committee came to the defence of Solar Harvest Ltd, in spite of existing evidence of the suffering of community members. Given the fact that the company's assistant manager is a grandson of the Tijo-Na (the customary chief), that the chairman of the central committee is a son of the Tijo-Na, and that the Tijo-Na was instrumental in having the company established in his traditional area, there might be conflicts of interest.

Of course, the above analysis does not imply any bad faith on the part of the company. Had it not been for the effects of the global economic crisis, the outlook might have been very different. Until it was deeply affected by that crisis, the company had created a reasonably good number of jobs and provided some important social amenities in the communities. Had it sustained the job creation and improved on its CSR demonstrated to date, its relations with the communities might have continued to be cordial. The MoU between Solar Harvest and MiDA also appears to be pro-smallholder farmers and the diversification into food crop production are measures that, if well planned and managed, could restore the good image of the company in local communities. The capital required by the company is, however, a key determinant of how successful these measures would be.

5. Conclusions and recommendations

The current wave of agricultural investment in Ghana is taking place in a context where legal and institutional frameworks at local and national levels are very weak. In this context, there is a real risk that deals concluded with customary authorities benefit only a minority. This could have negative implications for the livelihoods of most smallholder farmers who are the poorest of the poor. The nature of the impact of an agricultural investment seems to depend on the framing of the business model, as well as on factors affecting commercial viability. If the manner in which the business entity is structured in wealth creation takes into account the needs of communities and provides room for partnerships and equitable sharing of benefits and risks, the impacts on people's livelihoods are likely to be positive. Thus,

TABLE 12
Key features of the Solar Harvest Ltd business model

Ownership	The company is owned by Norwegian nationals. It has leased land for 25 years on the basis of a lease renewable for another 25-year term. As part of a separate, more recent development, the company will upgrade and expand irrigation facilities on land belonging to local farmers under a MoU with MiDA.
Voice	Decision-making lies in the hands of the management of Solar Harvest Ltd. The chain of command appears to be top-down. The land acquisition process was characterized by lack of consultation with local people – the deal was struck directly with customary authorities.
Risk	The company bears business risks, and so do the workers – the workforce has shrunk from 400 to 5 as a result of the global economic downturn. Community members who lost land now face livelihood challenges and even risk destitution.
Reward	Jobs have been created – 400 at peak, currently 5 and earning some 75 Ghanaian Cedi (less than US$50) per month as watchmen. The salaries of the manager and his assistant were not disclosed. Two small dams and a corn mill have been provided by the company at Kpachaa and one small dam each for the communities of Jimle and Jashee. The corn mill facility was not in working order at the time of the research in February 2012. The company's expectation of rewards from jatropha production has also suffered due to financial constraints, and the company has revised its business model towards greater diversification. The bulk of the rental sum for the first lease period of 25 years appeared to have been taken by the chiefs; community members who lost their land use rights got little if anything in compensation, as indicated by community members interviewed.

a business model's degree of inclusiveness will be high or low depending on how much control over ownership, voice, risk and reward is exercised by a single entity.

Attracting more FDI to the economy, especially to agriculture, is seen in Ghana as a key strategy to achieve the MDGs, while addressing the structural impediments the country is bound to face to become economically independent. Building on some of its natural competitive advantages, Ghana has also taken several FDI related reform measures to enhance its agribusiness climate. This effort has paid off, making Ghana the second fastest growing country behind Burkina Faso in terms of inflows and the third largest FDI stock-builder in West Africa. The benefits of these outcomes are unevenly distributed with little investments going to agriculture in terms of cash (1 percent) and projects (5 percent), compared to other subsectors such as manufacturing, building and construction, general trade and services. FDI originates from the country's long-standing partners such as France, United Kingdom and United States, but also from new and distant investors from Asia (China, India, Indonesia, Singapore and the Republic of Korea).All these partners are expressing renewed interests in available land acquisition or leasing opportunities to farm energy crops and fish while maintaining investments in traditional farming of cash and food crops. The disparity in FDI distribution remains true at the regional level where the Greater Accra region appears the largest FDI beneficiary.

Land tenure insecurity is hampering the remarkable effort being made to drive more FDI into the agriculture sector and food industry. Despite the plethora of institutions currently acting to address the land related issues in the country, Ghana has not yet succeeded in simplifying the process of acquiring or leasing land in a safe and transparent manner and this task needs to be confronted. Contrary to the perceived belief that land is abundantly and readily available for venture in any profitable economic activity in Ghana, the amount of land remaining by region is shrinking like washed leather, underscoring the sensitivity of land issues today across Ghana, especially for smallholders whose livelihoods are highly dependent on rural incomes. Consequently, there are no large tracts of land available now for sale or lease without the encumbrance of some sort of claim or ownership attached, which would suggest an urgent need to address the problem of land tenure insecurity that is prevalent in the country. Providing support to the mapping of the most favoured foreign investment areas and maintaining online land banks is worthwhile if one wants to see more FDI inflowing to Ghana's agriculture.

The impacts of FDI have been favourably felt in many areas across the Ghanaian economy in general, including its food sector where the case of the palm sector and the free zone scheme were extensively discussed to illustrate the economic, social and environmental contribution of FDI therein, as they relate to various income generation opportunities and risks created to help provide a livelihood for people. Foreign direct investment inflows to Ghana have helped ease various structural imbalances including the state of food insecurity, while raising the risks of seeing more and smaller land users dispossessed of the land they work to make ends meet, and also of seeing the country exposed to more climate change effects. Though encouraging, the results obtained are well below expectations in many other areas including the persistent agricultural yield gaps that the on-going sluggish rates of technology transfer have failed to address in key supply chains.

Joint-venturing appears as the most preferred mode of entry (78 percent of the agribusiness projects registered with GIPC) investors used to penetrate the Ghana agribusiness sector. The findings suggest that the palm oil sector is one of the most profitable agribusiness activities where good public policy coupled with a sustained financial support to the industry would make a big difference in the fight against poverty and food insecurity in the country. The review of the GOPDC business model suggests that there is room for interested foreign investors to invest responsibly in the agricultural sector while keeping in check the expected economic, social and ethical commitments. There exists a win-win solution to maximize the economic, social and environmental bottom lines in the agriculture sector while reducing the risks for all involved including the smallholders who are most likely to be dispossessed of their land. The review suggests the need to go beyond agribusiness models established so far (out-growers and contract farming schemes) to include more inventive agribusiness arrangements, viz. farmer ownership models whereby smallholders would be given a chance to use their lands for equity swap arrangements while also keeping a greater value of the processed commodities derived from what they produce along global food value chains.

Within the context of the data limitations faced, the results of the business model case studies suggest that the business model developed by the ITFC presents a high degree of inclusiveness. The company management undertakes operational decisions, but given its outgrower scheme, control over much of the land is in the hands of outgrowers. The existence of an outgrowers' association enables farmers to exercise a voice *vis-à-vis* the management of the company. Community perceptions of the initiative are overwhelmingly positive, and the study documented amenities and services provided by the company. The food security risks posed by ITFC are minimal given that outgrowers are encouraged to intercrop with groundnuts, and the company has not undertaken large-scale land acquisition for its operations. However, the study also highlighted concerns that aspects of the contractual arrangements appear to favour ITFC, especially in terms of price fixing and marketing of produce.

Both outgrowers and company face and share risks unique to their roles and the terms of their contractual relations. While outgrowers' risks are largely production related (for example, the vagaries of the weather), management bears the bulk of financial and marketing risks. In the case of benefits, employment as well as CSR activities – particularly in the areas of health and education – have resulted in cordial relations between the ITFC and the local communities. The outgrowers expect an annual income of US$2 000 after 14 years, when debts are fully repaid. This promises better incomes for outgrowers than alternative livelihood sources in the area. There are, however, risks that this expectation might not be realized given the lengthy time frame involved. Yet, outgrowers regard themselves as better off even now, compared to other farmers or labourers.

It should be noted that the ITFC has received extensive development aid support, particularly with regard to its outgrower scheme. This raises issues about the extent to which the experience can be replicated and scaled, since the need for development aid support might have been a possible motivation for ITFC to embark on the outgrower scheme, in addition to the costs of large-scale land acquisition.

The case of Solar Harvest Ltd is the story of a private company that acquired land on a large scale in several communities, for the cultivation of jatropha. While the ITFC has been around since 1999, Solar Harvest is a more recent project, launched in 2008 as part of the global increased interest in land investments in the global South. Its business model concentrates the control of land and other key assets of the company in the hands of management. The process of land acquisition resulted in tension in the communities and still remains contentious, due to the imbalance in the power of stakeholders in the negotiation process. The company's promises of jobs and other CSR activities were seriously affected by the global economic downturn. This has left the company in a state of financial crisis and thus nearly all of its 400 jobs at peak of operations have been lost. These developments have led to significant deterioration in relations between Solar Harvest Ltd and local communities. This is a serious risk to the investor. The burning of the jatropha plantation at Kpachaa may well have been accidental – but the possibility of an act of sabotage cannot be ruled out. The experience is a dire warning to the many companies developing similar projects in many parts of Africa. Of course, outcomes and community relations might have been different had it not been for the global economic downturn. But the model itself (acquisition of large areas of land, dependence on lending) made the investment vulnerable to these outcomes.

The two case studies analysed in this chapter are very different in nature – not only in focus areas of crop cultivation, but also as business models. While the ITFC focuses on organic mango cultivation through an outgrower scheme, has been in operation for a long time, and has received extensive development aid support; Solar Harvest Ltd focuses on jatropha as feedstock for biofuel and is a fairly new company, which has yet to make a harvest and has already been hit by a financial crisis. The ITFC can be seen as an example of best practice in agricultural investment. Solar Harvest Ltd, on the other hand, offers useful insights for the future of agricultural investments that require large-scale land acquisition in areas prone to tenure insecurity.

Agriculture is the engine of growth for the Ghanaian economy. The country currently aims to become a middle-income country by 2015, under the Food and Agriculture Sector Development Policy II (FASDEP II) and the Medium Term Agriculture Improvement Plan (2009-2015). Evidence from this study suggests that achieving this would require measures that promote serious private agricultural investments. However, given the high risk of food insecurity of some agricultural investments, the way forward lies in a balance between a value chain approach to agricultural development and food security concerns.

This approach would require developing country-level operational tools to implement the principles of responsible agricultural investments (RAI) developed by the Food and Agriculture Organization (FAO), the International Fund for Agricultural Development (IFAD), the United Nations Conference on Trade and Development (UNCTAD) and the World Bank. To achieve this requires political will and commitment to strengthening the legal and institutional capacities and frameworks for effective land governance and agricultural development from local to national level. In the case of Ghana, the following specific recommendations are made:

1. The Lands Commission, in consultation with traditional authorities and other stakeholders, should urgently speed up the process of finalizing the draft guidelines for large-scale land acquisitions for agricultural and other purposes, for these to serve as a tool for operationalizing the principles for responsible agricultural investment (RAI);
2. The capacity of traditional authorities and local communities needs to be strengthened to enable them undertake productive negotiations with investors through the development of local land governance structures such as the Customary Lands Secretariats;

3. The Civil Society Coalition on Land (CICOL), the District Assemblies and Customary Land Secretariats can play an important role by undertaking periodic public education and sensitization of communities on their land rights and how these can be protected;
4. The Ghana Investment Promotion Centre (GIPC) and the Environmental Protection Agency (EPA) need to collaborate more in enforcing compliance with standards and conditions and periodically monitoring compliance by investors;
5. The Constitutional provision that government should not interfere with the chieftaincy institution must be reviewed to enable some level of intervention, especially where the land rights of communities are usurped by a chief for personal gain;
6. The Lands Commission should disclose and publicize the contracts involving land acquisitions for the public to evaluate how transparent, accountable and equitable these transactions are to both present and future generations; and
7. Government ought to balance efforts to promote agricultural investments with the need to promote national food security.

References

A **Ackah, C.G.**, Aryeetey, E.B.D. & Aryeetey, E. 2009. *Global financial crisis discussion series-Paper 5: Ghana*. Overseas Development Institute (ODI), May 2009.

Acquah, P.C. 1995. *Natural resources management and sustainable development: The case of the gold sector in Ghana*. UNCTAD/COM/41, 15 August 1995.

ADB (Agricultural Development Bank). 2008. *Annual Report and Financial Statement 2008*. Ghana.

____2007. *Annual Report and Financial statements 2007*. Ghana African Banker Magazine. IC Publication 4th Quarter. October 2008, issue N°6. IC Publications.

African Business Magazine. December 2009, N°359. IC Publications.

African Economic Research Consortium. 2006. *Foreign Direct Investment in sub-Saharan Africa: Origins, targets, impact and potential*. S. Ibi Ajayi, ed. Nairobi.

Agbessi Dos-Santos, H., & Damon, M. 1999. *Manuel de nutrition africaine* , IPD-ACCT-KARTHALA, Octobre 1999.

Agro-Chemicals Report, Vol. III, N°2, April-June 2003: Commercial farms in South-East Asia.

Ahomka-Lindsay, R. (Chief Executive officer). 2008. Ghana Investment Promotion Center GIPC). PowerPoint presentation made at Ghana-Nigeria Business Summit, 6th October 2008.

Aryeetey, C., Barthel, F., Busse, M., Loehr, C. & Osei, R. 2009. *Empirical study on the determinants and pro-development impacts of foreign direct investment in Ghana*. HWWI (Hamburg Institute of International Economics, Germany), Research Programme World Economy and ISSER (Institute of Statistical, Social and Economic Research, Ghana), University of Ghana, Legon.

Aryeetey, E. University of Ghana ISSER, & Kanbur, R., Cornell University.2008. *The economy of Ghana, analytical perspectives on stability, growth and poverty*. J. Currey, ed. Accra: Woeli Publishing Services

Brobbery, S.A. 2008. *The Law of Chieftancy in Ghana*. Advanced Legal Publications, Accra.

ECOWAS. 2004. *The ECOWAS Trade Liberalization Scheme: Protocols and Regulations*. ECOWAS Executive Secretariat. Abuja.

EDJA (Editions Juridiques Africaines).2009. *Code des collectivités locales du Sénégal annoté* .

EGIS BCEOM International/Associated Consultants.2009. *Integrated Transport plan-Ghana, Report on commodities generation and transport in Ghana.* April 2009.

ESD (Environmentally Sustainable Development Proceedings).1993. Series N°2. *In*: Proceedings of the first annual international Conference on Environmentally Sustainable Development. World Bank, 30 September– 1 October 1993.

F&D (Finance and Development). Septembre 2008, 45(3). (Available at: http://www.imf.org/external/pubs/ft/fandd/2011/09/).

ISSER (Institute of Statistical, Social and Economic Research), 2000. *The State of the Ghanaian*

Economy in 1999. University of Ghana Legon.

______.2006. *The State of the Ghanaian Economy in 2005*. University of Ghana Legon.

______.2008. *The State of the Ghanaian Economy in 2008*, University of Ghana Legon

Jaeger, P., Accord Associates LLP. 2008. *Ghana report horticulture cluster strategic profile study, article I: Scoping Review*. Prepared for WB-SDN/AFTAR/ MOFA./EU-AAACP.

Koroma S., Mosoti V., Mutai H., Coulibaly A.E., & Iafrate, M. 2008. *Towards African Common Market for Agriculture Products*. FAO: Rome.

Lehmann, M.B. 1984. *The Dow Jones –Irwin Guide to using The Wall Street Journal*.

Levy, M.L., Ewenczyk, S., & James, R. 1989. *Comprendre l'information économique et Sociale: guide méthodologique* .. Collection J.Bremond, 2e édition augmentée. Paris : Hatier.

Liondjo, F. (International Consultant) .2001. *Strategies for the multilateral trade negotiations and implementation aspects of the WTO agreements- Ghana.*, Cluster 2, August 2001.

Malik, M. 2009. *International Investment Agreements, Best practices bulletin#1*. International Institute for Sustainable Development (IISD), August 2009.

Matsumoto-Izadifar, Y. 2008. *Making better use of agribusiness potential*, OECD (Organization for Economic Co-operation and Development). (Available at: . www.oecd.org/dev/publications/businessfordevelopment).

Mbekeani K. 2007. *The Role of Infrastructure in Determining Export Competitiveness*, (Available at : http://www.aercafrica.org/documents/export_supply_working_papers/KMbekeaniInfrastr8DB3B.pdf-Similar pages

Ministry of Lands and Forestry. 1999. *National land policy*, Accra, June 1999.

Ministry of Lands, Forestry and Mines. 2008. *The Ghana Land Bank Directory, Second Edition, 2008*. Ghana.

MOFA (Ministry of Food and Agriculture, Ghana).. 2009 Speech of the honorable Minister's remarks on the Occasion of the launch of the FAO/IEA Project: *Articulating and mainstreaming appropriate* Agricultural trade policies

_____.2008. *Agriculture in Ghana: Facts and Figures-2007*, Statistical Research and Information Department (SRID), Accra.

_____.1998. *Marketing Margins for Selected Food Commodities* Accra.

Monke, E.A. & Pearson, S.R. 1989. *The policy analysis matrix for Agricultural Development*, New York: Cornell University Press.

MTM (Hebdomadaire-Marchés tropicaux et méditerranéens). 1999. *Spécial Ghana, 54e année*. Vendredi 2 juillet 1999, N°2799.

Nathan Associates Inc. 2009. *Ghana: Economic performance assessment* , USAID, July 2009.

Ndiaye, T.M. 1992. Matières premières et droit international, Dakar: NEAS.

Ndoye, D. 1997.: *Le problème des biens immobiliers de la collectivité leboue de Dakar , questions historique, économique, sociale et juridique*. EDJA (Editions juridiques africaines). *Nkoranza District Assembly. 2006. Third Medium Term Plan -2006.Nkwanta.*

Norton, R.D. 2004. *Agricultural Development policy: concepts and experiences*, FAO & Wiley.

Olayemi, J.K. 1996. *Poverty, Food Security and Development in West Africa*, Paper presented at the Regional Conference on Governance and Development, Prospectus and for 21st Century West Africa, Accra.

Smaller, C. & Mann, H. 2009. *A thirst for distant lands: Foreign investment in Agricultural land and water*. International Institute for Sustainable Development (IISD), May. Manitoba: IISD

Technoserve. 2007. Feasibility Report of Biofuel Production in Ghana: *Assessing competitiveness of the Industry's value chain*, unpublished.

UNCTAD (United Nations Conference on Trade And Development).2009a. *World Investment Report- Transnational Corporations, Agricultural Production and Development*,

______.2009b. *World Investment Prospects Survey 2009-2011*,

______.2008. *World Investment Report- Transnational Corporations and the infrastructure challenge*,

______.2008. *World Investment Prospects Survey 2008-2010*,

______.2007. *World Investment Prospects Survey 2007-2009*.

______. *2006.* Review of Maritime Transport., UN: New York.

______. 2002. *Summary of the deliberations of the investment policy reviews,* UN, 18 December 2002.

_____.2000. *Tax incentives and foreign direct investment, a global survey*, ASIT Advisory Studies N°16. UN: New York and Geneva.

University of Ghana Legon. 2008. *Harvest and Post Harvest Losses Baseline Study*.

WTO (World Trade Organization). 2007. Trade policy review- Report by the secretariat Ghana, WT/TPRS/S/194, 17 December 2007

Mali:

Large-scale agricultural investments and inclusive business models[1]

1. Introduction

This chapter discusses trends, drivers, legal frameworks and case studies of agricultural investments in Mali. The purpose is to generate evidence on a range of different models for structuring agricultural investments, with a focus on models that hold promise for the inclusion of local farmers and communities.

The chapter analyses the context in which agricultural investments are taking place, particularly with regard to Mali's policy framework regulating land use and tenure and to the economic position of local producers; it analyses recent trends in agricultural investment and land acquisition in Mali; it discusses the design and implementation of different business models, focusing on case studies of two investment projects; and it develops conclusions and possible ways forward. The two case studies involve a discussion of two recent agricultural investments: a bio diesel project run by Mali Biocarburant SA in the Koulikoro Region, which provides an example of agricultural investment that does not involve land acquisitions and has made the inclusion of smallscale producers a central pillar of its business model; and a sugar cane plantation and processing facility run by the Markala Sugar Company (SOSUMAR) in the country's Office du Niger area, which is located in Ségou Region. This second case study provides an example of 'public – private – community' partnership.

The chapter draws on a review of the literature and of documentary evidence, including some contracts for agricultural investments, on interviews with key informants, and on fieldwork based on qualitative semistructured interviews. Interviews with informants based in Bamako helped frame the analysis and collect data on the two case studies. Informants included researchers, officials of public and semipublic agencies and private sector officers. Fieldwork focused on the two case studies. Field research was conducted in May 2011 in Koulikoro Region and in the Office du Niger area. During the field visits, collective and individual interviews were conducted with the various stakeholders, including investors, local producers, government administration, technical services and funding agencies. The remainder of the chapter is structured in three parts. The next section analyses the national context within which agricultural investments are taking place, which is characterized by widespread poverty, the existence of considerable agricultural development potential, weak public funding capacity in the sector, the urgent need for private investment – all within a confused land tenure situation. Section 3 reviews current trends in agricultural investment, discussing key players and their drivers, features as well as potential and actual impacts. Finally, Section 4 presents findings of the two case studies. The conclusion summarises key findings and suggests possible ways forward.

2. The national context

Gaining a proper grasp of the context of private investment in the agricultural sector in Mali requires a discussion of three critical issues. Firstly, the country has considerable agricultural development potential but, due to its extreme poverty and the decrease in Official Development Assistance (ODA), it faces difficulties in funding the agricultural sector. This explains the desire of

[1] This chapter is based on an original research report produced for FAO by Moussa Djiré with Amadou Kéita and Alfousseyni Diawara, International Institute for Environment and Development

public authorities to attract private investment. Secondly, complex and pluralistic land tenure systems and the limited ability of national legislation to ensure effective regulation of the tenure dimension of private agricultural investment raise important challenges for the government's ability to manage investment in agriculture and for the protection of local rights that may be affected by investment projects. Thirdly, legislation has been adopted to promote investment and regulate its social and environmental impacts, but the effectiveness of this legislation in establishing safeguards for local people and the environment has been questioned. The next few sections discuss these three aspects in greater depth.

2.1 A country with major agricultural potential but facing financing difficulties

Mali is a landlocked country in the heart of West Africa, with a surface area of around 1 240 000 km². Its population was 14 517 176 in 2009.[2] Mali shares around 7 200 km of borders with Algeria to the North, Niger to the East, Burkina Faso to the SouthEast, Côte d'Ivoire and Guinea to the South, Mauritania and Senegal to the West. Much of the country is relatively flat, with rolling plains and low plateaux.

With its very low but steadily rising Human Development Index, Mali is amongst the poorest countries in the world. Although the poverty rate fell over the period 20012006, it is still very high, with a national average of 47.4 percent in 2006. Geographical variations are substantial: the poverty rate is 20.1 percent in urban areas and 73 percent in rural areas (CSRP 20072011). The country was ranked 175th out of 187 on the HDI in 2011 (UNDP, 2011). This pattern of poverty, combined with certain cultural and historical features, has made Mali the source of major migration, particularly towards West, North and Central Africa, as well as Europe and America.

The structure of the Malian economy is characterized by a predominance of the primary and tertiary sectors, which accounted respectively for 36 percent and 35.6 percent of gross domestic product in 2009 and 2010.[3] This pattern was expected to remain stable in 2011, with the primary and tertiary sectors losing a little ground to the secondary sector. In 2010, growth in real GDP was held to the same level as 2009 (4.5 percent) but below initial forecasts.[4] Mali presents considerable agricultural, forestry and pastoral potential. Rural land is estimated to amount to 46.6 million hectares, including 12.2 million hectares of arable land, 30 million hectares of grazing land, 3.3 million hectares of wildlife reserves and 1.1 million hectares of forest reserves (Ministry of Agriculture, 2008). The country has vast areas suitable for development and irrigation (2.2 million hectares), substantial water resources (2 600 km of rivers), considerable biological diversity, substantial forest and wildlife resources and large numbers of diverse, adapted livestock (7.1 million cattle, 19 million sheep/goats, 0.6 million camels, 25 million poultry) (Ministry of Agriculture, 2006 and 2008).

Nevertheless, agricultural resources are unequally spread over the national territory, twothirds of which is desert. In addition, funding is crucial to the expansion of the agricultural sector but this is becoming increasingly problematic.

The modernization of agriculture is one of the three main objectives of the Rural Development Master Plan (SDDR), together with environmental protection and improved natural resource management. The Master Plan was adopted in 1992 and updated in 2000. The provisions of the Plan are reflected in various other official documents. They were taken up by the second President of the third Republic of Mali just after his election in 2002 and developed particularly within the Economic and Social Development Programme, which he outlined during the presidential election campaign in 2007 (Toumani

[2] Provisional results, 4th General Census of Population and Housing.

[3] www.africaneconomicoutlook.org/fr/countries/west-africa/mali/

[4] www.africaneconomicoutlook.org/fr/countries/west-africa/mali/

Touré, 2009). So, agricultural modernization is seen as a policy priority at the highest level of government.

The desire to modernize agriculture also lies at the heart of the Framework Law on Agriculture (LOA), which was adopted in 2006. In article 3, the LOA states that 'agricultural development policy shall be based on voluntaristic promotion of the modernization of family farming and agribusiness, to foster the emergence of a structured, competitive agroindustrial sector integrated within the sub-regional economy'. However, agricultural modernization involves a financial cost that the country cannot meet from its own resources. The Strategic Framework for Growth and Poverty Reduction (CSCRP) 20072011 estimates the cost of taking action in the agricultural sector at FCFA 153 648 000,000 (CSCRP 20072011, Annex III). At an approximate exchange rate of 1 US$ = 500 FCFA, this is equivalent to US$307 296 000.

A brief look at public investment in the Office du Niger area, which is today favoured by investors because of its enormous hydroagricultural potential, can give an idea of the resources required to pursue this agricultural modernization agenda, in particular as regards irrigation schemes.[5] The Office du Niger is one of the oldest irrigation schemes in West Africa. Set up in 1932 in the inner Niger delta, it was to become, according to the original plans, the main supplier of colonial France's textile industries, the rice bowl for West Africa and a place of technical and social innovation. The objectives were ambitious, with over a million hectares to be irrigated over a 50-year period. The major structures were designed and built to meet those objectives. Using existing backwaters and a dense network of irrigation and drainage canals, the scheme now covers more than 87 692 hectares. Irrigated lands are used to produce rice, vegetables and sugarcane (Dave, 2010).

Until recently, all the schemes in the ON area were funded by the public authorities. Between 1934 and 2009, the government developed a total area of 63 713 hectares, including 4 653 hectares supported as of 2000 through the Special Investment Budget (African Development Bank, 2010). To these must be added interventions supported by donors. Donor agencies initially funded only the rehabilitation of older schemes, but then went on to fund the creation of new ones. The main donors have been the Netherlands (20 595 hectares rehabilitated and 5 829 hectares constructed), the French Development Agency (5 540 hectares rehabilitated and 1 700 hectares constructed, together with another donor), the European Development Fund (3 650 hectares rehabilitated), the International Development Bank (700 hectares rehabilitated and 520 constructed), USAID (1 971 hectares new schemes, usually with the Office du Niger or other donors), German Development Cooperation (3 100 hectares rehabilitated and 800 hectares new).

Table 1 provides data on the substantial volume of funding expected to be required for new schemes in the Office du Niger area. It shows that the irrigation of 79 865 hectares planned for the period 20102020 requires an amount of FCFA 266 756 291 750 (US$533 512 584), i.e. an average of FCFA 3 340 000 (US$6 680) per hectare, excluding costs relating to feasibility and related studies. According to data from SEDIZON (2010), funding for planned extensions and studies to be conducted by a Libyan investor, Malibya, over the same period would amount to FCFA 85 750 million (US$171 500 000). These figures illustrate the challenges faced by a country like Mali in financing its plans to expand irrigation infrastructure as a basis for agricultural modernization at this scale.

It is for these reasons that the Malian Government has worked to promote private investment in agriculture. Given limited capital availability within Mali, foreign investment is expected to play a particularly important role. Private investment is seen as a source not only of capital, but also technology, knowhow, infrastructure development and market access, and potentially as a catalyst for economic development in rural areas. On the other hand, family farming is considered in public discourses as oldfashioned and incapable of ensuring food security.

[5] The name Office du Niger designates both the irrigation scheme area and the institution set up by the government to manage the scheme.

TABLE 1

Estimated cost for the implementation of the development and rehabilitation programme under the Development Master Plan for the ON area (SDDZON)

Nature of the work	Area concerned (ha)	Estimated cost FCFA
Total extension work (ha) 2010-2020	79 865	266 756 291 750
Total rehabilitation work (agricultural plots only, i.e. excluding irrigation and drainage networks) 2010-2020	2 695	11 927 000 000
Total rehabilitation and extension studies (plots including reconversion Sossé Sibila)	24 855 362 874	
Total studies and work	303 538 654 624	

Source: Office du Niger, 2010

The call for private investment was accompanied by a 'charm offensive' to attract investors, which included revising the investment code and setting up a National Investment Promotion Agency, a Presidential Investment Council and an international cooperation office within the Ministry of Agriculture, backed by an intensive advertising campaign.

The call for private investment did not fall on deaf ears. Stimulated by the international food crisis and the increased interest in biofuels, private investors rushed to get hold of Malian agricultural land, particularly in the Office du Niger area. Some of these deals were very large. By way of illustration, 100 000 hectares were allocated in a deal with Malibya, an enterprise of Libyan origin, and a similar land area was allocated to Huicoma, a Malian company. These land allocations alone exceed the total area of the irrigation development schemes established in the ON area since colonial times.

However, this scramble for land took place against a background of relative confusion, given the limitations characterising the legal, regulatory and institutional framework. It is to this topic that the next section turns.

2.2 A hybrid land tenure system

There are two main land tenure systems in Mali: customary systems deriving from ancestral traditions and local practice, on the one hand, and the formal system of written law established by the state, on the other.

Customary systems and local practice

Customary patterns of land access are still the most widespread in rural areas. Throughout history, major empires and kingdoms have flourished on the territory of Mali, shaping lifestyles, beliefs and patterns of access to land and natural resources. This historical legacy explains the great similarities that exist in the traditional organization of social and land relations, although land tenure regimes are still widely different as a result of specific historical, geographical and socio-cultural factors.

Relationships between individuals and social groups are organized according to principles like kinship; gerontocracy and the corollary principle of seniority, based on respect for the elders; the pre-eminence of indigenous communities, particularly as regards the exercise of local political power and access to land; and a gender hierarchy in which men take precedence over women.

These principles guide the organization and operation of village institutions and indeed the entire social and tenure structure in rural areas. However, generally speaking, their implementation varies depending on the agroecological zone concerned, the nature of production systems and especially social and historic factors.

Access to land in rural areas follows two essential patterns: intra-lineage access and interlineage access. The predominant method of access to land, common to all geographical zones, is intralineage access. This takes two main forms: inheritance and allocation of a portion of the lineage holdings to one family or individual belonging to the lineage. As land ownership is passed on within families, it is possible through inheritance not only to gain access to land but also to become its manager according to customary rules (GERSDA, 2007). Management was originally based not on ownership rights understood in the sense of individual private property rights, but on a set of rights (access, usage, offtake, exclusion, disposal, etc.) held collectively by the members of the lineage or family and allocated in various ways to the members of those groups.

Intralineage access patterns depend on the size of lineage landholdings and tenure issues in the area. In many families, there is an increasing trend towards splitting up lineage holdings following the enlargement and dismantling of family farms. As a result of various different factors, large families are breaking up and giving rise in various places to the emergence of nuclear families as customary holders of the land they work.

Interlineage access is organised around arrangements that transfer rights, permanently or temporarily outside the landholding lineage. These arrangements include gifts, loans, rental and, more rarely, sharecropping and sale. The latter three arrangements have developed recently as a result of the growing monetization of land relations. They tend to involve relationships between indigenous people and recent migrants, rather than between lineages within the same community. The various arrangements can be combined; the predominance of one or the other depends on local land relations and the economic stakes in the area concerned.

Despite the existence of principles common to the different customary systems, rules governing access to land vary according to local issues, social and historical factors. They are also profoundly influenced by dynamics concerning the design and implementation of national law enacted by the state.

Tenure systems under written law

Formal (written) law establishes various methods of access to land. The provisions of general legislation must be distinguished from the norms regulating particular areas such as irrigation schemes.

The Land and Property Code (*Code Domanial et Foncier*, CDF) is the piece of legislation that provides the foundation of national law governing tenure. As a general rule, Malian legislation follows the principle of state ownership of land. The state plays a leading role in land relations, and directly holds land as part of its public or private land estate. The latter category consists of land that has been explicitly registered as belonging to the state, but also land classified as 'vacant and unclaimed' and land held by virtue of customary rights (article 28 of the CDF). The CDF does protect these customary rights, requiring that their compulsory acquisition requires public purpose and payment of compensation (article 43). But ownership on these lands is vested with the state. So it is the state that has the legal authority to decide on and negotiate transactions affecting these lands.

Article 35 of the CDF states that the private land estate can be transferred in a number of ways, namely such as rural concessions, allocation, longterm lease, leasehold with the promise of sale or title deeds. In the case of rural concessions, for example, the public authority grants the concessionholder the right of temporary use of a piece of land to develop it on the terms set out in the concession deed and attached specifications. In a longterm lease, the lessor grants the lessee a longterm use right that can be mortgaged, against payment of an annual fee. Title deeds are regulated by article 169 of the CDF, which states that titles are permanent and cannot be challenged. A title deed is seen by the Malian courts as the sole starting point of all property rights at the time of registration.

While customary rights are formally recognized and protected by legislation, the procedures to establish and register them have still not been determined. This is because the necessary implementing decrees have not yet been adopted. This circumstance makes Mali's land legislation incomplete in important respects. Customary land holders that wish to formalise their rights can only do so through the procedure provided for rural concessions. This procedure is costly and cumbersome, and arguably not suited to recording customary rights.

In virtually every region of the country, there are schemes set up by the state or NGOs where tenure systems varied depending on the legal status of the scheme in question. Plots are typically allocated in these schemes on the basis of permits or usage agreements. This chapter focuses on the case of the Office du Niger, where various tenure systems coexist.

From its creation until the present day, the Office du Niger has undergone various changes which have resulted in a wide range of tenure arrangements. Pursuant to Decree No. 94-004 of 1994, the Office du Niger (ON) is a public industrial and commercial establishment (EPIC) responsible for managing land irrigated or irrigable through the Markala dam. Decree No. 96188/PRM of 1996 confirms the ON's control over not only the land which has been developed and equipped, but also the land located in undeveloped areas, i.e. irrigated land and land that is capable of irrigation by means of the Markala dam. Article 3 of the 1996 decree specifies that the remit of the ON management can extend to nonirrigable land if the government deems it appropriate. However, according to article 4, this land must, like land already developed and the surrounding protected areas, be registered in the name of the Malian state, which will bear the cost of clearing customary rights exercised over that land and all expenses connected with registration. Undeveloped land in the ON area, as in the other rural regions of the country, is in practice held by local communities and managed according to customary rules. Any intervention by the State in those areas would require prior negotiation with the customary holders.

The 1996 decree sets out the following mechanisms for accessing land: the Annual Usage Contract (***Contrat Annuel d'Exploitation*** - CAE), the Farming Permit (***Permis d'Exploitation Agricole*** - PEA) and the housing lease in irrigated areas; and the ordinary lease and emphytheutic lease in areas not yet irrigated. The last two methods are used for largescale investments, and are briefly discussed in Box 1.

In practice, recent large-scale land acquisitions in Mali and in the Office du Niger have involved use of contractual arrangements not explicitly mentioned in the legislation discussed so far.

The first such contractual arrangement is the investment agreement, or convention of establishment. This is a relatively recent practice in private investment in the agricultural sector, although a few rare examples can be found from the 1990s. These agreements reflect the investors' wish to obtain legal safeguards from the government concerning aspects capable of affecting the success of their investment. This mechanism has been used by large foreign or

BOX 1
Access to undeveloped Office du Niger land

- Ordinary lease: granted on undeveloped land for a maximum period of 30 years, renewable by express agreement between the parties. The lessees must develop irrigation infrastructure. Non-payment of rent and failure to maintain the hydraulic network will result in cancellation of the lease. No structure erected in connection with a lease can be destroyed if the contract is terminated.
- Emphytheutic lease: granted on undeveloped land for a period of 50 years, renewable by express agreement between the parties. On expiry of the long-term lease, the lessee leaves the infrastructure constructed by the project in place as it stands, without compensation from the Office du Niger. The lessee undertakes to develop the land within three years from the date of signature of the lease. This period may be renewed once, either tacitly or by express agreement between the parties. The leases are typically subject to conditionalities determined by the Office du Niger. The lessee bears the cost of developing the land and establishing the hydraulic network and all other facilities enabling the land to be used.

BOX 2
Content of agricultural investment agreements

The content of investment agreements is extremely diverse. In general, the agreements start with recitals setting out the background and purpose of the investment. They then establish the two parties' commitments, the terms for granting the land, access to water, use of mineral resources that might be discovered on the site, participation of third-party enterprises, assignment of rights deriving from the agreement and settlement of disputes. They specify the area made available and the duration of the agreement, noting that the government will make these areas available to the investor free of all legal encumbrances and all tenure rights. They also indicate any public easements that the state might impose on the land, as well as the investor's commitment to carry out studies, undertake the development and comply with all required legal formalities. The tax regime is also set out in line with the provisions of the investment code.

Details of the parties' commitments and even the nature of the institution signing the agreement on behalf of the government vary from one agreement to the other, as will be explained later. Much depends on the type of investment and the institutional entry point chosen by the investor. In most cases (with the important exception of the agreement with Malibya), the investment agreement specifies that the investor must conclude a separate agreement with the Office du Niger in order to implement the provisions of the investment agreement concerning land acquisition.

national investors like Illovo, a firm based in South Africa, the China Light Industrial Corporation for Foreign Economic and Technical Cooperation, Malibya, the Société des Moulins Modernes and a few others.

But despite all the political, strategic and legal interest of investment agreements, the transfer of land rights is actually implemented through a lease contract. In this respect, the investment agreement can be seen as some sort of letter of intent which cannot be implemented until the studies required by it have been carried out and its provisions have been operationalized, whether totally or partially, through a lease contract. In some cases, while the investment agreement covers a very large land area (e.g. 100 000 hectares for Malibya and 20 000 hectares for the Malian company GDCM), the lease contract may only cover a smaller area for which the feasibility and impact assessment studies have been conducted and the development plans submitted (e.g. an initial 25 000 hectares for Malibya and 7 400 hectares for GDCM).

The lease contract is signed by the Managing Director of the Office du Niger and by the investor. It specifies the nature of the lease (ordinary or emphytheutic), as well as its duration and the exact location of the land. The contract sets out the timeline for the development of the land (usually three years), the agreed land use and the terms and conditions, including the terms of access to water and payment of the water fee, conditions of land use, and terms for cancellation of the agreement, withdrawal of the plot and settlement of disputes. Once signed, the contract is registered at the land registry in Ségou.

Constraints on tenure security for rural producers

As a broad generalization, the national law regime regulating land tenure is ineffective. State law is modelled on the French legal tradition, rather than customary tenure systems. National and customary laws are governed by different and partly contrasting principles. In rural areas, two different systems of authority, the government and customary regimes, claim legitimacy. By placing customarily held land within the private land estate of the state, the CDF has undermined the security of tenure of the majority

BOX 3
Procedure for obtaining the lease

There are four stages in the procedure for obtaining a land lease from the Office du Niger:

1. Anyone wishing to obtain a lease from the Office du Niger must send an application to the Chief Executive Officer. In response, the prospecting investor will be invited to contact the technical department to discuss the project and identify an appropriate site for its implementation.
2. Following this and based on the findings of the technical department, the management of the Office du Niger will send the applicant a letter of intent so that the project can be set up.
3. The developer will then carry out the required studies, i.e. a feasibility study on the development project, including technical, socio-economic and financial assessments, and the environmental and social impact assessment. The technical studies must, inter alia, deal with the primary, secondary and tertiary irrigation and drainage infrastructure and the plot layout work to be done by the developer. These studies must be conducted within one year.
4. When and if the findings of these studies have been deemed positive and validated, the lease contract will be concluded with a schedule of conditions for developing the allocated plot.

of rural people, who have little decisionmaking or management power on their own land. The lack of the implementing decree required to regulate the procedure for recording customary rights makes it more difficult for rural people to have access to formal documentation for their land. Customary and statutory systems also coexist in irrigation schemes that are governed by special regimes under national law, such as the Office du Niger. Here, while irrigated land is accessed through the arrangements articulated in the 1996 management decree, undeveloped areas are effectively managed through customary systems. The different principles that inspire customary and statutory law create latent tensions that can easily explode as largescale investments enter the local arena.

Although the CDF establishes various measures intended to ensure the transparency of the procedure through which investors may access land under national law, these measures are, in practice, sometimes breached or sidestepped. The effect is to weaken the procedures, undermine the land rights of rural communities and affect the credibility and reliability of deeds issued under national law. In addition, official procedures are based on mechanisms that are unfamiliar and inaccessible for the majority of rural people, and have costs which exclude these people from land ownership (Djiré, 2007).

The effectiveness of national law and of government procedures is restricted by multiple factors, including barriers to rural communities' access to justice, incomplete and inappropriate legislation, and heavy bureaucracy.

2.3. Measures to promote investment and regulate its social and environmental impacts

A new investment code to promote private investment

In order to promote private investment, Mali has, like the other countries in the region, enacted a law determining conditions and procedures for both foreign and national private investment. First adopted in 1991 (Law No. 91-048 of 1991 and Decree No. 95-423/PRM laying down its implementing provisions) and subsequently considerably revised (especially in 2005), the Investment Code was drawn up under the aegis of international financial institutions well before the current wave of largescale land acquisitions.[6] It does not therefore take account of some of their specific aspects, despite its successive revisions.

The Investment Code defines investment broadly as the contribution of 'fixed assets and initial working capital in connection with a development project'. Despite the dryness of this definition, it does have the advantage of excluding exclusively commercial transactions (sale/purchase) from the scope of the Code. The latter operations are governed by the Commercial Code, together with the OHADA Treaty, which concerns the harmonization of business law in Africa. Also excluded from the application of the Investment Code are mining exploration and exploitation and petroleum exploration and exploitation which, although covered by investment agreements, are governed by the Mining Code and Petroleum Code, respectively, and their implementing provisions.

The Code sets out the mechanisms and provisions designed to promote investments, through legal and institutional arrangements which are attractive. It grants many benefits to investors, without discrimination, such as tax and financial advantages, or flexible hiring and firing terms. Industrial developments are encouraged through an increase in the duration of the exemption from the tax on industrial and commercial profits and from the business tax. Apart from equal treatment between national and foreign investors, the Investment Code offers several other safeguards, including the right to repatriate profits and salaries and recourse to international arbitration to settle disputes with the Malian Government. Finally, the Code guarantees protection of established rights, including through general stabilization clauses.

The Code sets no minimum investment threshold. The essential criterion for project eligibility is that the rate of added value must be 35 percent or more.

To ensure efficient enforcement of the Code's provisions, the state has reorganized the departments dealing with investment. At the government level, an Investment Ministry has been established, under the supervision of which the Investment Promotion Agency is tasked with increasing direct investment, particularly foreign direct investment. A onestop shop was set up in 2008 to deal with all administrative procedures relating to enterprise creation in respect of new investments, and to shorten the time taken to complete the formalities.

These advantages explain to a large extent why major national and foreign investors prefer to sign an investment agreement with the government before approaching the Office du Niger for a land lease. Indeed, the investment agreement triggers the application of the Investment Code. In addition, prior approval from the highest level of government authority, which is usually involved with the signing of investment agreements, can help to facilitate the allocation procedure.

Addressing social and environmental issues

Largescale investments typically raise important social and environmental issues. Parallel to the development of legislation to promote investment, the Malian Government has enacted legislation to manage social and environmental risks. While progress has been relatively slow with regard to social risks, environmental legislation has made important advances over the past two decades. The Malian Constitution of 25 February 1992 enshrines the right to a healthy environment as a human right. Similarly, it considers environmental protection as the common duty of citizens and the state. Indeed, article 15 of the Constitution provides that 'Everyone is entitled to a healthy environment. The protection and defence of the environment and the promotion of quality of life are the duty of everyone and the State'.

Reflecting these constitutional provisions but also under international pressure from environmentalists and ecologists, relevant regulations have gradually been put in place to ensure proper protection of the natural and human environment.

In 2001, basic legislation was enacted to combat pollution and nuisance (Law No. 01-020

[6] The 1991 Code repealed the first one adopted back in 1986 (Law No. 86-39/AN-RM of 8 March 1986) and is currently under review.

of 2001 on pollution and nuisance). According to article 3 of that law, any activity liable to harm the natural and human environment is subject to prior authorization from the Environment Minister based on an environmental impact study. Article 5 of the same law requires an environmental audit of any industrial, agricultural, mining, craft, business or transport activity, work or development that could be the source of environmental pollution, nuisance or degradation.

First adopted in 2003 to implement these provisions, the decree concerning the environmental impact assessment (Decree No. 03-594/P-RM of 2003) also deals with the social impacts of projects, although its title does not mention that element. The desire to ensure greater consideration of the impact of projects on people living in the area led the authorities to adopt a new decree (No. 08-346/P-RM of 2008). This decree places greater emphasis on the social impact of projects, and establishes the rules and procedures governing the environmental and social impact assessment (ESIA). Further minor amendments were made in 2009.

In principle, projects subject to an ESIA cannot begin implementation without an environmental permit issued by the Minister for the Environment. The permit would require the mitigation and compensation measures recommended by the ESIA. As part of the ESIA, the project developer must inform the local people, particularly those liable to be affected by the project. Also, a public consultation must be organized by the government representative or mayor in the project area to enable local people to voice their concerns. The ESIA must be accompanied by an environmental and social management and monitoring plan (ESMP). These provisions apply to all projects, including agricultural development projects, liable to have negative environmental and social impacts. In practice, the decrees regulating the management of the social and environmental impacts of investments face major problems in implementation, as will be discussed later.

It is in this context of inadequate legal safeguards for local interests, whether in law or in practice, that the recent wave of land acquisitions in Mali has taken place.

3. Trends in private agricultural investment and large-scale land acquisitions

3.1 A long tradition of land acquisition by urban elites

The current trend towards private agricultural investment began in peri-urban areas. It is not a new phenomenon, dating back to the colonial era and ever increasing urbanization. Many government officials and traders used their professional positions or their social relations (friendship, marriage ties, etc.) to acquire plots of land in villages not far from towns. This land became the subject of rural concessions and then, in some cases, title deeds (Djiré, 2007). The trend was encouraged at independence by the authorities of the first Republic, who advocated 'returning to the land' and, as a result, set about dividing land into lots and establishing rural concessions for the benefit of city dwellers, especially in the areas around Bamako.

Rampant urbanization since independence encouraged city dwellers to continue acquiring land in periurban areas. Claims that these activities aimed to set up modern farms served to provide them with social legitimacy. In reality, however, although some city dwellers did establish livestock farms (poultry or dairy cattle), most were interested in speculative land acquisitions; the land acquired would be divided up and sold in the form of housing lots (Kéita, 2010; Djiré, 2007).

Despite the absence of official statistics on this phenomenon, there are a few case studies available to help gauge its scale. A study conducted in 2005 in the rural municipality of Sanankoroba, 30 km from Bamako (Djiré, 2007), showed that, while the number of title deeds in the municipality had increased exponentially in recent years, the cumulative 268 title deeds issued until then by the land administration were distributed as follows: government officials (40.29 percent); the state itself (35.44 percent); enterprises (19.40 percent); private organizations (1.88 percent); smallscale farmers (1.49 percent); artisans (0.75 percent); retired people (0.37

percent); and students, undoubtedly acting on behalf of their parents who already held other deeds (0.37 percent). Given that land titling is a condition for the acquisition of land ownership in Mali, these figures show that Malian farmers are being excluded from (official) land ownership. Ownership of valuable lands is increasingly concentrated in the hands of public servants and entrepreneurs living in town. And as the capital city expanded outwards, some landowners began to divide up their land and sell lots for residential use. Plots of a few hectares covered by a single title deed sometimes gave rise to hundreds of lots and respective deeds (Djiré, 2007).

Another study, conducted in the rural municipality of Baguineda-Camp, 35 km from Bamako, showed that the land under the management of the Baguineda Irrigation Scheme Agency (*Office des Périmètres Irrigués de Baguineda - OPIB)* was the subject of almost 40 longterm leases held in the name of public servants, traders, army officers and private sector executives. In the floodplains of the OPIB, 900 out of 2700 contracts allocating plots for rice production were held by non-resident city dwellers (Kéita, 2003). The average size of these plots between 3 and 5 hectares, with a few plots reaching 10 hectares.

To some extent, these acquisitions of periurban and irrigated lands by local elites foreshadowed current trends in land acquisitions – albeit at a slower pace and covering smaller areas. Like the largescale land acquisitions that have attracted so much media attention, these smaller land deals can undermine the tenure security of local dwellers in rural and particularly periurban areas. Farmer organizations in Mali have called for this phenomenon to be taken into account in debates about 'land grabbing'.

3.2 A process that accelerated and diversified in the second half of the 2000s

Following the renewed interest in agriculture and the efforts of the Malian Government to attract investment, the trend described above has accelerated and expanded beyond periurban areas. The nature of the land acquirers has also changed, particularly with regard to the substantial involvement of foreign investors. The size of individual deals has increased exponentially, with some deals covering tens of thousands of hectares. The Office du Niger has become a favoured target for both national and foreign investors.

While the recent wave of land acquisitions affects the whole of Malian territory, the number and size of investments and acquisitions vary significantly from one area and region to another. In the absence of comprehensive information on developments across the national territory, the trends analysis focused on the Office du Niger (ON) area, where the most iconic cases can be found. The Office du Niger area hosts a major share of Mali's irrigation potential, and is considered to have attracted particularly intense investor interest. In addition to private investment, the ON area has witnessed considerable public investment schemes. It is worth briefly recalling key features of both types of investment.

Public investment schemes supporting by development partners include:

- Land allocations to regional organizations for irrigation development, with plots to be made available to citizens of the member countries: CENSAD (Community of SahelSaharan States) and UEMOA (West African Economic and Monetary Union);
- An experimental scheme centred on creating title deeds to be assigned to Malian individuals as part of a project funded by the International Finance Corporation (IFC);
- A scheme funded by the United States Government under the Millennium Challenge Account (MCA), also based on the issuance of individual title deeds.

Schemes involving partnerships with regional institutions have had mixed success. The first such scheme involved the Community of Sahel-Saharan States (CENSAD). This is a relatively new organization bringing together countries from Northern Africa and from the Sahel, and covering an area of 12 million km². CENSAD countries tend to suffer from food shortages and low incomes. At the 6th Conference of Leaders and Heads of

State of the Community of SaharaSahel States, held in Bamako in May 2004, the then President of Mali announced that 100 000 hectares of irrigable land would be made available to CENSAD in the Office du Niger area. The aim was 'to help meet the target of food security for all member countries'.

After several meetings of a steering committee set up by the Malian Government, a project document was prepared and submitted to CENSAD, together with a draft agreement (Ministry of Agriculture, 2005). The document estimates the total cost of the programme at FCFA 312 600 000 000 (US$625 200 000), at an average cost of FCFA 3 126 000 (US$6 252) per hectare.

Various sources suggest that, when the report's findings were presented to the following CENSAD Summit, some Heads of State saw the Malian proposal as a poisoned gift. For these reasons, the CENSAD scheme did not go ahead, though the project was taken up by a Libyan company (Box 4).

Another development scheme involving a regional integration organization has made more progress compared to the CENSAD experience. Following a similar logic to the failed CENSAD scheme but taking account of lessons learned through that experiment, the Malian Government offered an area of 11 288 hectares to the UEMOA under an agreement signed in April 2008, as part of a wider regional programme to develop Office du Niger land. The allocation covers two pieces of land located within the hydraulic scheme of the Sahel - Fala de Molodo canal: an area of 9 114 hectares in Kandiourou sector and an area of 2 174 hectares in Touraba sector.

The UEMOA project has three components:

- Infrastructure development (hydro-agricultural schemes and private developers' installations): the UEMOA project aims at establishing farms of varying sizes for nationals of the member countries;
- Upgrading existing schemes, which involves intensification of rice production, crop diversification and support measures;
- Programme organization and management to ensure effective project implementation. Under this component, UEMOA acts as contracting authority in developing the plots that will subsequently be allocated to private operators from the member states. Under this arrangement, UEMOA will cover the cost of the feasibility studies, together with the construction costs of installing the main irrigation and drainage networks and the internal and external road systems; while UEMOA nationals will cover costs for the secondary and tertiary irrigation and drainage networks, together with levelling and preparing the plots, with pre-financing from UEMOA.

The scheme is designed to be open to three kinds of farmers: indigenous smallscale farmers, who will be allocated small plots with a unit size of between 4 and 5 hectares; private farmers with adequate technical and financial capacity to farm plots with a unit size of 10 or 20 hectares; and major private investors capable of setting up agri-businesses, who can be allocated blocks of between 30 and 60 hectares. Malian beneficiaries may receive title deeds but non-nationals will have to make do with longterm leases. With around FCFA 19 million funding from the European Union, the work started on 18 September 2010 and should in principle be completed by the end of 2012.

Issuance of private land titles to individual farmers is a key feature of the UEMOA scheme. This idea was first introduced in the Office du Niger area by another development scheme, the Koumouna project, which was supported by the World Bank.

The Koumouna project bears the name of the place where the scheme is implemented. First funded by the World Bank under the National Rural Investment Programme (*Programme National d'Investissements Ruraux* - PNIR) in the early 2000s, the project is designed to test the impact of granting title deeds to small- and mediumscale farmers. The project covers an area of approximately 830 hectares (reduced at the end of the project to 444 hectares), which were divided into 130 three-hectare lots and a small number of larger lots. It is based on the assumption that land titles and farmers' participation in the investment will produce

BOX 4
The conclusions of the feasibility study for the CENSAD project

As CENSAD did not have the necessary expertise to analyse and react to the proposals from the Malian side, it sought assistance from the FAO to advance the project. A consultancy took place from 21 July – 12 August 2005. After visiting Rome, Tripoli, Bamako and the Office du Niger area, the consultant made some important observations and recommendations which cast doubt on the project's viability. First of all, the consultant's report confirmed that development of the land offered to CENSAD would require extending the hydraulic infrastructure of the Office du Niger. It also noted the need to enlarge the intake canal and the second to the necessity of funding ancillary infrastructure, particularly roads and social facilities (education and health).

The report then tackled issues relating to seasonal water availability, which could seriously threaten the profitability of commercial farms. Building the Fomi dam was seen as the only way to increase availability during the low-water period and to facilitate dry-season cropping in the CENSAD project area.
Finally, the consultant looked at production systems and economic considerations, noting that the reasons for high yields (an average of 6 tonnes paddy/hectare, with peaks of more than 8 tonnes/hectare) and low production costs in the ON area include the modest size of farms (an average of around 3 hectares), local farming techniques and almost exclusive use of animal traction for soil preparation. He observed, however, that the planned farms on the land made available to CENSAD would be run in a radically different way, with a preference for large-scale mechanization, despite there being no convincing evidence of its effectiveness under the operating conditions of the Office du Niger area. The consultant drew attention to the 30-year lease granted by the ON in 1998 to the Chinese company COVEC to set up a 1000 hectare experimental farm using large-scale mechanization. The experiment failed and the company rented the land out to small-scale producers who, because of the shortage of irrigated land, agreed to pay a higher rent than the water charges that farmers using state land must pay.

The report also mentioned the cost implications of large-scale mechanization, which would make the scheme very expensive. Finally, in the conclusions and recommendations, the consultant suggested beginning work as a trial on 10,000 hectares within the schemes covered by the Development Master Plan. The findings would be used to inform feasibility studies on the remaining areas.

Source: Aw, 2005.

greater security, motivation and a more rational approach to farming.

The ON management, PNIR and World Bank set up a committee to review applications. The results bear witness to the failure of the initiative. The committee was supposed to select candidates on the basis of criteria drawn up by the three organizations, but an initial session held in July 2005 only found one candidate who had met all the financial criteria. A new call for expressions of interest to make up the number was issued by the ON management in October 2006 and the stakeholders jointly drew up a new scoring grid. Of the 16 applications received, 11 were deemed admissible and 5 inadmissible (due to noncompliance with procedures, particularly failure to provide required documentation). Of the 11 admissible applications, 6 received low scores against indicators like solvency, track record and ability to pay a share of development costs. These six applications were therefore rejected pursuant to article 4(2) of Decision No. 05-0187/MA-SG of 2005, which regulates the operation of the Committee. Only five applicants have scores above minimum legal requirements and were thus approved. In effect, land allocations were made by default. Some of the beneficiaries have now begun farming.

Like the Koumouna experiment, a separate and more recent MCA-supported project is also built around the notion of introducing title deeds in the Office du Niger area. The project is the agricultural component of a substantial funding package granted by the United States to Mali, another component of which involves renovating Bamako airport. The objective of the agricultural component is to increase farmers' income in the project area (Alatona) through extending the hydroagricultural schemes, improving security of tenure, increasing the area under cultivation, livelihood diversification, and agricultural intensification. To this end, the MCA project involves developing irrigation infrastructure in Alatona and allocated irrigated plots to farmers. The project has obtained 22 441 hectares, which will then be divided into a large number of title deeds (ranging between 1 and 80 hectares each). Plots would be allocated to people from the area, who enjoy priority, and to farmers from elsewhere. In the latter case, open calls for applications are used for allocating blocks of 5 and 10 hectares for young graduates and rural people, and blocks of 30, 60, 90 and 120 hectares for commercial farms. Of the 5 200 hectares to be developed in Alatona, 1 000 hectares have already been developed and plots distributed to 200 new farmers. The project also includes activities in the fields of education, health and organizational capacity-building.

Publicly funded projects like the Koumouna pilot, the UEMOA project and the MCA project are designed to promote farmer entrepreneurship in the Office du Niger area. In recent years, the Office du Niger area has attracted a substantial number of private investors motivated by other concerns. Over the period 2004-2009, 871 267 hectares were allocated to investment projects, with the pace accelerating after 2007. These allocations were made either by the Office du Niger or by the central state, in the main to large investors, on a permanent (50 419 hectares) or provisional basis (820 848 hectares). They cover an area almost 10 times the size of the irrigation schemes set up since the creation of the ON in colonial times.

While much attention has focused on land acquisitions by foreign investors, 90 percent of the known applications have been submitted by national developers, even though nationals represent less than 50 percent of the total area allocated (Papazian, 2011). Although there are some large land applications from national investors, most of national players seek land areas below 50 hectares. A staggering 38 percent of all applications covers areas between 1 and 5 hectares. On the other hand, no foreign investor has acquired less than 500 hectares (Papazian, 2011).

Land allocations to Malian applicants include: farmers (individuals or groups) already settled in the ON area; farmers (individuals or groups) without farming permits who wish to settle in the ON area; and large private investors. The first group consists of farmers that are already settled in the area that hold a farming permit (PEA) from the ON, and that wish individually or collectively to expand their farms and obtain greater security of tenure by means of a lease contract. These people are mostly farmer representatives who sit on ON joint management bodies, ON zone representatives or local political or association leaders who were the first to be informed of this new opportunity to access land. While many make individual applications, others prefer to set up associations with friends and family. Examples of the latter include the Nièta de Phédié Association, Modibo Kimbiri de Dogofri Association, and land allocations made to the Samabalagnon and Dunkafa-Ton cooperatives. The second group includes people wishing to settle in the ON area but who, having failed to gain access to serviced plots, are applying for undeveloped land. They generally work seasonally on fields belonging to nonresident farmers or work on land sub-let by farmers holding large areas. In general, they access plots through associations and cooperatives. Large private investors are developers that mostly do not live in the area and whose main activity is not farming. Some of them even live outside the country. Like foreign investors, they apply for very large land areas. Just 10 of them hold a combined total of 50 percent of all the areas allocated to

Malians. Significant players include the Tomota Group (100 000 hectares) and the companies Yatassaye (20 000 hectares), Société Africaine de Production Agricole (20 000 hectares), CAMEC (20 000 hectares), SOCOGEM (20 000 hectares), Ndiaye et frères (15 000 hectares), Société des Moulins Modernes (7 400 hectares) and BMB Export (10 000 hectares).

Foreign investors are just as diverse a group as national investors. Following a classification developed by Papazian (2011), they include private investments through sovereign wealth funds, such as the Libyan company Malibya;[7] industrial groups (national and multinational) from the food processing and energy sectors, such as the Chinese investments Sukala and N.Sukala;[8] and foreign investors involved with public-private partnerships with the Malian Government. The latter category includes a large number of projects in which the Malian Government plays an active part through partnerships with the investor or the government of the investing country. This trend is illustrated by the case of PSM, which is one of the two case studies examined in this chapter and is discussed further below.

In line with Malian legislation, land allocations to these investors typically relate to land that has not yet been developed (i.e. irrigated) and is governed according to customary rules. However, in some cases, the state already has title deeds in respect of the areas concerned. Currently, land use is agro-pastoral, and the inhabitants include sedentary farmers, who grow cereals like millet, and transhumant herders. The arrival of private investors on this 'undeveloped' land often causes tension between investors and the local community.

Data from the Office du Niger (Office du Niger, 2009) suggests that developers are mainly interested in rice, oilseeds and sugarcane. Only 5.8 percent of the 871 267 hectares allocated is covered by a leasing contract. Of the 94.2 percent remaining, projects still at the 'letter of intent' stage account for 60 percent. So much land allocation is still covered by provisional instruments like letters of intent, rather than 'hard' lease contracts. Of the areas allocated under leases (which constitute 5.8 percent of total allocations), only 23 percent have actually been developed. So only a tiny percentage of total land allocations has actually been developed. Of the areas still subject to provisional allocation, 54 percent come under letters of intent where the deadline for conducting studies as a precondition for obtaining the lease and commencing farming has already expired (Papazian 2011; Djiré and Wambo, 2010). Under Malian legislation, these allocations should be cancelled. These observations suggest that developing agricultural land is not the main concern of most of the 'investors' active in the Office du Niger. In many cases, what is observed is speculative land acquisitions based on the recognition that high-value land is becoming scarce and will be of major financial and strategic importance in coming years.

It is therefore worth looking again at the institutional framework surrounding this race for land and analysing its effectiveness in the light of actual practice.

3.3. A legal and institutional framework under threat from current practice

A multitude of management and regulatory bodies and mechanisms

The Office du Niger, already briefly introduced, lies at the heart of the institutional arrangement and is responsible for managing the land allocated to the scheme. The ON has long been presented as 'a state within the state'. Although not entirely false, this assertion is gradually being challenged, particularly with the arrival of major private investors and the various donor-supported pilot projects being undertaken. A wide array of organizations has mandate to work on agricultural development in the ON area. Various central bodies act directly or indirectly upstream of the land allocation process and sometimes downstream through their decentralized branches in the field.

[7] For more details on the Malibya project, see Diallo and Mushinzimana (2009); Oakland Institute (2011); Adamczewski and Jamine (2011).

[8] On N-Sukala, see Papazian (2011).

For example, the Presidential Investment Council (CPI) and the Investment Promotion Agency (API) are mandated with increasing private investment, in agriculture and beyond. Established in 2003, the CPI is chaired by the Head of State and has foreign and national members representing major mining, industrial and financial groups, together with several ministries (Oakland Institute, 2011). The API was set up in 2005 to ensure greater private sector involvement in the national economy. Answering to the Ministry of Industry, Investment and Trade, the agency's task is to facilitate and increase direct, particularly foreign, investment. A one-stop shop was set up in 2008 to deal with all administrative procedures relating to enterprise creation in respect of new investments and shorten the time taken to complete the various phases. All applications for approval under the Investment Code and requests for prior authorization to set up businesses are, in principle, centralized at this one-stop office.

In addition, various central government departments are involved in managing investment in general and agricultural investment in particular. For a long time, the Office du Niger was answerable to the Ministry of Rural Development and, following an administrative restructuring, the Ministry of Agriculture. But after a ministerial reshuffle in 2009, responsibility for supervising the ON management was transferred to a new Secretary of State attached to the Prime Minister's office – the SEDIZON. Differently to the ministry responsible for rural development and then agriculture, SEDIZON is specialized in dealing with the Office du Niger, given the area's strategic importance. It is responsible for implementing the Sustainable Development Master Plan for the ON area, *Schéma Directeur de Développement Durable de la Zone Office du Niger* (SDDZON), which was adopted in December 2008. More fundamentally, the establishment of SEDIZON reflects the desire of the highest government authorities to bring decision-making power from the ON management, located in Ségou, back to the Malian capital (Papazian, 2011).

But various ministries remain involved with decisions affecting agricultural investments in the ON. The Ministry of Housing, Land-Use and Town Planning deals with granting title deeds when this procedure is required, as well as registering lease contracts at the Ségou land and property register. It also handles the compulsory taking of local land rights and is involved in resettlement operations. The Ministry of the Environment play a part in environmental impact studies and in environmental monitoring, and issues environmental permits. The Minister of Finance manages the tax benefits granted by the Investment Code. Ministries responsible for water, energy, agriculture and other may also be involved in preparing and/or monitoring projects, e.g. by sitting on the validation committee for an ESIA report or the technical committee which supervises and monitors leases.

Outside Bamako, several institutions play a key role in the governance of agricultural investments. The main of these is the Office du Niger. As discussed, this is a 'public industrial and commercial establishment' (EPIC) endowed with legal personality and financial autonomy. Set up in 1932 to develop irrigation in the Niger River delta, it was restructured in 1994. The ON has its head office in Ségou, not far from the dam in Markala that feeds the irrigation scheme. The Office du Niger area is divided into six production zones under autonomous management, where activities are carried out according to plans and programmes approved by the board of directors. Several joint management committees with representatives from the ON management and from farmers assist the ON in managing the land, water and infrastructure and in settling disputes.

Challenges for land governances

Despite this complex institutional set up, major shortcomings affect the ability of the Malian state to manage large agricultural investments. There is much diversity of institutional entry points (the authority that negotiates the contract, for instance) and of form and content of the agreements concluded between investors and state. Manifest gaps between law and practice in the process of implementing contractual arrangements have been documented. Generally speaking, legal requirements on managing the environmental and social impacts of investment projects are often sidestepped or ignored. 'Letters of intent' and even actual land leases are given out in the absence of strategic planning. It is useful to discuss these challenges in greater depth.

The first striking feature of the various agricultural investment contracts signed by the Malian Government concerned is the diversity of entry points chosen by investors (Cotula, 2011). In theory, the process for obtaining a lease involves an application to the ON management, followed by a 'letter of intent' and then a lease contract. This process is followed by most Malian investors (with a few exceptions). But large foreign investors mostly rely on 'investment agreements' (or 'conventions of establishment') with the central state, which effectively take the place of the 'letter of intent'. Moreover, different contracts have been signed by different government agencies. For example, the agreement with Malibya Agriculture was signed by the Minister of Agriculture, the agreement concerning the Markala Sugar Project (PSM) was signed by the Minister of Industry and Trade, while the N-Sukala contract was signed by the Minister of Housing, Land-Use and Town Planning. Another contract with GDCM was concluded by SEDIZON. And the allocation of 100 000 hectares to the Tomota group was not the subject of any agreement with the central government, despite the large land area concerned.

As a consequence of this situation, the ON management tends to be faced with a fait accompli. Based on the contract signed by the central government, the ON is legally required to do everything it can to meet the various commitments undertaken by the state (Cotula, 2011). Also, signature of the lease contract by the ON management should, in principle, be preceded by validation by an ON lease committee on the basis of a final discussion between the various ON officials and preparation of a schedule of conditions clarifying various aspects of the contract, particularly the investor's obligations. This committee was set up at the Office du Niger at the end of 2007 as a result of the large numbers of applications, but is apparently not yet operational. Therefore, existing lease contracts were signed directly by the CEO of the Office du Niger, with no prior assessment by the committee.

Furthermore, while the structure of the investment agreements and lease contracts is more or less the same for all private investors, there are major differences in their content, particularly as regards the tenure rights allocated to the investor, land and water fees, and various other important aspects of the contract. In other words, the content of the contracts can vary in important respects from one project to another depending on the institutional entry point chosen by the investor.

For example, while the Malibya agreement provides for a long-term lease free of charge, the agreement concerning the Markala Sugar Project (PSM) project involves a long-term lease for much of the land area, and transfer of land ownership for the land where the processing facility will be located (857 hectares), with land fees being determined and deemed to be an in-kind contribution from the Malian state in exchange for an equity stake in the project. Similar considerations apply to water rights and fees, which are mentioned in all the investment agreements and lease contracts. According to the Malibya agreement, for example, Mali undertakes to give Malibya any necessary permits to use the water from the Macina canal and underground water as per the project's economic feasibility study. It also undertakes to 'permit Malibya to use the quantity of water needed, without restriction, during the rainy season' and to 'provide the necessary quantity of water for less water-dependent crops' from the Macina canal during the low-water period (the authors' translation). The water fee is set at FCFA 2 470 per hectare/year for sprinkler irrigation and FCFA 67 000 per hectare/year for gravity irrigation. The latter is the same amount paid by small-scale farmers on plots developed by the state. The same amounts appear in the investment agreement with another company, M3 SA. Setting the amount of the water fee in the investment agreement seems to conflict with the provisions of the ON management decree, which states that this amount should be set by an Order adopted by the Minister of the Agriculture. The desire to bring contracts into line with national law may explain why the lease contract with M3 SA, unlike the investment agreement, does not fix the water fee, and merely refers to an order to be adopted by the line ministry supervising the Office du Niger.

Finally, although largely ignored in most contracts, the issue of resettlement is an important aspect of the agreement with M3 SA. According to the clauses of its agreement with the government, that company undertakes amongst other things to put forward a plan for resettlement of displaced people, where appropriate, and to propose an operating model which includes resident communities in the project. Furthermore, as per standard legal practice, the land concerned is allocated free of any legal encumbrances preventing its use. However, the agreement stipulates that 'if the allocated land encompasses sensitive areas such as villages, sacred places, transhumance routes or fields, its use is subject to the compensation provisions in force'. This provision does not appear in many earlier contracts, though compensation is in any case required by national law.

Challenges for strategic planning and scrutiny of investment proposals and projects

In the absence of effective co-ordination between the various agencies, particularly the ministries concerned and between the latter and the ON, not only are contracts concluded with different content but also the effectiveness of strategic planning arrangements is called into question. The central government has allocated land to foreign investors, while the Office du Niger was allocating land to national investors. As indicated above, 871 267 hectares were allocated between 2004 and 2009 alone, vastly exceeding the extension target of 120 000 hectares by 2020 set in the SDDZON.

The effectiveness of the screening carried out by the ON management is also dubious, given the large number of letters of intent issued and the failure to carry out the required studies for many of these. It seems that the Office du Niger does not properly consider the track record or the technical and financial capacity of the applicant before issuing the letter of intent. Because of the lack of transparent selection criteria, in effect anyone can file an application and receive a letter of intent. Among other things, this procedure allows applications from developers that are far removed from the field and from agriculture, even in the absence of an ability to carry out the necessary studies (Papazian, 2011). Multiple sources in the field indicate that land allocations tend to be influenced by subjective considerations, such as links between national developers and local or national decision-makers, or relationships between foreign developers and the state. These circumstances explain the low rate of conversion of letters of intent into leases, and the recurring failure of lessees to develop the land allocated to them. Legal requirements concerning the deadlines for carrying out feasibility studies (within one year from issuance of a letter of intent) and developing the land (within three years from the lease) are not being respected or properly monitored.

The land governance challenges raised by these dynamics have been recognized to some extent by the ON and by the government. This is reflected in the recent establishment of a new Secretary of State, attached to the Prime Minister's office, responsible for the integrated development of the Office du Niger area (SEDIZON). It is also reflected in the initiation of a revision of the ON management decree, and in the cancellation in 2010 of many letters of intent for which investors had not complied with requirements to carry out feasibility studies within an agreed timeframe. The decision to cancel letters of intent affected 224 219 hectares.

Challenges for monitoring compliance with social and environmental standards

Investment projects in the ON area all have social and environmental impacts and are therefore subject to ESIA (environmental and social impact assessment) requirements. However, compliance with these provisions is uneven and the degree of compliance usually depends on the origin of the funding.

As ON personnel were not very familiar with the relevant procedures, they tended not to apply them. As a result, the ESIA did not form a direct part of the formalities prior to obtaining several leases. Some developers obtained their lease contracts without having carried out any

ESIAs. The mass influx of foreign investors eager to obtain thousands of hectares brought the issue into sharper focus. These investors were applying for large areas of land used by farmers and transhumant herders. In several cases, construction works began without any prior ESIA and sometimes even before the lease contract had been signed with the ON. Some foreign developers considered an investment agreement signed with the central government to have sufficient legal authority to authorize commencement of operations, and saw signature of the contract with the ON as 'just one more administrative stage' (Papazian, 2011). During the fieldwork for this chapter, several people commented that, when ESIAs do take place, it is very rare for the proper procedures to be followed.

Another key issue that large agricultural investments must deal with is payment of compensation for affected communities and with resettlement if local communities are displaced. For example, Libyan company Malibya reportedly began construction of the road and canal not only without any prior ESIA but also without taking any account of existing land uses in the project site. Traditionally, the area of Macina is used for transhumant herding. A local convention and development scheme for the agro-pastoral areas supported since 2006 by the German cooperation was trumped by the implementation of the project. Temporary camps were apparently destroyed and transhumance routes obstructed along 7 km in Kolongo municipality (Brondeau, 2011).

N-Sukala and Tomota also began to clear the land without any public consultation or preliminary ESIA. Tomota cleared around 1,400 hectares in the same way as Malibya and with the same effects. In the area of Bewani, the land cleared by N-Sukala belonged to the local villages and was used for grazing, firewood collection and dry cereal cropping. Local people were not adequately informed through the public consultation required by the decree concerning the environmental and social impact studies. They received no prior compensation. The same happened in the area of Sanamandougou, where local people originally opposed the M3 SA project before giving up after confrontations with the police which were followed by arrests and various promises made by the developer (Papazian 2011, Oakland Institute, 2011; Diallo and Mushinzimana, 2009; Adamczewski and Jamine, 2011; and data collected during fieldwork).

Conversely, as will be discussed later, the operational guidelines of the African Development Bank, similar in content to those of the World Bank, were applied in respect of the environmental impact study and resettlement plan in connection with the PSM. Various provisions favouring the local people were put forward as support measures.

In the absence of clear national guidelines in respect of the displacement and resettlement of affected communities, everything depends on the goodwill of the developer and any requirements imposed by lenders.

A fundamental problem lies in the government's commitment to make land available 'free of all legal encumbrances and tenure rights'. As already mentioned, the land leased is usually outside the irrigated perimeter – investors are allocated undeveloped land for them to build irrigation infrastructure. In these areas, local communities exercise rights, whether customary or not, to use the land for cereal farming or for livestock grazing. Although the ON management decree affirms the monopoly of the Office du Niger over any land that can be irrigated from the Markala dam, the legal status of this undeveloped land falls into two categories: i) land that has already been registered, with title deeds transferred to the Office du Niger; and ii) land that has not yet been registered and over which resident communities exercise customary rights. Registration of this latter land category requires the taking of customary rights and compensation for the holders, following the spirit and letter of the CDF, the management decree and the ESIA decree. This is a task for the government, but government agencies are not always in a position to perform this task to standard. By 2002, only 199 046 hectares of ON land had been formally registered with the state, mainly within the irrigation schemes (according to the 2002 ON Framework Agreement). So the vast majority of ON land, and even more so of undeveloped land in particular, has not yet been registered with the state.

Nor is the allocated of land that has already been registered with public authorities problem-free. In some cases, the state has registered the land in its name without having carried out all the required field investigations, particularly the identification and compensation of people who exercise rights over the land concerned. In other cases, the state registered the land long ago (whether or not following the proper procedures), but then left the land fallow, so that it was settled by communities who eventually came to consider themselves as the legitimate owners. In this latter case, even if local groups have no legally recognized ownership rights or even customary rights over that land, it is difficult for political, social and humanitarian reasons to evict them without compensation.

In principle, the state must cover the cost of compensating communities holding customary rights. But according to the provisions of the ESIA decree, payment of compensation is a matter for the project developer. This seems contrary to the letter of the management decree concerning the land allocated to the Office du Niger. In practice, the issue is handled on a case-by-case basis depending on the project.

In the case of the Malibya and PSM projects, the Malian Government is responsible for compensation. As regards the N-Sukala project, on the other hand, the Chinese side undertakes to 'cover costs related to information, removal and resettlement of villages and PAP' (article 7 of the contract). Negotiations with Tomota are still ongoing. In this last case, the Malian Government refuses to pay compensation, arguing that the developer has not undertaken any work in the area, while Tomota is also refusing to pay, arguing that other projects have received state funding to cover the cost of compensation (Interview with a ON official in Ségou; cf also Papazian, 2011). In this sense, the agreement with M3 SA stipulates that the company must take responsibility for paying compensation.

Equity concerns and the soundness of policy choices

Apart from problems related to compliance with legislation, private investment in the ON area also raises issues of equity and soundness of the policy choices made. Large private investors were initially allowed to come in without any concern for small-scale farmers. As Benoît Dave points out, there are some 25 000 family farms in the area, with the size of their small plots averaging 3.7 hectares (Dave, 2010). These farms are becoming smaller and smaller, as witnessed by the fact that the average area worked per family has been divided by three in the space of twenty-five years, so that it amounts to only 3.14 hectares (Bélière et al., 2003).

These farmers do not own the land. They rent it free of charge but must pay an annual charge for maintenance of the irrigation system. Failure to pay is sanctioned with eviction. These farms face many problems, which Benoît Dave mainly attributes to the shortage of land. According to that author, 56 percent of family farms have less than 3 hectares of irrigated land, the minimum size considered necessary for rice farming in the Office du Niger (Dave, 2010). This percentage is rising, because many farms split up as a result of inheritance or family conflicts, or because over-indebted farmers are obliged to sell some of their fields, even though the practice is forbidden by national law. Conversely, farmers have no real possibility of obtaining further land: there are few new schemes for small-scale farmers and the land tends to be allocated to public servants, traders or new farmers. Moreover, access to credit is beyond the reach of many small-scale producers, who are therefore unable to develop irrigation infrastructure themselves.

Against this background, allocating thousands of hectares to private investors without making any provision for a substantial increase in areas allocated to family farms is bound to raise equity issues and compound the concerns voiced by small-scale farmers that they will end up working as farm labourers in the near future.

In addition, questions have to be raised about the relevance of the policy choices made. With a few exceptions, the contracts and agreements for large investments in the ON area make no reference to the end market for the projects' produce. For example, the recitals of the agreement between Malibya and the Government of Mali quote food self-sufficiency as one of its objectives, but the contract makes no mention of the destination of the produce. How can a project contribute to food self-sufficiency if produce is sold on export markets?

Similarly, the key issue of whether enough water will be available in the longer term against the cumulative number of approved projects, raised by various studies (Schuttrumphand et al., 2008; Oaklahand Institute, 2011), has not gone away. In addition to water issues, the feasibility study for the CENSAD project (Aw, 2005) also highlighted several important issues going beyond the specific project and affecting the entire ON area. These issues include the importance accorded to large mechanized farms which are unsuited to rice production in the area, land use changes, and inclusion of small-scale farmers.

4. Case studies of inclusive investment models

While much attention in earlier research has focused on the more worrying experiences with agricultural investments in Mali, this study deliberately focused on two experiences that are widely seen as being part of better practice. One such experience is a sophisticated public–private–community partnership involving a sugar cane plantation and processing facility in the ON area – the Markala Sugar Project (PSM). This project involves the establishment of a 14 123-hectare sugar cane plantation and of a processing plant for the production of sugar, ethanol, and electricity. The plantation would involve a combination of estate production and outgrower schemes. Involvement of a multilateral lender involved application of international social and environmental standards. The second experience examined is the work of Mali Biocarburant SA (MBSA) in the Koulikoro Region, which is outside the ON area. This experience involves the production of bio diesel for the national market. The company has invested in a processing facility, and sources jatropha nuts from local farmers on the basis of contract farming. The farmers are organized in a cooperative that has an equity stake in the Malian subsidiary of the company, and thus representation on the company board. While the PSM involves land acquisition, albeit in the form of an interlocked set of joint ventures, MBSA has not acquired any land for farming – it sources its entire produce from family farmers.

Besides using different models and being implemented at different scales, the two experiences are also at different stages of implementation: the MBSA experience is relatively advanced and lends itself to an analysis of preliminary outcomes, whereas the PSM is still at the stage of fundraising and testing varieties.

Beyond these differences, the two models share a common concern about taking the interests of the local communities into account. This chapter discusses advantages and disadvantages of the two models. Given the major differences between the two experiences, the intention here is not to carry out a comparative study. Also, limited access to data means that the analysis is inevitably preliminary and incomplete.

4.1. The Markala Sugar Project (PSM) – A public-private-community partnership model

Originally designed as a public-private partnership (PPP) and later expanded to a tripartite public-private-community partnership, the Markala Sugar Project is unlike most private investments in the ON area, because of the way it has been set up and the support it received from the African Development Bank (ADB). The project is led by Illovo Sugar, a South Africa-based sugar company, which is in turn controlled by a British company.

The project has two components: a farming component involving the establishment of a 14 123-hectare sugar cane plantation with sprinkler irrigation, designed to produce 1.48 million tonnes of sugar cane per year; and an industrial component involving the establishment of a processing plant for the production of 190 000 tonnes of sugar and 15 million litres of ethanol per year, together with cogeneration of 30 MW of electricity. The plantation is divided in two separate zones. In Zone A, water abstraction will be from the existing Costes Ongoïba canal, while the second zone, Zone C, will be irrigated from the existing Macina canal. The chosen irrigation method is by sprinkler (central pivot system) and, according to the project documents, this choice was essentially guided by a concern to save water. The first phase of the agricultural component will include clearing and preparing the land for the sugarcane plantations. The natural vegetation will have to be cleared and the arable land currently used for cereal production, together with the grazing land, will be converted into sugarcane plantations. The second operational activity in this component will be the installation and management of 200 irrigation pivots, together with construction of the other plantation infrastructure such as access roads, primary, secondary and tertiary canals. The land between the pivots will represent around 1 000 irrigated hectares available to the local communities. It will be used to grow vegetables and generate income, ensuring the food security of an area known for its very low level of food production.

The project is located in the Office du Niger area to the north-east of the town of Ségou, the capital of the fourth administrative and economic region of Mali. It falls within Title Deed No. 2215 in Ségou District. With a total area of 111 377.46 hectares, this land was registered in the name of the state and the deed was issued on 23 June 2004. Within this, the land earmarked for the SOSUMAR project was split into two parts, with one to be transferred in full ownership to SOSUMAR, which is the company leading the industrial component of the project, and another to be given on a 50-year renewable lease to CaneCo, the public-sector company leading the agricultural component. The company SoSuMar is a joint venture between the Malian Government and Illovo, whereas CaneCo will be owned 90 percent by the Malian Government and 10 percent by SoSuMar, the industrial company. But, according to explanations provided by an Illovo official, "Illovo would have no economic interest whatsoever in the public sector company CaneCo. It is true that SoSuMar would own 10 percent of CaneCo, but SoSuMar has waived its rights in perpetuity to receive any dividend or profit share from CaneCo. The shareholding was purely symbolic".

Context: The shortfall in sugar production

The PSM reflects the desire at the highest level of government to promote the agro-industrial sector. The fundamental objective of the project, according to the project documentation, is to achieve self-sufficiency in sugar, to export surplus production to neighbouring countries, and to reduce rural poverty through irrigated agro-industrial agriculture.

Annual sugar consumption in Mali is estimated at 155 000 tonnes. There are currently only two sugar production plants in Mali, both located in the Office du Niger area (in Siribala and Dougabougou) and both operated by Sukala SA, a company in which a Chinese company, the China Light Industrial Corporation for Foreign Economic and Technical Cooperation, has a 60 percent capital stake. The country's current annual output provided by Sukala SA's plants is around 35 000 tonnes. As a result, 120 000 tonnes of sugar need to be imported to meet consumer demand. This situation, especially during the month of Ramadan when sugar consumption increases substantially, forces the Malian Government to grant customs duty exemptions to sugar importers, a drain on the public purse, to avoid vertiginous price rises.

Project history: From the feasibility study to the involvement of the ADB

Before coming to the conclusion that sugar was beginning to turn 'sour', the Malian Government had launched initiatives designed to meet national demand through local production. This was the background to the first feasibility study on sugar production undertaken in 2001, with funding from USAID. The study was conducted by the Schaffer & Associates LLC International (SAIL) group.

The findings of the study confirmed the potential for setting up an irrigation scheme capable of supplying very good quality sugarcane, together with a processing plant with a production capacity of more than 170 000 tonnes of sugar per year. The study recommended implementing the project in partnership with an experienced operator from the sugar industry.

As a result, in 2003, the Malian Government organized a round table in Bamako for investors in the sugar sector. The aim was to present the project to them and seek expressions of interest. In the end, South African company Illovo Sugar was chosen as the strategic partner for the project. Following various missions and complementary studies conducted by Illovo, the partnership was formalized in an agreement signed on 27 June 2007 between the Government of Mali, Illovo Group Holdings Limited and Schaffer & Associates International LLC.

The agreement required Mali to take part in fundraising efforts. So various institutions, including the African Development Bank, were invited to contribute funding. The bank responded positively to the invitation, and its participation induced the project to comply with ABD requirements on social and environmental standards.

The agreement of 27 June 2007, a very complex, technical document, leaves little space for the local community. Following opposition to the project from some villages and following the completion of the ESIA resettlement plan prepared according to the ADB's operational guidelines, important changes were made to project design to address this shortcoming.[9]According to Malian environmental legislation, the PSM is classified as a 'Category 1' project, subject to an in-depth ESIA and to the preparation of an ESMP. The ADB also considered the PSM as a project requiring preparation of a detailed ESIA. The ESIA reports on the PSM were therefore subject not only to national law requirements, but also to the African Development Bank's environmental and social assessment procedures. In addition, in application of the bank's policy on involuntary displacement of local people, a detailed Poverty Reduction Programme and Resettlement Action Plan (PRP or RAP) had to be developed based on a broad development perspective. Documentation produced by the developer in these regards in May 2009 was accepted by the ADB's project assessment committee.

Technical and financial partners (TFPs) working in the Office du Niger area made both general and specific comments on early versions of the ESIA prepared for the Board of Directors of the African Development Bank. These comments concerned matters such as environmental and social provisions and primary infrastructure, with particular reservations expressed in relation to the issue of water availability. Water was the subject of a further study conducted in 2010.

Following these various initiatives, the loan agreement between the ADB and Malian Government was signed in Bamako in June 2011. Under the agreement, the ADB is to contribute an amount of €65 million (around FCFA 43 billion) towards the financing the two major project components – FCFA 23 billion for the agricultural component and 20 billion for the industrial component – against the total cost of FCFA 275 billion (US$560 million).[10]This makes the Markala sugar project (PSM) the first PPP development project in the agro-industrial sector to be approved for African Development Bank funding.

Project partners and their motivation

The project brings together diverse players having different motivations. The Malian State is, as noted above, mainly concerned about the country's sugar supply and poverty alleviation. It sees the project as a good opportunity to solve this problem and to create employment, as well as to make foreign currency savings by importing less sugar and, in general, to promote socio-economic development in the area.

Illovo Group Holdings Limited (IGHL) is, as its name implies, a holding company. The

[9] Commenting this affirmation, Illovo's representative notes that it would only be fair to note that it was always the stated intention of Malian Government that the entirety of the earnings from the public sector component of the project would be used for purposes of poverty alleviation within the region and across Mali. The later defined transfer of 40 percent of the cane growing area into the direct ownership of the relocated people (RAP) and out-growers (PAP) was just implementation of the original concept, but not a change in the overall purpose i.e. poverty alleviation. While one cannot contest this affirmation, it is also indisputable that the 2007 agreement does not refer to poverty alleviation.

[10] It is useful to note that in a memorandum dated November 2011, the expected cost of the project had risen to €488 million ($634 millions).

Illovo Sugar group is a South African company and leading sugar producer in Africa. IGHL is registered in Mauritius and has subsidiaries in six African countries. The company is listed on the Johannesburg stock exchange. Illovo Sugar is majority owned by Associated British Foods Ltd, which owns 51 percent of its capital through British Sugar. Participation in the project will undoubtedly enable Illovo to achieve its stated objective of increasing its African sugar production by 50 percent over a five-year period.

Schaffer & Associates International LLC (SAIL) is a private corporation based in the United States which provides international research, management and support services for agro-industrial, energy and infrastructure projects. It is not common for this kind of company to be a direct shareholder in a project where it has carried out the feasibility study. It is understood that SAIL bought shares in the project company SoSuMAR at the request of the Malian Government, who wanted to encourage it in this way to continue and increase its involvement in the project and convince potential investors of the project's benefits.

Differently to the first three players, the African Development Bank (ADB) is not a party to the original agreement. It became involved in funding the project at the Malian Government's request, in line with its mission to fund development activities in Africa. More specifically, the Bank wanted to test public-private partnership funding in the agro-industrial field. The project was put together in two complementary stages, as can be seen from an analysis of the original agreement of 27 June 2007 and the essential contributions to project design made with the ESIA and the resettlement action plan (RAP).

The project set-up

A limited company registered in Mali was set up to implement the industrial component. Named SOSUMAR (Markala Sugar Company), its primary purpose was to build and operate a new sugar cane processing plant in Mali; to produce, market and sell sugar and its derivatives (molasses, ethanol, biomass, etc.); and to provide services for CaneCo, the second company to be set up as part of the PSM.

Article 3.1.3 of the 2007 investment agreement provides that the majority of SOSUMAR's capital will be held by a strategic private foreign investor. On the date of signature of the agreement, shareholders in SOSUMAR were the Malian Chamber of Commerce and Industry, IGHL, SAIL and an individual. At that time, IGHL held a minority share in SOSUMAR. However, IGHL had a purchase option to acquire the majority of the company's capital. It also made commitments concerning the future supply of technical services to SOSUMAR and CaneCo.

According to article 3.1.5 of the 2007 investment agreement, SOSUMAR is to set up CaneCo, the primary aims of which will be to establish plantations to produce sugar cane exclusively for the plant managed by SOSUMAR. Shareholders in CaneCo will be SOSUMAR (10 percent of capital) and the Malian Government (90 percent), although, as reported by an official from Illovo, SoSuMar has waived its rights in perpetuity to any dividends or profit share arising from CaneCo. So the Malian Government controls the company running the farming component of the venture, with control over the industrial component being in the hands of the investor once financing has been secured.

While CaneCo had not yet been established on the date of signature of the 2007 agreement, it was meant to approve the rights, benefits and commitments pertaining to it and be able to demand their enforcement in its favour. Similarly, although SOSUMAR is not party to the agreement, it can accept and take advantage of the rights, benefits and commitments pertaining to it in the agreement.

The financing structure of the project is quite complex. As regards funding of the necessary working capital, it is specified that: SOSUMAR will endeavour to ensure that CaneCo's working capital requirements are met by means of loans that CaneCo and SOSUMAR may conclude with lenders; and IGHL will endeavour to ensure that SOSUMAR's working capital requirements, including the amounts needed for CaneCo, are met by means of loans that SOSUMAR may conclude with lenders.

As regards the soft loans required for the project, article 6.3.2 of the 2007 contract stresses the government's responsibility to obtain them in order in its turn to provide sufficient financing to SOSUMAR and CaneCo to ensure that the project is fully funded. However, according to article 6.3.2.2, if the Malian Government is not successful in obtaining the entire amount of funding required, it will attempt together with IGHL to make up the shortfall by means of loans from financial institutions or other sources.

CaneCo, which will initially be incorporated with the minimum capital required for registration, is to be set up by the government and SOSUMAR, which will be its sole shareholders. The share capital of CaneCo is then to be increased and the Malian Government will take a 90 percent stake in the company's capital by means of a contribution in kind consisting of a long-term lease granted to CaneCo on the land allocated to that company, with an agreed value of FCFA 2 050 000 000. Following this, SOSUMAR is to pay cash for a 10 percent stake in CaneCo's share capital.

Under the 2007 agreement, SOSUMAR has the following obligations:

- To build and run the plant and provide technical support to CaneCo pursuant to a technical services agreement to be concluded;
- To build the plant in such a way that it has capacity to crush 7 680 tonnes of sugar cane per day and produce high-quality sugar in line with market requirements;
- To supply the Malian Government with six-monthly reports during the construction period on the progress of the work, staff training and any difficulties encountered;
- To employ at least 5 000 people in SOSUMAR and CaneCo activities when the latter have reached full production capacity.

Although it is not directly a party to the agreement, SOSUMAR declares in article 8.2 of the contract that it expect to create 7 200 jobs for the project and to produce 195,000 tonnes of sugar and 15 million litres of ethanol per year. It also states that the estimated cost of the industrial facility is US$167 million and the estimated total cost of the agricultural facility is US$150 million. Again according to SOSUMAR estimates, the date at which the plant should be able to commence crushing sugar cane is 1 December 2009.

For its part, IGHL undertakes to supply the necessary technical services for the efficient operation of SOSUMAR and CaneCo, and to provide SOSUMAR with its expertise to enable the latter to achieve the project objectives in terms of job creation, training and establishment of a drinking water and electricity supply for the benefit of other users.

SAIL undertakes to 'do everything necessary and required to ensure that the suspensive conditions mentioned in Article 6.2 are fulfilled' (the authors' translation). Amongst other things, these conditions relate to:

- The signatories' commitment to make every effort to facilitate signature by 31 December 2007 of the subsidiary agreements enabling the other conditions of the project to be fulfilled;
- Negotiation of the financing arrangements for SOSUMAR and CaneCo on the terms and conditions accepted by mutual agreement between the parties and all the project funders;
- Signature of a shareholders' agreement between the government, SOSUMAR and the other company shareholders, on the terms agreed between them, together with signature of the annexes to that agreement; and
- Establishment and registration of CaneCo and signature of the deed of incorporation and articles of association.

Given that the establishment of SOSUMAR is the responsibility of IGHL, according to the agreement, SAIL's role can be interpreted as supporting the process and play the role of facilitator in relations between the project partners.

The Malian Government undertook to contribute FCFA 1 500 000,000 towards the share capital of SOSUMAR, in the form of an assignment to the company of a title deed

CHART 1
PSM organizational chart

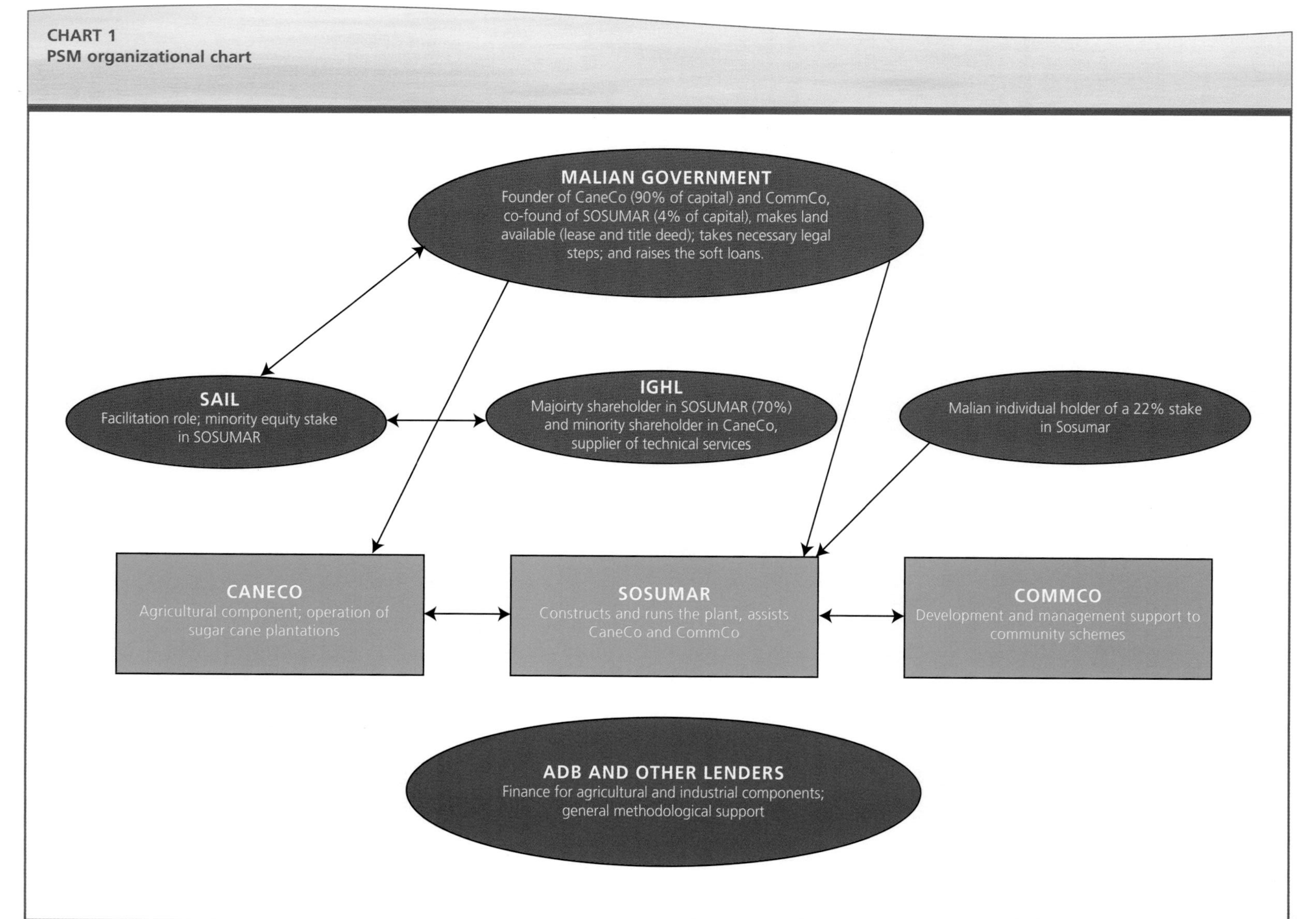

covering 857 hectares of land, plus a long-term lease on 134 hectares of land to be identified by SOSUMAR in the Markala area. The plant and related infrastructure would be built on the land covered by the title deed. This transfer of ownership and the granting of the lease represent the contributions in kind, in two instalments, of the Malian Government, which is not required to make any direct cash contribution.

In addition, article 12.6.1 of the 2007 contract stresses that if a future extension of the project requires additional funding, the government undertakes to grant SOSUMAR, under a long-term lease, an option giving it the exclusive right to occupy and use the additional land for fifteen years. The amount of the charge for the duration of this lease will be capitalized in SOSUMAR and represent payment of the Malian Government›s subscription to the SOSUMAR capital increase. The project extension may include contiguous or non-contiguous land to be chosen by SOSUMAR within the zones marked as Zone A, Zone B and Zone C on a map annexed to the 2007 agreement. SOSUMAR undertakes to exercise or renounce the option to extend the project area within 15 years from signature of the agreement (article 12.6.3). The government warrants that, over the same period (15 years from signature of the agreement) and until SOSUMAR has exercised its rights in relation to the extension, the land concerned will not be used in any way that could compromise SOSUMAR›s rights and the planned use. The lease covering the project extension, like the one granted to CaneCo, will be for a renewable term of 50 years. Legitimate doubts may be raised about the value of granting an investor option rights over a 'land reserve' in an area that is subject to heavy pressure on land. Should the investor decide not to exercise this option, Mali would sustain significant opportunity costs.

According to the 2007 investment agreement, the Malian Government is to make a 90 percent capital contribution in kind to CaneCo, in the form of a long-term lease on the land granted to CaneCo with a value of FCFA 2 050 000,000. Once the lease has been signed and registered, SOSUMAR will pay cash for new shares in CaneCo, becoming a 10 percent shareholder in that company.

This renewable 50-year long lease will, according to article 13.2 of the 2007 agreement, cover 19 254 hectares of land, on which CaneCo will conduct its agricultural operations. The terms of the lease contract will, inter alia, authorise the company 'to use its rights over the land as surety' to obtain future loans to develop its activities (authors' translation). Similarly, the terms grant CaneCo the right to sublet some of the leased land to other sugarcane producers, including SOSUMAR.

Following changes to project design made following the ESIA and RAP, in addition to SOSUMAR and CaneCO a third entity, 'CommCo', will be set up to develop 5 600 hectares to be allocated to the local communities. The area will be developed as an outgrower scheme, though this component has not yet been initiated. One part of this area will be used to grow cereals and vegetables and the other to produce sugarcane. Plots will be specifically allocated to women. The creation of this entity provides the PSM with its community dimension in addition to the original public-private partnership model.

In article 18.1 of the 2007 agreement, the Malian Government acknowledges the need to adopt legal measures to protect the national sugar market. It undertakes to preserve sugar›s status as a sensitive product and take the necessary steps to protect the national market as detailed in Annex D to the agreement. However, both parties acknowledge that UEMOA regulations could prevent enforcement of all these stipulations. If implementation of any provision of Annexe D would mean a breach of UEMOA norms by the Malian Government, the government undertakes amongst other things to attempt to exert a positive influence and obtain permission from other UEMOA member states to take steps to protect and preserve the status of sugar as a sensitive product.

Concerns about water

A key issue in the contract relates to commitments entered by the Government of Mali with regard to water. According to article 15.1, the Malian Government undertakes to ensure that SOSUMAR and CaneCo each have access at all times to a water supply for the needs of their respective operations. In addition to this overall commitment, the government warrants that it will do everything necessary to ensure that the two companies are granted full rights of access and extraction with regard to water from the Macina canal or any other canal bringing water from the river and/or another source of water close to the site, at an initial maximum rate of 20 m3/s every day of the year (increased to 35 m3/s if the site is extended as mentioned above).

The government is also to make every effort to ensure that the water charge is set at a rate that will not affect project viability or exceed the price paid by other major consumers of agricultural water in the Markala area. The rate must take account of the proportional length of the canals used by the project and reflect the comparative efficiency of using central pivot irrigation. Notwithstanding these clauses, however, if drought results in inadequate flow in the river Niger to meet domestic water demand and the requirements of international treaties, an emergency system of concerted water management will be implemented.

According to articles 15.2.1 and 15.2.2, once the minimum flow requirements laid down in international treaties have been met, the requirements of SOSUMAR and CaneCo on the one hand and Sukala on the other will be met in proportion to their respective areas of sugarcane plantations, 'with absolute priority rights over the quantities of water available in the Office du Niger scheme' (the authors' translation). So SOSUMAR and Sukala have priority access to water in the event of drought – for example, vis-à-vis other agro industrial developments and local farmers. Article 15.2.2 goes on to stipulate that the absolute priority rights will apply up to the total maximum agreed requirements of Sukala, SOSUMAR and CaneCo 'before any other user can be supplied with water by the Office du Niger. This disposition can be considered as discrimination which is not in favour of food crops. It carries prejudice to the other users of the Niger River, particularly in the area of Office du Niger.

Social and environmental standards

SOSUMAR and CaneCo warrant and give an undertaking to the government that they will do everything in their power to comply with environmental legislation applicable to the project. According to article 22.3, SOSUMAR, CaneCo and IGHL must each be classified in the agricultural category and sector as regards the employment and social security requirements laid down in the respective laws. The government will facilitate the conclusion of a collective agreement once a trade union has been set up. Each of the companies agrees to observe and comply with all laws and regulations applicable to labour and employment.

However, the contract also features a very broad stabilization clause. When not properly formulated, broad stabilization clauses can raise concerns about the ability of the government to improve social and environmental standards over project duration (see Cotula, 2011). According to article 7.3, the Malian Government warrants that no law can nullify the agreement or any one of its terms, or cause it or any one of its provisions to cease to have effect. To this end, the terms of the agreement will continue to be applicable and enforceable and 'take precedence over any new law enacted after signature of the agreement, the enforcement of which might affect the continuation of the project or cause the agreement or any one of its provisions to cease to have effect'. So the contract prevails over national law. In article 4.4, the Malian Government warrants that it will do everything necessary to ensure that the provisions of the agreement 'shall bind the government and local authorities and all other authorities or government or similar bodies in Mali...'. More specifically, article 4.4.1 states that 'the provisions of clauses 12 to 15 of the agreement [concerning the land and water rights aspects of the project] bind and shall bind the Office du Niger' and undertakes to do everything necessary to ensure that the latter complies fully with those clauses.

However, article 7.5 of the agreement stipulates that, if the Government of Mali adopts any measures more favourable to SOSUMAR and/ or CaneCo and/or their shareholders, the latter may individually or collectively adopt the more favourable arrangements provided that they adopt them in their entirety.

Final remarks about project design

Overall, IGHL and SAIL have been able to negotiate very favourable clauses for SOSUMAR and CaneCo, as regards to land and water rights as well as the stabilization of the provisions of the agreement. This makes the 2007 investment agreement look like a 'classic' contract where emphasis is on providing the company with legal rights and with safeguards for the protection of its investment. However, project design underwent significant changes at the financing stage, particularly following the involvement of the ADB. The arguably one-sided nature of some of the provisions in the contract was partly rebalanced through the changes induced by the ESIA and the RAP. Among other things, changes in project design involve the planned establishment of an outgrower scheme for 5 600 hectares of the plantation land, and of a third entity, 'CommCo', to complement SOSUMAR and CaneCo and run the above mentioned outgrower scheme. At least in its design, the project has therefore evolved from a straight PPP to a more innovative public-private-community partnership11.

The socio-economic outcomes of the project

The project is expected to become fully operational in 2017. It is far too early to assess its livelihood impacts on the ground. However, it is possible to outline a few considerations based on the ESIA and on the authors' fieldwork.

The overall catchment area of the project encompasses the territory of a total of six rural municipalities, with a population of some 156 000 inhabitants (African Development Bank, 2009). According to the ESIA, the population of 64 localities (1 718 households) will be directly concerned by the major negative effects of the PSM, including people subject to physical displacement. The latter come from 23 hamlets comprising 127 households (1 644 people), while around 4 294 other households will be indirectly affected.

There are not enough health centres in the project area and living and working conditions are extremely precarious. Given the heavy dependence on the river Niger and Macina canal as sources of water supply for the population of certain villages and hamlets, waterborne diseases are extremely prevalent in these municipalities.

The area›s economy is essentially based on the primary sector, which accounts for more than 90 percent of economic activities. Cropping (46.1 percent) and herding represent the major sources of livelihood. Although cereals occupy more than 95 percent of the cultivated area, yields are quite low and this means that PSM area has a considerable deficit in cereal production in relation to consumption patterns in the Ségou Region.

Alongside these two main activities, communities undertake secondary activities to meet their economic needs. Women in the PSM area engage in gathering and vegetable growing, which represent their major sources of income and make a substantial contribution towards meeting family needs. Herding and small-scale trading, engaged in by 2.02 percent and 1.75 percent of the population respectively, are just as important sources of income for certain families. Both men and women engage in small-scale trading and craft activities.

Fishing is practised in the River Niger and the irrigation canals of the Office du Niger, but income from this source is falling constantly due to the reduced fish stocks in the river. Analysis of average annual household income structure shows that cropping (83 percent) is the primary source of cash income, followed by herding (12 percent); remittances (3 percent) come third, followed by non-agricultural activities (2 percent). Wages and rents make a negligible financial contribution to household income.

According to the ESIA report, the main impacts of the agricultural component of the project will be the loss of community land; the introduction of monocropping, which will bring about an irreversible loss of fauna and flora; and risks of soil erosion and proliferation of grain-

[11] It is important to underline that CaneCo would be, economically speaking, 100 percent owned by the Government of Mali, for the sole purpose of poverty alleviation. Most of the indicated safeguard clauses, therefore, were about ensuring the viability of the project in the interests of the Government of Mali as well as for the investors in the private sector component.

eating birds. All community sources of income will be affected and there will be potential disruption to ecosystem balance.

The impacts of the industrial component are wide-ranging, affecting the air, soil and water and health and safety. However, the ESIA found that the industrial optimization practices proposed by the developer, consisting of water saving, cogeneration of energy, composting, wastewater treatment, emission control and monitoring of performance indicators during production, should help to reduce these negative impacts.

During the construction phase, the main impacts discussed in the ESIA include massive loss of vegetation cover when laying out the pivots and setting up the plant, psychological disturbance induced by displacement and the destruction and reconstruction of homes, loss of immediate cash income due to the halt in economic activity during the displacement and resettlement period and, finally, the loss of socio-economic infrastructure.

The production phase is expected to cause a massive influx of foreign seasonal or permanent workers into the area. The arrival of large numbers of foreigners, most of whom will be single men, is likely to result in the emergence of new habits and changes in behaviour. There will also be a high risk of increases in sexually transmitted diseases such as HIV/AIDS, together with a high risk of industrial accident in the sugarcane fields and plant, or during operation of machines introduced downstream for new economic activities (metalworking, mills, rice hullers, threshers, etc.). It is also likely that the increased population will cause local prices to rise sharply and encourage inflation.

In addition to these findings of the ESIA report, the agro-economic study estimates that the PSM will affect cropping, grazing and fishing areas. It is likely that some of these losses will be offset over time through new income-generation opportunities created by the project, together with the introduction of services such as electricity, schools and preventive health.

Nevertheless, community food production is expected to fall at least during the transition period, i.e. the time from actual occupation of the land to develop the sugarcane plantations and purchase of the cane by the plant until effective implementation of the poverty reduction project which is to improve cereal production (African Development Bank, 2009).

The ESIA also estimates that the sugarcane plantations will cause the destruction of several woody species of considerable economic and social value to the local community.

Analysis of the water management situation included in the ESIA shows that users› water needs in the dry season could only be met without major difficulty through measures to increase water availability (namely, construction of the Fomi dam upstream in Guinea), and that palliative measures would need to be put in place pending construction of the dam.

The environmental and social management plan and resettlement scheme include relevant measures to mitigate these negative impacts. According to the project documents, positive impacts at national level are expected to include currency savings of over FCFA 31 billion per year as a result of reduced sugar imports. Similarly, the project should generate around FCFA 4 billion in tax revenue for the national budget; promote income-generating activities and benefit 20 000 people through the introduction of economic activities directly or indirectly connected with sugar production; promote entrepreneurship; and establish favourable conditions for the development of small and medium enterprises.

From the social perspective, the project could help to reduce seasonal migration from rural areas and regional, national and international emigration as a result of the creation of local job opportunities; reinforcement of existing infrastructure; promotion of the local area; self-sufficiency in energy; and local development.

The PSM hopes to contribute to qualitative and quantitative changes in the agricultural sector through the introduction of mechanization, security of tenure, training and access to the means of production. For example, irrigation pivots could be transferred to local people who would operate them and sell the sugarcane produced to SOSUMAR. In addition to SOSUMAR and CaneCo, a third company, CommCo, will be set up by the state for the benefit of local producers.

As a result, activities to implement the PSM at local level could, if carried out as planned, offer the affected communities an opportunity to improve their livelihoods. An increase in income is expected, especially for women, given that some activities such as planting and weeding the sugarcane fields will be mainly done by women. There could also be an expansion of retail and wholesale trade.

Hoped-for positive impacts on health reflected in the ESIA include the opportunity for local people to take advantage of the new health infrastructure, which will be partly funded by the project in connection with the planned development of facilities.

The compensation measures planned under the RAP go beyond legal requirements under national law. A community development programme will be set up to fight poverty. Support measures are planned, including capacity-building in respect of intensive production for rural producers (rice and vegetable farmers, foresters, herders and fishermen) in the affected areas, to compensate for the other losses caused by the project.

In line with ADB policies in respect of involuntary resettlement, the project has involved the affected people (PAP) in designing the resettlement scheme. The aim of this scheme was 'to ensure that compensation measures, the choice of resettlement sites, development plans and service provision take account of their needs, priorities and development aspirations'. With a view to raising awareness amongst the PAP and helping SOSUMAR to put the scheme together, local government bodies in Ségou Region set up a local technical committee to help preparation of the resettlement action plan (*Comité Technique Local d'Appui à l'élaboration du Plan d'Action pour la Réinstallation des Populations* - CTLA). According to project officials, the CTLA was able to organize consultations during which local people could express their concerns. It is fair to note that some of the villages expressed their opposition to the project during the consultations.

As a result, the project is expected to only relocate fewer than 100 people. People involved with economic activities incompatible with sugarcane production can be resettled at their own request. A community development programme was set up to enhance RAP activities. Among other things, the project will rebuild PAP housing entirely in conventional, more durable materials, to enable them to re-establish and improve their living standards. The project will also allocate either rice or sugarcane fields, at their choice, to people who have lost their arable land. Grazing areas will be relocated to two rangelands located 54 km and 56 km respectively from the most remote places in the PSM area.

The community development programme should have positive consequences for employment and generate business opportunities. It is to be accompanied by a Poverty Reduction Programme for people directly affected by the Markala sugar project (PRP). The programme will run for 10 years and should help 6 012 households in 85 localities in the project's catchment area to pursue or commence economic activities. The programme's objectives could be described as ambitious, insofar as it will work in a wide variety of fields, including cropping, herding, fisheries, forestry, agro-forestry, conservation, product packaging and processing, energy, education, water, health, transport infrastructure and income-generating activities (Djiré and Wambo, 2010).

As project implementation has now started, albeit still on a limited scale, it is possible to start tracking outcomes on the ground. In the village of Welentiguila, for example, 70 hectares of land have been taken to set up sugarcane nurseries. Field research suggested that compensation paid to villagers (at a rate of FCFA 50 000/hectare/year), coupled with the wages earned by farm labourers working in those nurseries, have resulted in relative income growth in the area. However, compensation amounts are not regular incomes forever, and longer-term impacts remain to be seen.

The positive features of the venture do not mean that the project has had no local opposition. According to the authors' fieldwork, during the local consultations some villagers did not want to be relocated or become sugarcane growers and expressed fierce opposition to the project. Evidence also suggests, however, that this opposition was partly related to local political and clan struggles. Two major lineages in the area have been at loggerheads since the colonial era. As the municipal council of Sansanding is headed by a member of one of the two competing families, members of the other family have stirred up their allies against the project, arguing that the council, which is in favour of the project, had 'sold off' community land cheaply to foreigners. During the field research, some of the people who have been interviewed raised doubts about the project's ability to carry out its planned activities. These doubts have been fed by delays in project implementation.

Advantages and limitations of the project

It is still far too early to assess the socio-economic impacts of the PSM. Certainly, the project has gone a long way towards taking account of community interests in project design and implementation. In the project area, there is now a major investment project with important development components in places where there had been virtually no alternatives. If the measures recommended by the various studies carried out at project design stage are fully implemented, they could make a substantial contribution to socio-economic development in the area. However, the project has suffered major delays and has also met stiff opposition from some villages. There are also questions about the fairness of some important clauses included in the 2007 investment agreement. Only more implementation time will enable a more comprehensive assessment of the social, economic and environmental outcomes of this project.

4.2 A private-community partnership: The case of Mali biocarburant SA

The second case study examined by this chapter concerns a partnership between a company and a cooperative of family farmers. The venture is led by the company Mali Biocarburant SA (MBSA). Differently to SOSUMAR, the project is located outside the ON area, and is implemented in the Koulikoro Region. The project involves the production of bio diesel from jatropha for the national market. The company has invested in a processing facility, and sources jatropha nuts from local farmers on the basis of contract farming. So the project does not involve land acquisition for farming purposes. The farmers are organized in a cooperative that has an equity stake in the Malian subsidiary of the company, and thus representation on the company board. This section outlines the context of the biofuel sector in Mali, the history of the project, key features of the business model, the implementation of the business venture and the its early outcomes, advantages and limitations.

The institutional context for biofuels in Mali

The steep rise and instability of oil prices on the international market, combined with environmental concerns, have stimulated new interest in biofuels throughout the world. An agro-pastoral country heavily dependent on oil imports to meet its energy needs, Mali has caught the fever and has been exploring production of several biofuel feedstocks, including jatropha.

Even before the widespread interest in biofuels, jatropha was already known in Mali under the local name 'bagani' and was used as live hedging. Over the period 1990-2000, the German cooperation supported projects to plant jatropha, with the nuts being used to produce oil to power mills and generators in several villages in the Koulikoro Region. In its quest for alternative energy sources, the government became interested in the sector. Two ministries were initially involved: the Ministry of Mines, Energy and Water and the Ministry of Agriculture. The Ministry of Environment and Sanitation became involved in 2006. Within the Ministry of Mines, Energy and Water, CNESOLER (the Malian solar power and renewable energy centre) has always been responsible for research programmes relating to biofuels (mainly jatropha). The centre runs the national jatropha energy programme (*Programme National de Valorisation Energétique du Pourghère* - PNVEP). As part of this programme, CNESOLER has promoted biofuel supply chain development for local rural use, for example through the Kéléya project (Geres, 2007). It is also worth mentioning the Malian agency for domestic energy development and rural electrification (*Agence Malienne pour le Développement de l'Energie Domestique et l'Electrification Rurale* - AMADER), which is a public administration body. AMADER's primary task is to manage and monitor domestic energy consumption and develop access to electricity in rural and peri-urban areas. AMADER runs a rural electrification programme, through which it funds and grants electrification concessions to private operators. Many such operators have installed generators and are now confronted with a rise in diesel prices that cannot be passed on to rural customers because of their low purchasing power (Geres, 2007). AMADER is closely following biofuel developments, but this is seen a long-term solution that cannot be relied on to address the short-term shortages faced by operators.

The Ministry of Agriculture leads a multi-year programme known as the jatropha sector support project (*Projet d'Appui de la Filière Pourghère* - PADEP), which started in 2008. Also, the Rural Economics Institute, which is a public technological, scientific and cultural institution run under the auspices of the Ministry of Agriculture, provides services to the various projects. This institute also carries out research on jatropha.

Finally, a national biofuels development agency (*Agence nationale pour le développement des biocarburants* - ANADEB) was set up in March 2009 with the mandate of promoting biofuels. Within this context of policy and institutional support for the development of the biofuel sector, and in the absence of significant public funding to promote operational projects, several private initiatives have been started, including both development projects and business ventures. Mali Biocarburant SA (MBSA) is a prime example of the latter.

Origin of the initiative

MSBA is the result of a not entirely accidental encounter between a private company and local producers in Koulikoro Region who, against a backdrop of energy crisis and renewed interest in biofuels, were looking for a partnership. Koulikoro is the second administrative region of Mali, straddling the Sudanian and Sahelian agro-climatic zones (Western Sahel). Millet, maize and sesame form the mainstay of its agro-pastoral economy.

The project developer and MBSA manager, a Dutch researcher and agro-economist, has worked in Africa for a long time, initially in East Africa (five years) and then in Mali (four years), focusing on the development of value chains. According to his own account, he has always been interested in setting up a 'win-win enterprise' in which both farmers and the investor would benefit. This concern led him to study various investment models adopted in both East and South Africa. He found that none of these models ensured genuine producer representation or provided them with worthwhile benefits. He concluded that only a model where producers have an equity stake in the business and where mechanisms exist to ensure a transparent relationship between the parties can ensure such a win-win.

This thinking fed directly into the concept of MBSA. The Koulikoro Region of Mali appeared promising, as the Dutch development agency SNV had been working there for a while. Technical and socio-economic studies were carried out, leading to the establishment of MBSA as a company in February 2007. According to the MBSA manager, the studies found that the production of the main staple crop, millet, in the Koulikoro Region did not provide farmers with adequate income and could not always even keep them fed all year round (50 percent of the households covered by the studies were unable to feed themselves throughout the year). It was thought that, because growing jatropha does not, in principle, require great effort, combining it with food crops could help to bridge the gap. The Malian partners were initially the Koulikoro Chamber of Agriculture, and then the local union of jatropha producers' cooperatives in Koulikoro (*Union Locale des Sociétés Coopératives de Producteurs de Pourghère de Koulikoro*, ULSPP).

While the Dutch developer was nurturing these ideas, two farmer leaders in the region, who were also teachers approaching retirement, were wondering what activities they could undertake once they left teaching. Having found out about jatropha seed processing in other parts of the country, they had begun trialling the crop.

Indeed, according to the president of the ULSPP, he had already planted 5 hectares of jatropha before the MBSA initiative got under way. It was at this time that the Dutch developer got in touch through SNV with the Regional Chamber of Agriculture to present the project. At the time, one of the two teachers was the vice-president of the Regional Chamber of Agriculture. The Chamber of Agriculture had a fund available, which was provided by the Royal Dutch Embassy in Mali in connection with the Koulikoro Rural Development Programme. The Chamber used this fund to finance the preparation of the business plan and the purchase of processing equipment. It also assisted with the establishment of several cooperatives, including the ULSPP cooperative union.

The ULSPP was set up on 9 February 2007 and registered in Koulikoro. It currently has 15 cooperative society members in Koulikoro District, five cooperative society and four group members in Dioila District and one cooperative society member in Kolokani District, covering a total of 2 500 producers comprising 500 women and 2 000 men (ULSPP, no date). Also in 2007, MBSA was established as a company and registered at the company register, with the production and marketing of jatropha oil and its by-products as its primary purpose.

Jatropha planting started in the rural municipalities of Dinandougou, Doumba, Koula, Meguetan and Sirakorola, all in Koulikoro District, in 2007, and expanded in 2008 to cover the municipalities of Tougouni, Tienfala and Nyamina in Koulikoro and Dioila Districts.

Project design and business strategy

Having started with a relatively simple initial structure, MBSA is now turning into a transnational enterprise, with several international public and private partners and activities in both Mali and Burkina Faso.

The initial shareholders in MBSA were the Dutch developer, who was also the managing director of the company, the Dutch Royal Tropical Institute (KIT); the Dutch railway company pension fund Spoorwegen Pensioenfonds (SPF); and the private companies Power Packs Plus (PPP) and Interagro. Together, these shareholders held 80 percent of the company's capital, while the ULSPP held the remaining 20 percent. So from the beginning, the union of farmer cooperatives held a significant equity stake in the company. As per standard practice, the company is run by annual general shareholders' meetings; a board of directors comprising representatives of the various shareholders; and a general manager.

The cooperatives produce jatropha seeds that the cooperatives union buys and then sells on to MBSA. MBSA processes the seeds to produce bio diesel and sells the end product. In the original set-up, the ULSPP was responsible for extracting the oil from the seeds, while MBSA was to process the oil to produce bio diesel; but given the union's difficulties in performing processing, the company now does all processing. The purchase price of the jatropha seeds is set by mutual agreement between the ULSPP and MBSA.

According to the MBSA manager, the bio diesel produced is sold to HUICOMA and Grands Moulins du Mali, two industrial enterprises based in Koulikoro, and to the 'dourounis' (public transport minibuses) in the town. The company is also canvassing for business from Air France and other enterprises interested in biofuels. Apart from bio diesel, MBSA also produces glycerine, which is used by a women's cooperative belonging to the ULSPP to produce soap.

In addition to income from sales of bio diesel, MBSA generates revenues through the carbon credit market. For example, MBSA signed a contract for carbon credits with KIA Motors Netherlands. 80 percent of all revenues from carbon credits are to be passed on to the member cooperatives in the form of equipment. Technical support is also provided to farmers by MBSA and the ULSPP with the aid of the government technical services and extension workers trained for this purpose.

While the key features of this initial set-up remain valid to this day, some important changes have occurred as a result of a corporate restructuring in 2011. Indeed, the first few years of operation revealed that the model had some limitations: jatropha production was not sufficient to supply the processing plant; and some Dutch shareholders began to ask questions about the sustainability of the scheme. The risk of side-selling, whereby farmers receive training and technical support from the company and then sell to other buyers offering higher prices, was seen as a particular concern. According to the MBSA manager, governance challenges within the ULSPP heightened the partners' concerns. Meanwhile, the company was initiating operations in Burkina Faso, thereby losing its original exclusive focus on Mali's Koulikoro Region.

In light of these considerations, the company was restructured along the following lines:

- Mali Biocarburant SA was transformed into a holding company controlling two subsidiaries, one in Mali (Koulikoro Biocarburant) and one in Burkina Faso (Faso Biocarburant);
- Two foundations were established - Fondation Mali Biocarburant in Mali and Fondation Faso Biocarburant in Burkina Faso;
- The equity stake held by ULSPP in MBSA was converted into shares of Koulikoro Biocarburant – in other words, ULSPP now holds shares in the local subsidiary, not in MBSA itself;
- Measures were adopted to clarify relations between the producer cooperatives, the MBSA subsidiaries and the foundations.

As a result of this corporate restructuring, the set-up is now as follows. At the centre of the venture is the holding company, MBSA Holding. Its shareholders are KIT (48 percent), SPF (30 percent), PPP (12 percent), the company's manager (9 percent) and Interagro (1 percent). MBSA Holding finances the subsidiaries and facilitates funding of the foundations. It owns the processing facilities. MBSA runs operations in both Burkina Faso and Mali through the two national subsidiaries. Activities in Burkina Faso are beyond the scope of this research. In Mali, activities are led by Koulikoro Biocarburant. Ownership of this subsidiary is as follows: MBSA Holding 79 percent; ULSPP 20 percent; and a Koulikoro Biocarburant executive 1 percent. Koulikoro Biocarburant purchases the jatropha seeds from producers, extracts the oil to produce bio diesel and markets the product. Farmers produce the jatropha seeds and sell them to the ULSPP, which they are members of, and which sells the seeds on to Koulikoro Biocarburant. Producers also receive support from the Koulikoro Regional Chamber of Agriculture and from the government's technical services.

The Fondation Mali Biocarburant was established as an association under Malian law in 2010. Its registered office is in Bamako. The foundation is a non-profit organization. It members are MBSA, which holds the presidency, TFT (Trees for Travel), KIA Motors and the ULSPP, together with two other jatropha cooperatives – the Bagani Nafabo Ton cooperative of Kita and the Ouéléssébougou jatropha producers' cooperative. The foundation supervises producers and helps the farmer cooperatives to integrate jatropha in production systems without compromising food security. According to information on the MBSA website, the foundation is in direct contact with producers to encourage them to obtain equipment and training through farmer field schools. The foundation is responsible for managing carbon credit revenues, most of which are used for the operational activities it conducts, with the balance allocated to the producer cooperatives in the form of equipment.

Both MBSA and Fondation Malibiocarburant have various partners and funding sources. The foundation has received financing from the KIA Motors Company, which is linked to the enterprise by carbon credit contracts negotiated before corporate the restructuring; by TFT, which acted as an intermediary between KIA Motors and the foundation, of which it is also a member; and through development aid funding, for example from USAID. Similarly, in addition to contributions from the shareholders identified above, MBSA has received investment subsidies from the Dutch Ministry of Cooperation and loan sureties or long-term loans granted by KIT and the French Development Agency (AFD).

Finally, as part of its efforts to promote sustainable development of the jatropha production chain, MBSA has pursued assiduous cooperation with various research institutes.

Early outcomes, advantages and limitations

Although the venture has been running since 2007, it is still too early to assess its longer-term outcomes. Important positive contributions are already visible. MBSA has created an entire jatropha value chain in Koulikoro Region where previously none existed. It has catalyzed the organization of rural producers. Today, the ULSPP has 2 500 members. The company has established an industrial jatropha oil production unit and a soap factory using the glycerine obtained during processing. The soap factory is managed by a women's group. Fifty-five permanent jobs have been created, and a large number of farmers have received support through the farmer field schools and producer training. As a result of the publicity generated around the biofuels sector and the success of the farmer field schools, the number of producers is growing year on year. For example, the aggregate area planted with jatropha for 2009 was forecast to be 1 000 hectares; but by the end of that season, 2 028 hectares were under cultivation, more than double the initial target. The end-of-season figure for 2010, the latest available to the study, was 2 020 hectares against a forecast of 2 000 hectares. Despite this slowdown, several cooperatives have submitted applications for ULSPP membership.

The company has paid particular attention to gender, by actively promoting women›s participation in the process. The iterative approach to the business, reflected in the various changes made to the original set-up, and the extensive collaboration with research institutions underlie a genuine commitment to learning and innovation. The venture also contributes to realising Mali's aspiration to tap into renewable energy sources and reduce its reliance on imported oil. Planting activities and the carbon credit scheme can provide a contribution to mitigating climate change, and Jatropha has been shown to produce benefits in terms of soil improvement and regeneration. The innovative nature of MBSA's business model has attracted considerable international interest, as reflected in the company's international partnerships and in direct mentions in several international research reports and United Nations documents.

Challenges have also emerged, however, which highlight the difficulties of making company-community partnerships work on the ground even in the presence of innovative and committed private sector players. A first such challenge concerns the limited success of the venture in raising income for rural people. As it might be recalled, this objective was an important consideration at project design stage. The assumption was that, when inter-cropped with food crops, jatropha farming can help farmers to increase their income without compromising food security. The project developers expected that, between 2009 and 2014, more than 20 000 farmers would earn an aggregate income of around €5 million, with estimated extra income of between €1.14 and €1.90 each per day. In addition, producers would receive dividends from the profits made by the company, from their equity participation in the company, and revenue from the sale of carbon credits.

While the project is still at an early stage and 2014 is still a long way away, and while production has not yet reached full capacity, the authors' fieldwork suggests that producers are beginning to become impatient. First of all, productivity is hampered by several factors. One of the underlying assumptions in jatropha production projects in general is that the plant can be grown successfully on marginal land with limited water. But in the arid or semi-arid zones of Koulikoro Region, producers need to water the crop during the dry season. Lack of equipment is also a constraint on productivity, and so are the white termites which destroy the seedlings in many areas and seriously compromise production. Some of the producers interviewed complained about the poor quality of the seeds used, though the authors could not verify these statements.

Pricing is another factor that adversely affects the project's ability to increase local incomes. Given the production costs of bio diesel, farmers can sell jatropha seeds at a very low price (FCFA 50/kg). In comparison, other crops suitable to the local ecology would offer higher returns. Sesame, for example, is sold at FCFA 300/kg in the region.

As the company has not yet turned a profit, no dividends have been paid to shareholders so far. The revenues received from the carbon credit scheme have been used to dig wells and provide basic equipment such as carts or tanks for a small number of farmers, selected on the basis of their production. Some farmer field schools have also been set up. In the longer term, these activities may result in higher productivity and hence incomes, but did not provide a direct contribution to raising those incomes in the shorter term and at scale. Some of the beneficiary producers expressed dissatisfaction about the support received. Some producers claim to have received carts with no animals to draw them. More fundamentally, some farmers felt that these interventions did not address one of the most critical issues, white termites. Farmers do not have the means to purchase the insecticides necessary to deal with this problem. According to the MBSA manager, a solution is now being developed to address the termites problem. But many of the producers interviewed for this research appeared disillusioned. Given the income challenges faced by the farmers, some of the carbon credit revenue might have perhaps been distributed directly to producers, rather than invested in equipment.

Another challenge that emerged during the fieldwork relates to the functioning of the institutional set-up. Although the venture reflects a partnership between a company and local communities, the fact remains that the interests of the two parties do not always coincide. Avenues for communication and negotiation are therefore critical. Both MBSA and ULSPP have a general meeting of members / shareholders, a board of directors and management staff. These bodies do periodically hold their statutory meetings. But there appear to be problems in communications among the multiple stakeholders. Communication challenges between the company management and ULSPP management have emerged. For example, the terms for awarding equipment funded from carbon credit revenue do not appear to be fully understood by producers and have caused disagreements between MBSA manager and the ULSPP leader. ULSPP officials also felt that MBSA management had taken some decisions without prior consultation of the ULSPP. An example cited was the decision to post extension workers to the villages, which was apparently taken without the knowledge of the ULSPP – though the MBSA manager disputes this point. The Union opposed this move until the details could be worked out together with company management. The extent of the communication challenges between MBSA and ULSPP management is illustrated by the minutes of an extraordinary meeting held on 13 December 2010, which the research team had access to. Even at that advanced project stage, the minutes reflect discussions about the size of the ULSPP›s equity share in Koulikoro Biocarburant, a matter which one would have expected to be very clear by then. There also appear to be communication challenges in relations between ULSPP management and its members. When interviewed, several members of the Union did not seem to be well informed about the business venture. Finally, some observers and development practitioners interviewed during the study felt that extending the business to Burkina Faso despite the challenges faced by the Malian enterprise reflected a desire of MBSA management to tap into the many public funding streams in the green energy sector. In response to this statement, the MBSA manager explains that the decision to invest in Burkina Faso was based on other factors: Farmers from Burkina Faso visited MBSA operations in Koulikoro since December 2007 and requested the MBSA management to visit their fields and set up a daughter company in Burkina Faso. After a due diligence as well as a feasibility study of Burkina faso, MBSA decided to set up a daughter company and one of the main arguments was the price of diesel. In Mali, the price of diesel is subsidized by the government resulting in a low price (now 630 FCFA/litre) while the price in Burkina is close to 800 FCFA/litre. Therefore, MBSA management anticipated an opportunity where they could pay a higher price for jatropha and still profitably sell biodiesel.

Despite these limitations, MBSA still offers much potential. The cooperatives union has plenty of competent members who have gradually built their negotiating and management skills and whose commitment guarantees that the process will be taken further. The manager is a socially committed businessman who retains a marked research bent and is continually looking for innovations. The model›s chances of success are boosted by the manager's in-depth knowledge of biofuel marketing circuits and networks and of the carbon markets. The manager's ability to mobilise the numerous partnerships described above holds promise for MBSA's ability to overcome its challenges. And despite their disillusionment, producers say they are ready to continue involvement with the venture because they have put in a great deal of effort from which they are still hoping to get benefits.

5. Conclusions and recommendations

This chapter has discussed trends, drivers, legal frameworks and two case studies of agricultural investments in Mali. The country has great potential for agricultural, forestry and pastoral production. Faced with major challenges in mobilising the resources required to finance an ambitious agricultural modernization strategy, the Malian Government has made concerted efforts to attract private (and particularly foreign) investment in agriculture. But the ensuing wave of large-scale agricultural investments is taking place in a national context that still appears ill-prepared to ensure that benefits are maximized and risks properly managed. For example, legislation adopted to manage the social and environmental impacts of large-scale investments has faced major implementation challenges.

Even more importantly, the recent wave of large-scale land acquisitions for agricultural investments has taken place in a land tenure context characterized by growing conflict and major governance challenges. In Mali, land tenure is governed by two main systems: the formal system under written law established by the state and customary systems that are most widespread in rural areas but differ from place to place. There are bridges between the two systems, for example when holders or acquirers of customary rights undertake formalization procedures provided by national law.

Despite efforts to legislate in ways that take account of the diversity of contexts and tenure patterns, many provisions of national law are incomplete, ineffective and out of touch with the local socio-economic reality, particularly in rural areas. Some national law norms are so ambiguous that they lead to confusion, resulting in conflicts and abuses, and in the ensuing tenure insecurity and poor land governance.

On the ground, multiple pressures are exacerbating competition for valuable lands and increasing the number and intensity of land conflicts between communities and the state, and between different communities. These pressures also have a negative influence on the quality of land governance, creating fertile ground for land speculation and corruption, abuses of all kinds and insecurity of tenure for the most disadvantaged groups.

While the recent wave of land acquisitions affects the whole of Malian territory, the number and size of investments and acquisitions vary significantly from one area and region to another. In the absence of comprehensive information on developments across the national territory, the trends analysis focused on the Office du Niger (ON) area, where the most iconic cases can be found. The Office du Niger area hosts a major share of Mali's irrigation potential, and is considered to have attracted particularly intense investor interest.

Given the diversity of the types of investments and farms in the Office du Niger area, this area can be seen as a laboratory where various forms of tenure can be tested, and a breeding ground for the country's future land policy. Two main categories of agricultural investment can be identified, each with several subcategories: (i) public investments made by the state with or without support from donor agencies; and (ii) private investments made by large-scale investors, whether national or foreign, with or without state involvement, and private investments made by small-scale private investors or farmer groups.

Until recently, all schemes in the ON area were publicly funded. Following the global food and financial crisis and the related renewed interest in private agricultural investment, together with the biofuels boom, the Office du Niger area has become a favourite target for private investment. Over the period 2004-2009, 871,267 hectares were allocated to investment projects, with the pace accelerating after 2007. These allocations were made either by the Office du Niger or by the central state, in the main to large investors, on a permanent (50 419 hectares) or provisional basis (820 848 hectares). They cover an area almost 10 times the size of the irrigation schemes set up since the creation of the ON in colonial times.

There is much diversity of institutional entry points (the authority that negotiates the contract, for instance) and of form and content of the agreements concluded between investors and state. Manifest gaps between law and practice in the process of implementing contractual arrangements have also been documented. Generally speaking, legal requirements on managing the environmental and social impacts of investment projects are often sidestepped or ignored. 'Letters of intent' and even actual land leases are given out in the absence of strategic planning. The size of some large land allocations, compared to the neighbouring areas allocated to family farmers, raises serious equity concerns.

The land governance challenges raised by these dynamics have been recognized to some extent by the ON and by the government. This is reflected in the recent establishment of a new Secretary of State, attached to the Prime Minister's office, responsible for the integrated development of the Office du Niger area (SEDIZON, from the French name of the institution). It is also reflected in the initiation of a revision of the ON management decree, and in the cancellation of a number of letters of intent for which investors had not complied with requirements to carry out feasibility studies within an agreed timeframe.

In addition to these recent developments, ongoing initiatives related to the implementation of the Framework Law on Agriculture (LOA) and of the deliberations of the 'Etats Généraux du Foncier' (Malian Land Tenure Congress, EDF) offer opportunities to improve land governance in the Office du Niger area and beyond.

While much attention in earlier research has focused on the more worrying experiences with agricultural investments in Mali, this study deliberately focused on two experiences that are widely recognized as part of better practice. One such experience is a sophisticated public – private – community partnership involving a sugar cane plantation and processing facility in the ON area – the Markala Sugar Project (PSM). This project has two components: a farming component involving the establishment of a 14 123-hectare sugar cane plantation with sprinkler irrigation, designed to produce 1.48 million tonnes of sugar cane per year; and an industrial component involving the establishment of a processing plant for the production of 190 000 tonnes of sugar and 15 million litres of ethanol per year, together with cogeneration of 30 MW of electricity. The plantation would involve a combination of estate production and outgrower schemes. Involvement of a multilateral lender involved application of international social and environmental standards. An ambitious development programme accompanies the investment.

The second experience studied is the work of Mali Biocarburant SA (MBSA) in the Koulikoro Region. This experience involves the production of bio diesel for the national market. The company has invested in a processing facility, and sources jatropha seeds from local farmers on the basis of contract farming. In other words, the venture does not involve land acquisition for farming purposes. Farmers intercrop jatropha with food crops. So although based on promoting a cash crop, the venture is not, in principle, detrimental to food security. The farmers are organized in a cooperative that has an equity stake in the Malian subsidiary of the company, and thus representation on the company board.

Both projects are based on innovative institutional designs. Both have gone a long way towards promoting inclusion of local farmers and consideration of social and environmental issues. While both projects are still at an early stage, they both have strong potential to benefit local groups through development opportunities. In the case of MBSA, the venture provides a potential source of additional income for smallholders. The profit-sharing principle on which this experience is based should help to reduce poverty in the medium to longer term. The project also offers opportunities for combating soil erosion. Similarly, SOSUMAR is an ambitious project that can bring multiple developmental benefits – from job creation to development of processing capacity, from opportunities for smallholders and local businesses through to improved energy access.

Both projects also present major challenges, however. In the case of SOSUMAR, for example, some clauses in the contract with the Malian Government appear to disproportionately favour the investor to the detriment of the others actors of the area. Yet, it is fair to note that the safeguard clauses in the Convention of 2007 aim, in general, to protect the viability of the overall project, not just the foreign investor. Also, opposition from part of the local community and the slow pace of implementation provide cause for concern. In the case of MBSA, problems in communication lines between the company, the management of the farmer cooperative and cooperative members, as well as difficulties in agricultural production, raise challenges for the inclusiveness and sustainability of the venture. The two experiences show that even where inclusiveness is integrated in the design of the business model, making it work in practice is riddled with difficulties, and positive outcomes cannot be taken for granted.

For a country like Mali, the renewed interest in agricultural investment presents important opportunities but also major risks. It is critical to tackle the challenges affecting the governance of land relations at both local and national level. Measures must be taken to fill the gaps in the governance of land tenure and agricultural investments. Steps need to be taken to accelerate the implementation of the land tenure provisions of the Agricultural Orientation Law (LOA). This law requires the government to develop a rural land policy to secure local land rights. Steps are also needed to strengthen institutional arrangements to monitor and ensure compliance with existing legislation. This applies particularly to regulations concerning environmental and social impact assessments and management plans. Finally, there is a need to strengthen the mechanisms to promote accountability in decision making affecting land relations. At the national level, the government has experimented with the 'espace d'interpellation démocratique' – a forum that enables civil society and citizens at large to bring concerns to the government and hold decision makers to account. Similar arrangements can be developed in relation to institutions involved with land governance at the local level – from local government bodies to the Office du Niger, through to deconcentrated government departments.

In addition to measures to improve the governance of land in general, several important steps can be taken to specifically address issues linked to large-scale land acquisitions. Land allocations should be subject to the free, prior and informed consent of local landholders. This would require going beyond current consultation requirements already included in legislation regulating impact assessment studies. Investment contracts with companies should also make it very clear that any land acquisition requires the consent of local landholders. There is a need for a coherent and comprehensive policy on agricultural investment, bringing together scattered provisions from different policies and laws. National policy should set land area size ceilings on land acquisitions. The duration of land leases, which is currently standardized (30 and 50 years, renewable, in the Office du Niger), should be tailored to the economics of investment projects, including based on nature of the economic activity and land area size. While local landholders at best obtain one-off compensation for their losses, thought should be given to arrangements for ensuring equity participations for local landholders, so as to enable ongoing sharing of project benefits. Land allocations above a certain size should be approved by parliament, and all contracts should be published. The capacity of government agencies to negotiate contracts with investors should be strengthened.

More fundamentally, there is a need to look at a wider range of models of agricultural investment. Family farmers have shown they can invest. In the Office du Niger area, there are experiences of cooperatives acquiring land for their members. For example, Association Niéta has obtained a lease for about 300 hectares that will benefit some 100 farmers. Smallholder farmers account for the bulk of agricultural production in the Office du Niger area. Yet their landholdings are shrinking with demographic growth, and their tenure is insecure. National farmer associations are developing tools to enable family farmers to have access to leases (i.e. the same type of contracts that are granted to large investors) for new land areas. They are also providing legal support to their members whose land rights are being threatened. These efforts deserve to be supported.

References

Adamczewski A., and Jamine, J-Y. 2011. 'Investisseurs libyens', in *Le Monde Diplomatique*, September.

African Development Bank, 2010, Etude relative à l'établissement d'un bilan des ressources en eau au droit de la zone de l'Office du Niger, Interim report, March, Annex 1.

African Development Bank, 2009, Markala sugar project, Executive Summary of the Action Plan for the Resettlement of the Population.

African Development Bank, 2009, Markala sugar project, Executive Summary of the Environmental and Social Impact Assessment.

AW, D., 2005, Rapport de Mission de "L'étude préliminaire pour la mise en valeur des 100 000 ha de terres dans la zone ON, mises à la disposition de la CENSAD par le Gouvernement du Mali», FAO, August.

Bélière, J.-F., Coulibaly, Y., Keita, A., Sanogo, M.K., 2003, «Caractérisation des exploitations agricoles de la zone de l'Office du Niger en 2000», Ségou, URDOC/ON Nyeta Conseils

Brondeau, F., 2011, "L'agrobusiness à l'assaut des terres irriguées de l'Office du Niger (Mali)",*Cahiers Agricultures*, 20(1-2).

Club du Sahel et de l'Afrique de l'Ouest/ OCDE, 2011, Investissements et régulation des transactions foncières de grande envergure en Afrique de l'Ouest, Synthèse rapport de recherche, http://farmlandgrab.org/post/view/19599.

Cotula, L., 2011, *Land deals in Africa: What is in the contracts*? IIED, London.

Coulibaly, C., 2006, "L'Office du Niger en question. 1902-2002: Cent ans de vicissitudes", Les cahiers de Mande Bukari, No. 5.

Dave, B., 2010, *Mali, Office du Niger. Le mouvement paysan peut il faire reculer l'agro-business ? Interview in "Dynamiques Paysannes",* No. 2.

Djiré M., 2007, Les paysans maliens exclus de la proprité foncière ; IIED, dossier N° 144

Djiré, M., and Wambo, A., 2010, Investissements agricoles et régulation des transactions foncières de grande envergure en Afrique de l›Ouest, Draft Synthèse rapport de recherche, Club du Sahel, OCDE.

Diallo A., and Mushinzimana G., 2009, *Foreign Direct Investment (FDI) in land in Mali,* GTZ.

Keita, A., 2003, Le phénomène des citadins paysans au Mali (Stratégie de création d'exploitation agricole moderne ou de spéculation foncière ?), Rapport de Recherche CLAIMS, Bamako.

Ministère de l'Environnement et de l'Assainissement, Partenariat pour le Développement du Droit et des Institutions de Gestion de l'Environnement en Afrique (PADELIA-Mali), Recueil de textes en droit de l'environnement au Mali, T.1, Textes nationaux régissant l'environnement et les ressources naturelles, Bamako, 2007. Ministry of Agriculture, 2008, Donors Round Table, 'Mali: Orientations stratégiques et priorités d'investissement pour un développement agricole efficient et une croissance accélérée', Bamako.

MLAFU, 2010, Recueil des textes législatifs et réglementaires du Mali.

Office du Niger, 2010, "Situation des attributions", Ségou, Office du Niger.

Papazian, H., 2011, "Les investissements fonciers au Mali: état des lieux et perspectives des acquisitions foncières à grande échelle en zone Office du Niger», Dissertation, Institut National Polytechnique, Ecole Nationale Supérieure Agronomique de Toulouse.

Schuttrumph, and Bookkers, T., with **Sangaré, A.**, 2008, Analyse du potentiel d'irrigation lors de la saison sèche dans la zone de l'Office du Niger, Rapport étude, KFW.

ULSPP, undated, Note d'information sur l'union pourghere de Koulikoro, ronéotypé.

Senegal:

Assessing the nature, extent and impact of FDI in Senegal's agriculture[1]

1. Introduction

Despite the numerous attempts made since the early 1960s to design the right set of agricultural development policies that could successfully set Senegal on a higher paced economic growth path, the country still remains dependent on foreign agricultural supplies to meet its food security needs. However, results are not yet visible to the naked eye. Almost 50 percent of the population remains officially unemployed, more than half the country lives in poverty with GNP growth for 2009 expected to reach just 1.5 percent. Furthermore, the country relies on imports for 70 percent of its food needs – a rate higher than most countries in sub-Saharan Africa[2]. The dependency rates could even go higher, to reach 90 percent when considering the case of key staple food items such as rice. It is perhaps worth recalling here the dilemma Senegal authorities faced during the 2008 global food crisis when the prices of many food items skyrocketed in international markets as a result of cutbacks on food supplies from its key Asian food exporters (Thailand, Pakistan, etc.).

To overcome these weaknesses and reverse these alarming economic growth trends just mentioned, Senegal must swiftly reform key sectors of its economy starting with its agriculture, in order to achieve the highest economic growth rates possible – in the order of 7 percent or more per annum. In particular, to achieve the over-reaching goal of feeding its 13 million people, and especially the 2.5 million people currently known to be undernourished, Senegal must modernize its agriculture, placing more emphasis now on the reforming of its agribusiness starting with the land related issues and best ways for addressing long-term sustainability issues. This chapter reviews selected initiatives including various efforts made in the country recently to attract more FDI to help reduce the impacts of these weaknesses drawing as much as possible from foreign resources.

As is the case with other countries in Africa, Senegal's lending decisions appear severely biased against agriculture due to the high risks associated with existing segmented gaps along the value-chain of many commodities. Agricultural lending in Senegal currently accounts for less than 5 percent of the total portfolio of the financial sector. There are almost no loans for capital investment. Loans to farmers, fishermen and herdsmen are very rare, totalling less than 1 percent of all loans. Indeed, the agriculture sector is desperately in need of credit: thus, foreign investments are vital. This is so despite the robustness of the Senegalese financial sector[3].

Senegal's leaders are well aware of the country's persistent vulnerabilities to poverty, food insecurity and lack of general economic competitiveness. In the wake of the 2008 food crisis, the government responded with the OANA

[1] This chapter is based on an original research report produced for FAO's Regional Office for Africa by Adama Ekberg Coulibaly, consultant

[2] Unless identified otherwise, statistics cited in this chapter are drawn from a number of sources, including the Economic Intelligence Unit (EIU) country profile (2008), the CIA's online world fact book (2009), the OECD's Africa Economic Outlook (2009), various studies of the United Nations, and other external publications, which themselves draw most of their data from international sources or the Senegalese Government's own institutions.

[3] Total deposits in the financial sector are about US$3.8 billion with the total loan portfolio standing at about US$3.4 million according to central bank figures covering the first quarter of 2009. Deposits grew 13 percent year by year from the first quarter of 2008, while loans grew at 11 percent.

programme[4], an ambitious agricultural growth-led scheme whose timely implementation requires injecting, in addition to the scarce public funds available, a sustained massive finance the country does not have. For the period of October 2008 to 2010 for example, the GOANA set goals for itself to produce 2 000 000 tonnes of maize, 3 000 000 tonnes of cassava, 500 000 tonnes of paddy rice and 2 000 000 tonnes of other cereals (millet, sorghum and fonio), 400 000 000 litres of milk and 600 000 tonnes of meat. In this connection, CFA 116.8 billion [5] were forecast to meet the agricultural production subsidies needed for the 2005/2006 to 2008/2009 growing seasons alone. It remains to be seen how the massive financing needed for these schemes can be met, and revive key subsectors such as sunflowers, sesame, bissap, potato and agro inputs (WTO, 2009).

2. Trends in levels of FDI inflows to Senegal

Senegal has not fared well with respect to other competing destinations in West Africa when it comes to attracting FDI and building FDI stock effectively. The country accounted on average for US$317 million of FDI inflows to West African economies in 2005-2008, which is only 1.9 percent of the total. This is lower than the FDI inflow levels registered in countries such as Ghana (5.8 percent). The same can be said with regards to FDI stocks achieved in Senegal (1 percent) relative to Ghana (4.5 percent) and Côte d'Ivoire (6.4 percent) over the same period, suggesting a much lower level of gross fixed capital formation therein.

Table 1 shows the major FDI projects as annually registered by economic sectors in Senegal over the period of 2003–2009. A total of 160 major FDI projects were registered with APIX from 2003 to 2009, representing an amount of CFA 165 759 billion. These investments were unevenly distributed with agriculture accounting for CFA 58 206 billion (35.1 percent of the total) or 58 registered projects (36.2 percent). Agriculture is followed by agro-processing and fisheries, which appear among the most attractive subsectors to foreign investors in Senegal accounting for 36.2, 31.25 and 25 percent of the total registered projects respectively. The annually reported numbers (Table 2), show fisheries as a subsector in the economy with a clearly declining trend in foreign interests. As for the wood industry, this sector has remained at a standstill because of the depleted state of natural resources.

FDI inflows to Senegal have traditionally and essentially originated from France which provided about 90 percent of the total until the year 2000 when Senegal started diversifying its FDI sources. Today, the major Senegal FDI providers include the Arab states, Malaysia, China and selected African countries such as Mali. It should be noted that France has been losing ground to the group of other investors just mentioned with its share in FDI inflows now representing no more than 50 percent of the total[6].

Some studies have provided a thorough review of the Senegalese economy in general and the conditions and opportunities for doing business in Senegal's agriculture in particular, suggesting the need to address several underlying, structural bottlenecks known to negatively affect its overall competitiveness. The present chapter draws as much as possible from these results to underscore some of the key vulnerable areas, such as:

- **High illiteracy rate.** Fundamentally, Senegal's rates of illiteracy – around 70 percent of women and nearly 50 percent of men – are among the highest in the world. The country's effective relegation of women

[4] The GOANA was launched on 18 April 2008 by Senegal's head of state with the view of achieving "food sovereignty, in line with the objectives announced in the Agriculture-Forestry-Livestock framework Law (Loi d'orientation agro-sylvo-pastorale – LOASP) passed in 2004 (APIX, 2009). See also article " La GOANA est plus qu'une rupture" 6 May 2008. at www.lesoleil.sn, on 12 April 2009.

[5] US$1= 494 FCFA (2005–2008).

[6] Note sur l'évolution de l'investissement prive au Sénégal, APIX. 2009.

TABLE 1
Annual FDI projects registered with APIX by sector in Senegal, in millions of FCFA, 2003–2009

Activities	No. of Projects	No. of Projects (%)	Total Investments 2003-2009	Total Investments 2003-2009 (%)	2003	2004	2005	2007	2008	2009
Agriculture	58	36.2	58 207	35.1	10 503.0	3 621.0	11 852.4	14 846.0	12 395.1	4 989.0
Agro-Alimentaire	10	6.25	1 031	0.6	81.4	356.5	264. 9		94 .5	233.8
Agro-Industrie	40	25	80 082	48.3	27 643.7	2 984.8	16 072.7	15 244.7	14 190.9	3 944.7
Bois	3	1.8	447	0.2			73.0	200.0		174.0
Elévage	9	5.6	3 372	2.0	515.0	148.2		633.0	1 926.0	150.0
Pêche	40	25	22 620	13.6	1 868.3	5 286.2	6 689.6	4 558.0	3 699.2	518.2
Grand total	160	100	165 759	100						

Source: APIX

TABLE 2
Annual FDI projects registered with APIX by sector in Senegal, in millions of FCFA, 2003–2009

Target sub sectors	No. of Projects	Total Investments millions of CFA	2003	2004	2005	2007	2008	2009
Agriculture	58	58 207	8	6	14	9	15	6
Agro-Alimentaire	10	1 031	1	3	2		1	3
Agro-Industrie	40	80 082	5	8	14	3	6	4
Bois	3	447			1	1		1
Elevage	9	3 372	2	1		3	2	1
Peche	40	22 620	4	4	15	9	5	3
Total general	160	165 759						

Source: APIX

to unskilled, unhealthy work means that it is ignoring the productive capacity of more than half of its population.

- **Poor basic infrastructures.** Senegal's infrastructures - particularly the agricultural sector, is entirely inadequate. The fact that trucks cannot transport goods without undue police interference and on the-road "shakedowns" almost dooms any hope of competitive trade.
- **Unreliable power supply.** The unreliable source of electricity and source of water supply or affordable irrigation systems remains among the most severe impediments to foreign investments in agriculture. Even where the national grid had reached pack houses and farm gates, electricity costs appear to constitute 50 to 70 percent of companies' operating costs. Where the grid did exist, it is reported not as reliable as hoped. The risks also remain high of producers sometimes losing their entire crops when their irrigation pumps go down for the day as a result of lack of power.
- **Unfriendly tax system.** The Senegal business environment is seen as unfriendly when it comes to assessing its tax

administration and tax regime relative to other West African competing FDI destinations. In general terms, taxes in Senegal are considered very high by the business community. Its tax authorities are found to be uncooperative in helping to meet the requirements of the country's tax laws. For example, it takes about a month and a half more for businessmen to process the myriad tax related issues (Table 3) in Senegal compared to their counterparts in the region (Table 4). The only penalty assessed for improper tax filing and payment in Senegal is a fine. No one has been known to go to jail for tax fraud or corruption; in short, the entire tax system must be changed to rebuild confidence with taxpayers[7].

- There is no affordable agricultural insurance scheme in place to safeguard investors against the risks involved with subsistence agriculture, which is the most dominant mode of agricultural production in the country.
- **Unrest in Casamance.** Investors may perceive the no war-no peace environment caused by the rebels in Casamance as a risk to stability.
- **Land insecurity.** The uncertainty of title and access to land is a particular disincentive to investment, particularly in the country's agriculture sector.
- **Inflexible labour code.** The inflexibility of the labour code in Senegal is notorious among agribusiness owners. The relative difficulty in terminating contracts for example, results in a situation where it often makes more sense for business owners to hire employees on a short term, without contracts than to hire contracted employees.
- **Corruption.** According to a report by USAID[8], Senegalese businessmen, civil society and government all reported high levels of corruption. It should be noted however that Senegal has a relatively strong showing for the region.
- **Ambiguous competition policies.** The government must also confront the concerns of several stakeholders who have been reporting persistent anticompetitive practices known to plague the business climate in several key food subsectors (sugar, rice, wheat flour, etc.).

[7] USAID, 2009, p. 89.

[8] USAID (United State Agency International Development). 2009. "AgCLIR: Senegal- Commercial legal and institutional reform diagnostic of Senegal's agriculture sector". September 2009.

TABLE 3
Taxation in Senegal that imposes the greatest administrative burden

Tax items	Tax base	Number of declarations per year	Estimated time required (hours)
Corporate income tax	Taxable profit	3	120
Value-Added Tax (VAT)	Sales value	12	450
Payroll tax	Payrolls	12	96
Retirement contribution	Payrolls	12	96
Social Security contribution	Payrolls	12	96
Total			666

Source: Economy Watch, Economics & Investing Reports
http//www.economywatch.com/doing-business/paying-taxes.html

TABLE 4
Comparative tax burden at country, Africa and OECD levels, 2009

Items	Senegal	Africa	OECD
Payments(number)	59	38	13
Time(hours)	666	312	211
Profit tax (per cent)	14.8	21.2	17.5
Labour tax and contributions (per cent)	24.1	13.2	24.4
Other taxes (percent)	7	32	3.4
Total tax rate (per cent profit)	46	66.7	45.3

Source: World Bank Doing Business Website;
http://www.doingbusiness.org/ExploreEconomies/economyid=164

TABLE 5
Additional tax burden on business in Senegal

Tax	Tax base	Tax rate
Tax on interest	Interest income	15%
Advertising Tax	Value of advertising	Variable
Stamp duty on contracts	Number of contracts	CFA2 000 (fixed)
Tax on insurance contracts	Insurance premium	Variable
Tax on unimproved property	Rental value of land	5%
Tax on improved property	Rental value of property	5%
Business tax	Rental value of business premises	Different rates
Vehicle tax	Engine capacity	CFA50 000 (avg)
Fuel tax	Fuel costs	included in fuel price

Source: Economy Watch, Economics &investing Reports
http//www.economywatch.com/doing-business/paying-taxes.html

3. Institutional, regulatory and policy framework for investments in Senegal

Senegal has set for itself, since 2007 the objective of becoming an emerging country within 25 years. This is the main thrust guiding the Accelerated Growth Strategy (AGS) that Senegal hopes to achieve with the delivery of an average annual economic growth rate of 7–8 percent and a public and a private investment rate of 30 percent of GDP annually[9]. The following describes some of the key FDI related measures the Government of Senegal (GoS) has implemented, starting with the establishment of key institutions with mandates to carry them out.

[9] The two components of the AGS are the creation of an international class business environment, and the promotion of growth-inducing clusters (agriculture and agro-industry, marine and aquaculture products; textile-clothing; ITC and teleservices, tourism, culture industries, craft products).

3.1 Office of the Accelerated Growth Strategy (AGS)

The government strategy towards agriculture is led by the Great Agriculture Offensive for Food and Abundance (GOANA) as well as by the AGS and selected key institutions in charge of their implementation: the Ministry of Agriculture, Office of the AGS and APIX.

3.2 Agency for the promotion of investments and major projects (APIX)

As for attracting foreign investments to start a business in Senegal, the central institution is the Agency for the Promotion of Investments and major projects (APIX). This agency is tasked to address the following:

- Improve Senegalese business environment;
- Promote Senegal as an investment location;
- Research and identify national and foreign investors;
- Follow up on investment contacts and project evaluations.

To achieve these goals, the Government of Senegal (GoS) has offered a number of incentives to target potential investors[10] which can be consulted at the APIX website[11]. These are included in the Senegal's investment code, a key national law passed in 2004 but amended several times since then.

3.3 The Senegal investment code

As outlined on the APIX web site, this code specifies tax and customs exemptions according to the size of the investment, classification of the investor and location. There are measures specifically designed to encourage potential agribusinesses to engage in business in Senegal. Noteworthy examples from the investment code include:

- Customs exemptions (3 years);
- VAT suspension (3 years);
- Tax credits
  - 40 percent eligible investment;
  - Five years;
  - 50 percent taxable profit.
- Exemptions extend to five or eight years if certain conditions of employment generation and distance from Dakar are met. This means that investments outside of Dakar receive longer periods of exemption from taxes than those carried out in the Dakar greater zone. This is a feature shared with terms met in the Ghanaian incentive package to direct more new investors towards the inland investment areas most in need of jobs or projects.

There are also incentives intended to target those who wish to invest specifically in agriculture in general sense. In order to qualify for the following, a company must show that it will export at least 80 percent of its output:

- Unlimited recruitment of expatriate workers;
- Duty-free importation of capital goods;
- Exemption of customs duties on vehicles ;
- Exemption from various taxes, including the land tax and income tax on dividends;
- Exemption for registering or modifying registration of corporate documents; and
- Unlimited transfer of monies in and out of the country, subject to respecting the restrictions arising from the UEMOA money laundering Act.

These features are very appealing but not very different from what investors get also from other competing FDI destinations such as Ghana, Côte d'Ivoire and Nigeria in addition to the advantages these countries have in terms of larger market size, better infrastructures, more conducive agribusiness environment, more robust free zone package, to name but a few. Aside

[10] The list of all possible incentives is fairly complex and can be found at www.investinsenegal.com. Note: in French only. There are incentives for large firms, for making investments above a certain level, for using local inputs, and for locating in the less industrialized regions of the country.

[11] www.investinsenegal.com

from the unquestionable geographic advantage Senegal presents as a result of its proximity to the key European and American sea and airports of entry, the country has no choice but to be more aggressive towards investors contemplating investing in West Africa. Overhauling the Senegal free zone programme is an example of a good area to start reforming.

3.4 Senegal special economic zone

In terms of free zone institution buildings, the GoS also instituted an authoritative body for the administration of all special economic zones within Senegal, to deal with business licenses and registration thereof. This Authority is set to operate as a one stop-shop for the formation, registration and licensing of companies. A wide range of services, including the provision of buildings are also made available to companies operating within the zone, including the delivery of working and residence permits to foreign nationals. The Authority enjoys the powers of municipalities while serving as the Delegate of the Prime Minister and all ministers within the special economic zones. However, Senegal has not been particularly proactive about the development and use of special economic zones, such as the industrial areas that benefit from concentrated government services (utilities, licensing, customs, etc.), and tax incentives for export oriented companies of the kind successfully experienced in Ghana. Steps taken more recently have included moves to replace its free trade zone initiatives with the Enterprise Zone Franche d'Exportation (EZFE) scheme with aims at reducing taxes and offering duty free imports to companies located within the zones.

Apart from the case of the old Dakar Free Industrial Zone (ZFID) which is virtually inactive[12] now, it is perhaps worth recalling the US$800 million Agreement which the GoS signed in 2007 with Jafza International of Dubai with the view of establishing, constructing and running the Dakar Integrated Special Economic Zone (DISEZ), a special economic zone for sanctioned investments outside of Dakar. This project was set at the time of its signing to be opened in 2010, with hopes for creating jobs as many as 30 000 posts. These outcomes barely match the results Ghana has harnessed in this area (Table 39) considering the long list of free zone companies, whose impacts have been felt in various areas of the Ghana economy as discussed in the chapter on Ghana.

3.5 Investment treaties

As discussed earlier for the case of Ghana, Senegal has also signed a number of investment treaties with countries which have demonstrated interest in developing business relationships with West Africa. These include a bilateral investment treaty with the United States, signed and ratified by the US Congress in 1990. This treaty provided for "most favoured nations" treatment for investors, internationally recognized standards of compensation in the event of expropriation, free transfer of capital and profits, and procedures for dispute settlement.

Senegal has also signed similar agreements for protection of investments with several other countries namely France, Switzerland, Denmark, Finland, Spain, Italy, Netherlands and Japan. The country is also member of the World Trade Organization (WTO), the African Organization of Intellectual Property (OAPI) and the World Intellectual Property Organization (WIPO) suggesting that it is adequately equipped with the legal instruments used to address the concerns of foreign investors in areas as sensitive as the protection of capital investments and intellectual property rights. As discussed earlier, these legal devices are important safeguards to minimize the risks for all involved be they from investors' host or home countries, but mirror very much the kinds of incentives other FDI competing destinations usually offer. This would imply that Senegal needs to go beyond just overhauling its investment package to make a difference. The country should adopt an integrated policy approach that comprises not only agricultural investment and investment policies, but also other crucial policy areas (good governance, infrastructures, competition, trade, R&D, land

[12] stopped issuing new licenses in 1999 although firms located there will continue to receive benefits until 2016.

and water policy), to tackle simultaneously the core underlying factors hindering its global competitiveness. This view is taken now to shed some more light on Senegal land-related issues starting with the role played by the key land-related institutions established to address them.

3.6 Senegal General Tax and Land Authority or Direction Générale des Impôts et des Domaines (DGID)

The DGID is one of the most influential land-related divisions of the Ministry of Economy and Finance but is in charge primarily for administrating the country's tax policy. It is also the agency responsible for the collection of tax revenue and the administration of all national land classified as national patrimony (Domaine National (DN)). The question of the DN is one of the more complex subject areas to discuss in the context of Senegal; suggesting the need to briefly describe how land available in the national patrimony or DN can be identified, negotiated, assessed and registered safely from the perspectives of interested foreign investors.

3.7 Land offices with conflicting responsibilities at local levels

Senegal has established a land office in each of its ten regions, which contains three agencies with a mandate to report to the DGID: the Cadaster, the Conservation Foncier (CF) and the Bureau des Domaines (BD). These are branches of the national State, not the local government.

Cadaster

The cadaster is tasked to survey and map land and to determine property boundaries. It is not concerned with land allocation, or with the issuing of licenses. It does have responsibilities for the subdivision of land (morcellement), which is becoming more and more of a challenge in a country such as Senegal where population is growing at faster rates.

Conservation Foncier (CF)

The Conservation Foncier is the office responsible for the registration of land, including privately owned land and apartments. The CF is supposed to authorize morcellements (land dividing operations) before the Cadaster can survey and finalize them.

Bureau des Domaines

The Bureau de Domain (BD) is the office responsible for land zoning, and for the direct administration of State lands. Investors who want a zoning change must apply here.

3.8 Local government entities set to enforce Senegal 1996 local government code

A new local government code was issued in 1996 to provide a basis for the authorities recognized in the regions, municipalities and rural communities to serve as seats of local government. A wide range of traditionally stated state competencies – including the management and use of state, public and government lands – was transferred to selected local government institutions in the context of the government decentralization scheme, but without the required accompanying financial resources. This scheme has resulted in the creation of a three-tier system with central, regional and local levels which do not match the existing national government five levels of administrative operations: national, regional, departmental, sub-prefectural, and village. This seems complicated, especially when it comes to sorting out land-related disputes. In practice as will be seen shortly, land conflicts between the different branches and levels may arise. However, in rural communities where most agricultural production takes place, the local authorities tend to be much stronger than the State. It is the rural councils which exercise the greatest degree of power at this level, while state representatives such as sub-prefects and village chiefs are little more than record-keeping bureaucrats and tax collectors.

3.9 The special power of rural councils to allocate and withdraw available rural lands

Land is allocated by the Rural Councils (RC), free of charge to beneficiaries who must live in the

TABLE 6
Major steps to be followed when allocating land by a rural council in Senegal, 2009

STEPS ACTION	
1.	Seeking for an appropriate project site in line with the project objectives and the scope of the project;
2.	Meeting with the President of the concerned Rural Community (PRC);
3.	Submitting an official request to the concerned Rural Community (CRC) ;
4.	Setting up the Survey Land Commission (SLC) ;
5.	Calling up the Rural Council in the case of the approval and meeting of the RC for deliberations;
6.	Locating of landmarks;
7.	Getting a copy of the deliberation certificate signed by the PRC and the Prefect or the Sub Prefect;
8.	- Payment of the compensations to existing land occupants; - Payment of landmark fees if the vote is favourable;
9.	Carrying out the land mapping operations by a Land Surveyor;
10.	Starting of the land preparation works.
LAND REGISTRATION - KEY IMPLEMENTING INSTITUTIONS IN THE CONTEXT OF SENEGAL	
•	Court
•	Government of Senegal
•	Rural councils
•	Land offices, including Cadastre, Conservation, Foncier and Bureau de Domain

Source: APIX

rural communities and be able to use the land productively. Any natural or legal person who is allocated a plot receives a means of production for an indeterminate period. When they die, their heirs are allocated the land, provided they can put it to productive use[13]. Table 6 illustrates the major steps to be followed by potential land investors seeking to work through the land policy process of identifying, acquiring and registering land potentially available for economic uses at the rural levels. As can be noted below, there are ten major steps investors have to go through with additional ones ignored for brevity. In step no. 9 for example, it is expected that the land map be signed by the land registry (Cadastre), where tasks carried out at that level are done by a private land surveyor instead of a public one. In short, there are too many land-related steps and institutions involved suggesting the need streamline the process. Furthermore, to carry out the land registration process smoothly, surveyors must be available in number and quality. This is a frequent obstacle, due to the cancellation of training programmes producing this kind of skilled labour since 1986. In the absence of surveyors, "surveys" are sometimes carried out by completely untrained individuals, resulting in "maps" that have no legal force but are used anyway. And where available, surveyors sometimes lack the equipment required to carry out their jobs in a professional manner. Consequently, projects such as PAMOCA[14] which has attempted to tackle some of the prerequisites to resolving land insecurity concerns should be

[13] Under the 1964 law, "productive use " is understood as defined for each region by order of the Prefect, but this never seems to have been done. So the question of what constitutes "mise en valeur" is left to the discretion of the local councils.

[14] An EU and ADB funded land project designed to develop a national map, cross-indexed with both satellite images and with local records, provide training and equipment to cadaster offices across Senegal, and assist the government in policy formation. See description of the project at www.devex.com/projects and www.aps.sn/aps.php.

promoted. Unfortunately, PAMOCA was due to close in December 2009 with no prospects for extension. Nor is any donor involved in a follow-on project[15].

3.10 Senegal's current land policy

The legal framework for land use in Senegal is complex and includes provisions enacted in the colonial era as well as several laws enacted during this decade. In this connection, the most important of Senegal's land laws is the 1964 law on the "domain national" (DN), which abrogated customary land tenure and nationalized most of the land. Under this law, the State is the exclusive trustee of the country's land and is responsible for its management. As described, the DN is subdivided into four categories of land areas, namely:

- the urban zones (zones urbaines);
- the zones for special uses (zones classées);
- the development zones (zones pionières), which remain under the control of the state; and
- the zones for agricultural production (zones de terroir) which are by far the most important for agriculture production.

National domain land accounts for about 95 percent of the country's total territory, and most of this is termed *zones de terroir*. Most of these were rural lands held under customary regimes following national independence, but with the post-independence legislative framework they are now covered by the common law regime of the national land law. In accordance with the laws, territorial lands include all lands that a rural community needs for housing, farming, livestock rearing and woodlands, and their possible expansion. The boundaries of each of the territories under review are determined by decree, and are administered by the Rural Council (RC), usually established under the oversight of the local sub-prefect, the State local representative. These coincide with the boundaries of the rural community, and the land within them is regarded as a space for development, not as a legal and economic asset. As such, the land belongs to no-one, does not form part of any estate and consequently has no rightful owner, since if everybody owns everything then no one owns anything. This administrative power has given the rural councils the authority to allocate and withdraw land and to monitor its use at local levels. In theory, the Rural Council can withdraw plots for two reasons: to sanction noncompliance with the conditions of allocation, particularly the productive use requirement, in which case the land is withdrawn without compensation; or in the interests of the community, in which case the landholder should be allocated a similar plot if practicable - although this is not possible in most rural communities.

From the above, three separate regimes are now emerging to govern land ownership in Senegal.

i. private property, a legacy of the colonial system that exists mainly in urban areas and has grown rapidly due to urban sprawl and modern economic activities;

ii. public ownership, which gives the state the option to allow local government to own or use land assets, consisting mainly of land attached to public buildings or communal amenities that do not include agricultural lands; and

iii. rural lands, most of which are covered by the national land law and enforceable through official state institutions.

This system of land ownership and control as outlined is almost universally recognized as a constraint to agricultural modernization, suggesting that the GoS pay special attention to these concerns and take measures to swiftly reform the country land policy.

3.11 FDI related land conflicts in Senegal courts

As expected, this system of land management is not without fuelling disputes. It is no wonder that Senegal's court system receives

[15] USAID, 2009.

a great many land cases every year. In Dakar for example, land disputes were estimated to account for about 20 percent of the court's workload, while it was estimated that St Louis receives "between a third and half", and Kaolack, "about half". In these courts, the common types of cases include boundary disputes, arguments over rights to use lands, disputes arising from decisions of the rural councils, and disputes over who has the right to use lands. These are very common, as the rural council decisions are sometimes at odds with the law. These are sometimes not written down or grounded on any legal devices of irrefutable legal quality. In this connection, there are no established best legal practices or framework yet to guide for example the drafting of land model contracts – be it for sales or leasing – involving rural councils and foreign investors. It is striking indeed to see how "easy" it is for a foreign investor to formally arrange the alienation of a large tract of land of the order of 5 000 hectares without conceding much to the other party.

Furthermore, the land-related decisions mentioned earlier may also be delivered by a single council member, or a group of the council members, rather than the Council as a whole. Overcoming these impediments calls for strengthening technically and financially the capacities of all land-related entities starting with the rural councils.

The Senegalese legal system is recognized as technically competent to handle the kinds of land disputes mentioned earlier, but it does not seem to be working for the poor farmers. Like in Ghana, the courts are reported to be very slow, not very consistent or sufficiently reliable to allow everyone, especially the poor, to access its services. These difficulties are attributable to the high fees, costs and physical distances that the poor can rarely afford.

As dysfunctional as it may seem, Senegal's current land system has the merit of preventing a switch to a system of pure fee, simple ownership without adequate safeguards for the poor, which would have accelerated the concentration of the most fertile lands in the hands of wealthier farmers, entrepreneurs and land dealers. Such a system would have resulted in turning the poor farmers into landless tenants and sharecroppers, or part of a great drift into the cities.

It is also striking to see how many of the offices or agencies described here are for the most parts underfunded, understaffed and underequipped suggesting the need to also take a serious look at problems of financing. Without adequate resources it remains to be seen how these key entities can meet their mandates, whatever these may be. Unsurprisingly, taking on land-related duties including land registration is one the slowest operations on record, given the numerous processing steps involved (Table 6); a known source of stress for all those who have attempted them.

4. FDI impacts on Senegal's agriculture – some economic, social and environmental impacts

Senegal has been confronted with a serious structural current account imbalance problem which has worsened over the past ten years to reach levels of over US$1 billion per annum. In the same vein, there has also been a structural deterioration of the balance of trade over the same period as a result of the escalating spending registered for import items, reaching about US$5.17 billion in 2008. In this regard, it is encouraging to see the upward trend performance in direct investments as shown in the financial operations account from 2001 to 2008. This illustrates a positive contribution made by foreign investors in helping Senegal to improve its overall balance of payments. The balance of trade provides another good illustration of these contributions.

4.1 FDI and trade balance related effects

As will be seen, Senegal has been running a structural trade imbalance over the past ten years ending up with US$2 919 million in 2008 that is, up 586 percent from 2001 underscoring the crucial importance of any export earnings opportunities experienced elsewhere in the economy. Foreign investors have participated actively in the agriculture production with some producing primarily for export. Horticulture is very illustrative of this orientation whereby companies such as GDS and SOCAS are among the best known, non-traditional product exporters. Until recently, in this area, Senegal had gained ground by increasing significantly its exports of non-traditional products in the period 2000–2005, thereby contributing to the increase of export earnings over the period under review. As shown in Figure 1, the country saw its tomato exports growing on average 40 percent per annum over the last ten years to reach about 18 000 tonnes in 2008/2009. In the same vein, the volume of mango exports grew remarkably well – almost ninefold from 1999/2000 to 2008/2009. This situation was very encouraging in helping the country to diversify away from its traditional export products such as groundnuts and fisheries.

There is little doubt that the tax revenues from the various production, processing, exporting and sales activities over a wide spectrum of investors involved with the food industry contributed to some extent towards improving of the levels of structural imbalances shown below in the government fiscal accounts (Table 7).

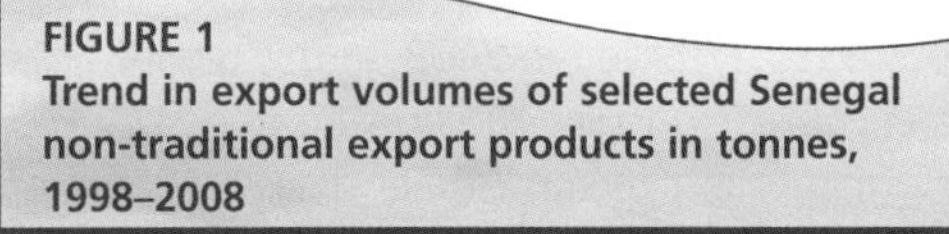

FIGURE 1
Trend in export volumes of selected Senegal non-traditional export products in tonnes, 1998–2008

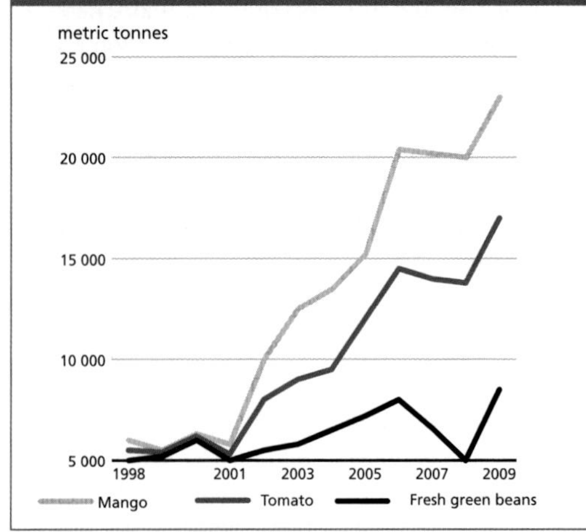

Source: Direction del l'Horticulture, MInistere agriculture, Senegal

4.2 Some economic impacts of FDI in Senegal's agricultural sector, including food security

There is a fair number of FDI related companies that are actively involved with the Senegal food industry. The review of the portfolio of activities of these companies suggests a wide range of agricultural outputs and services being carried out, and to which FDI have been contributing positively in the supply of some high quality fresh and processed food products in the country. In the same vein, FDI related companies have created large number of jobs suggesting an improvement in the levels of incomes distributed, as well as improved purchasing power on the part of concerned stakeholders. Assuming that these jobs did really materialize across the various food supply chains, the participation of FDI in Senegal's agricultural production system can be said to be positive. These improvements are the natural result of the wide diversity of operations and the high export orientation of many of foreign investors involved.

In the case of the tomato industry (discussed below), it is the only sector known to be operating on the basis of mutually beneficial arrangements for all involved. The industry has been very successful in attracting several billions of FCFA in foreign investments to increase the production levels of fresh as well as processed tomato. There is enough supporting evidence provided by this industry to demonstrate how beneficial this orientation has been in raising and stabilizing incomes earned at the farm-gate or in exports. No doubt the resulting increase in the purchasing power of the various tomato stakeholders involved – including the State – has meant greater demand for available food products, hence improved levels in food accessibility in the food industry.

Furthermore, investing in one hectare of land in the rural areas of Senegal with the view of producing high quality fresh tomatoes to supply nearby tomato processing plants of the kind operating in the Senegal River Valley is proving beneficial in terms of levels of income distributed along the tomato value chain. This shows tomato production related activities as a profitable

TABLE 7
Government revenues and Senegal fiscal balance, % of GDP, 2005–2011

Indicators	2005	2006	2007	2008	2009	2010c	2011c
Central government revenue	20.9	21.4	23.4b	22.5	20.0	21.7	22.5
Central government expenditure	23.6	27.4	27.2b	27.5	24.6	25.8	25.1
Central government balance	-2.7	-6.1	-3.8b	-5.0	-4.7	-4.1	-2.6

Source: APIX

venture, not only for the foreign investor concerned but also for the wide range of other stakeholders (producers, service providers, bank, state institutions, etc.) involved with this industry. It should be also be remembered that the tomato is the only food industry in Senegal whereby such clearly defined working arrangements – including the setting of guaranteed farm-gate prices – have been successfully reached among the food industry participants, contributing to ensure a win-win outcome for all involved.

4.3 Some FDI related agro-processing level impacts

Studies have shown that FDI inflowing to Senegal is essentially directed towards the formal, modern sectors, including that of agroindustry. As the following figures will show, there are also some signs of progress in the country's agro-processing subsector as a result of the active participation of FDI in this area. In the first seven months of 2009, for example, agro-processors experienced a significant improvement in production levels, faring about 9 percent year on year, in contrast with most other industrial sectors. Among the food subsectors showing the strongest trends in growth, are vegetable oil production, cereal and sugar processing. These encouraging trends were partly attributable to the excellent agricultural season registered in Senegal in 2008/2009, particularly for groundnuts and cereals.

The recent developments mentioned previously are encouraging; however, it should be remembered that key activities such as vegetable and groundnut oil production and exports are still performing below levels reached in 2006. Taking the agricultural sector broadly into account, the overall impact of FDI inflows on the sector so far registered is rated as below expectations. An exception is the subsector driven by small farming agriculture which has demonstrated some real increase in contribution to GDP formation (Figure 2), while key subsectors such as fishing, livestock and hunting, industrial farm and export oriented farming, forestry and logging have not been contributing sufficiently to the GDP.

Putting this into a better perspective, the entire modern sector in Senegal, which currently best captures the wealth-creation activities of foreign investors, (all sources considered), accounted for only US$1 631.416 million in value addition in 2008. This is comparable to the individual achievements of selected emerging bankers in terms of their net 2008 banking product[16] : The Standard bank group of South Africa (US$6 504 million); the First Bank of Nigeria (US$1 547 million). Figure 2 clearly shows the key contribution which the domestically oriented small food producers make to GDP relative to other sectors. This figure also shows how limited the impacts of industrial and export oriented activities have been on the country's economic output, despite all the FDI incentives provided. These suggest that Senegal should thoroughly review the activities of established foreign investors or established agribusiness segments, gauging carefully in due context what each investor or agribusiness venture is really bringing relative to what is it is costing in the balance to the country.

[16] Standard bank group of South Africa (US$6 504 million), First Bank of Nigeria (US$1 547 million) and Ecobank international, Togo (US$826 million). See Les 200 premières banques africaines. Classement exclusive 2008. Jeune Afrique. Hors série no. 22. octobre 2009.

FIGURE 2
Contribution of selected economic activities to Senegal GDP, 2005–2009, in CFA billions – nominal terms

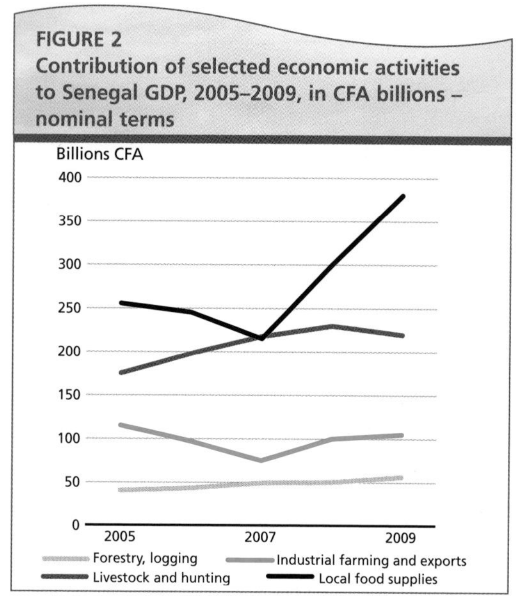

Source: APIX

FIGURE 3
Levels of technical inefficiency in selected subsectors in Senegal

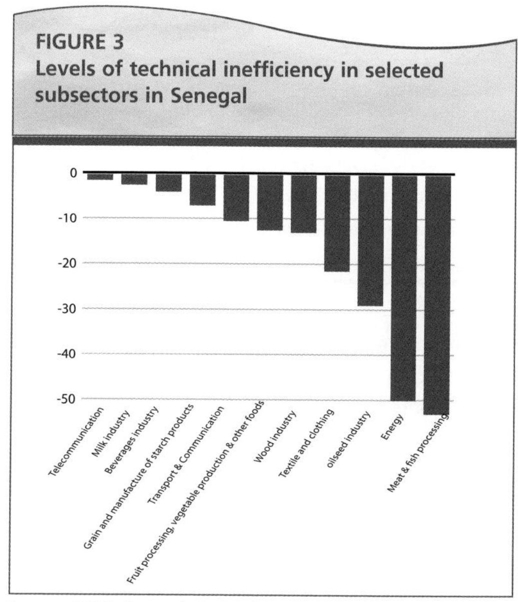

Source: DPEE, Senegal, September 2009
Note: (Estimated % below optimal outputs, given particular labour and capital inputs)

The relatively low performance just discussed could well be attributed to some structural difficulties or technical inefficiencies reported as negatively affecting the entire Senegal economy. Drawing from the results of a recent study in Senegal, fish and meat production investors were estimated to be losing 52 percent of their optimal outputs due to some technical inefficiency. As shown in Figure 3, key subsectors such as energy, textiles and clothing, and grain processing appear to be operating also at 50 percent, 21.1 percent and 7 percent respectively, below potential owing to technical inefficiencies.

This poorly rated performance could provide some explanations of the limited FDI related impacts on the agricultural sector as registered so far in several key strategic areas including technology transfer and R&D which are discussed in the following section.

Technology transfer and innovation in Senegal's tomato industry

Technological progress is crucial for agricultural development. Studies have shown that improvements in agricultural productivity were closely linked to policies towards and investments in agricultural research & development (Alston, Pardey and Smith, 1999). Agricultural development through innovation is vital for reducing rural poverty as the green revolution has shown for the case of several emerging Asian economies. As in Asia, an increase in the levels of research and development spending in support of more innovation, greater productivity and increased profitability can make a big difference in rendering an economy competitive, contributing more to exports and less to imports. This has not yet happened in Senegal's agriculture and food industry; the tomato sector is a case in point.

As the following figure 4 will show, productivity levels for many Senegalese farm products are encouraging but still below achievable levels attained elsewhere. In the tomato industry, for example, which is known to offer the best working, farm industry operational framework, yields grew only 1 percent on average over 1998/1999–2007/2008 to reach 21 tonnes per hectare in 2007/2008[17] suggesting that the increase registered in tomato production is essentially attributable to expansion of land areas allocated to the crop (Figure 5). There has been a

[17] Down 40 percent from the pick level of 35 tonnes per hectare achieved in 2005/2006..

FIGURE 4
Yield performance of selected farm products in Senegal, 1998–2008 average, in tonnes/hectare

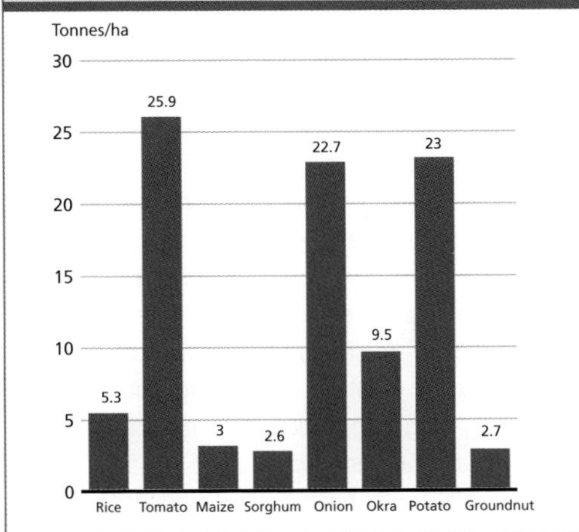

Source: Agence Nationale de la Statistique et de la Demographie (ANSD), 2009 see wwww.ands.sn

FIGURE 5
Averaged tomato yields achieved in 2000–2005 in selected countries, tonnes/hectare

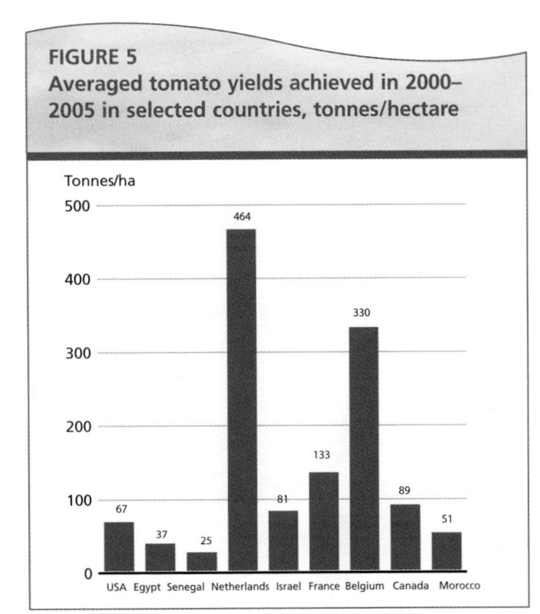

Source: FAOSTAT- www.fao.org

limited technology transfer and related spillover effects from SOCAS to downstream farming operators including the smallholders in this industry, calling for participants to take a serious look at the issues involved.

4.4 The social and environmental impacts of FDI on Senegal's tomato industry

The following scenario illustrates a number of positive social and pro-poor development opportunities created in local economies as a result of the presence of FDI in the tomato industry, while drawing attention to some of the accompanying and emerging risks attached.

Shaping a win-win contract farming model in Senegal – the case of SOCAS

Since 1995, SOCAS has been successful in implementing a contract farming scheme which has made the tomato industry stand out from the rest of the country's agricultural landscape. As can be seen in Box 1, tomato concentrate is produced for local outlets and exports in Senegal by SOCAS out of fresh tomato grown essentially in the Senegal River Valley, and on the basis of a unique contract farming arrangement[18] between SOCAS and some 12 000 tomato farms. As discussed earlier, this arrangement has contributed to a rapid increase of the production of fresh as well as processed tomato, thereby raising incomes, and improving the social and economic conditions of all the stakeholders involved, including the small tomato producers[19]and the State. This

[18] This "sucess story" is essentially due to the close collaboration between the tomato producers and the processor. The partnership started with the setting up of a consultation framework in 1995 - the National Comittee for Concertation of the Tomato Industry (or Comité National de Concertation de la Filière Tomate Industrielle (Cncfti)). This industry-wide, decision-making body included the tomato producers, the processor (SOCAS) who are the most active members, the representatives of suppliers, traders, consumers, and the State services. The Committee intervenes, among others, in the negotiation and the management of available agricultural credits. La Société d'aménagement et d'exploitation des terres du Delta (SAED) du fleuve Sénégal is tasked to serve as the secretariat. SOCAS works with producers on the basis of a contract under which the payment terms as well as farmgate pricing of fresh tomatoes are outlined.

[19] Groupements d'interêts économiques (GIE).

FIGURE 6
Cultivated land areas of major food products in Senegal, 1998–2008 average

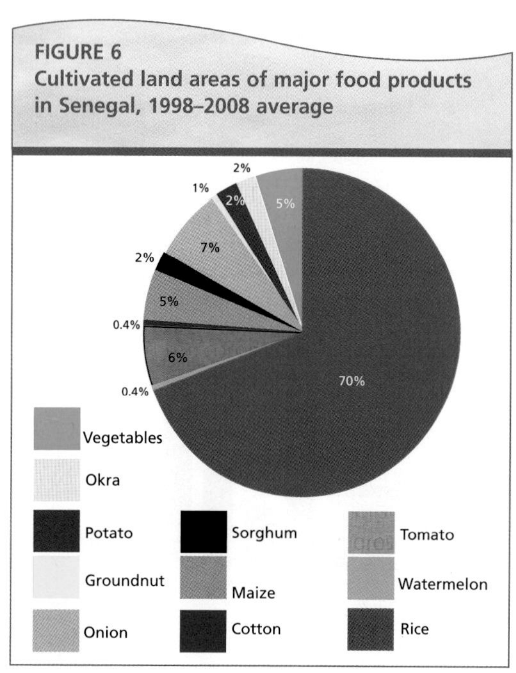

Source: ANSD, 2009 , see www.ansd.sn

framework helped the small producers to produce tomatoes in spite of scarce agricultural credit resources. In addition, the arrangement is believed to have brought about a number of other benefits, especially to famers. These include the improvements in rural income generation opportunities and its stability; and some reduction in tomato post-harvest losses. Indeed, unlike the GDS agribusiness model whose working terms are discussed later in Box 1, SOCAS did not deliberately cut past the tomato smallholders in an effort to vertically integrate its food supply chain and meet the stringent food safety standards now emerging in global food markets.

In 2008/2009, the tomato industry is estimated to have produced 80 000 tonnes of produce, underscoring the importance of income levels injected accordingly in the local economies. However, it should be remembered that with these output levels, SOCAS is believed to be processing beyond its absorption and processing capacities. These developments underscore the need for increasing investments in new processing capacities, thereby creating new jobs while further reducing the level of post-harvest losses, and consequently, the farmgate poverty levels of thousands of people whose livelihoods are dependent on quality tomato outputs. Drawing from available data as shown above, Senegal's tomato production accounted for 6 percent of the cultivated land areas allocated to major food products in the country. Depriving small producers of access to those lands – be it through land acquisition deals, land leasing or land alienation of any sort – would have some lasting and devastating social and economic implications for the local communities concerned, suggesting that the FDI and land reform related issues discussed extensively earlier need to be swiftly confronted. These remarks remain true for millions of other small producers involved in the cultivation of rice, maize, sorghum, onion, groundnut, cotton, potato, fruits and vegetables.

The agribusiness model whose advantages include the contract farming arrangements discussed above is not the rule but rather an exception. Since this institutional arrangement is a very common feature of agriculture in other regions of Africa, including some of the Senegal's neighbours, the lack bears some examination. The following discussion focuses on the GDS agribusiness model and on why this kind of farming arrangement should be discouraged. There is a great need to build more trusting agribusiness relationships with the small farming communities in order to ensure more pro-poor development outcomes in the longer run.

Building lasting agribusiness legacies in Senegal agriculture – the case of GDS

Unlike many other countries in Africa, particularly Kenya and Ghana, very little contract farming has occurred in Senegal. The main difference between these locations seems to be that in Senegal, most export agribusiness such as GDS (see Box 2) prefer to grow their own crops under strict quality control procedures, especially when product quality and food safety standards are paramount.

For seasonal export products that are readily available in local markets, such as mangoes, it is simpler and easier for exporters to make spot purchases from reliable producers to fill their product requirements, rather than contracting with small growers. But for others, many agribusinesses established in Senegal consider

BOX 1
Building responsible agribusiness models – the case of SOCAS in Senegal

In 1965, a French subsidiary company of the SENTENAC Group started testing the production of fresh tomato in the Senegal river valley with the view to producing fresh tomato to meet the growing demand of Senegal in tomato paste. The group has since then been manufacturing wheat, maize and millet flour, animal feeds (poultry and livestock), and other related food preparations distributed in the West African region including Senegal. It is composed of two agro-industrial subsidiaries, namely les Moulins SENTENAC and SOCAS (Société de Conserves Alimentaires au Senegal) Limited with the latter starting tomato paste production in 1970 with a first tomato processing plant with the capacity of 200 tonnes/day of fresh tomatoes and a pilot-farm capable of producing 3 to 4 000 tonnes of fresh tomato per season.

Since then, SOCAS has aimed at producing concentrate, juice and dried tomatoes out of the fresh tomato produced locally in Savoigne and Dagana near Saint-Louis, in the Senegal river delta where land and water are abundant. As initially intended, the company planned to meet local demand by 1987 with the outputs provided on a contract basis with about 12 000 tomato farmers, on condition that both the company and the State committed fully to provide the necessary technical assistance free of charge to the tomato famers and protection to the local industry respectively. Under this arrangement, Socas bought at guaranteed prices its entire needs in quality fresh tomatoes which have been successfully grown from the 2 685 hectares cultivated on average from 1998-2007 by selected small farmers now well organized in farms organizations. This make the only industry where there is an established board representative of all the industry interest who sat every year to discuss the entire issues facing the industry before setting the fresh tomato farm gate price along with terms of remuneration of all relevant services providers.

SOCAS employs about 300 permanent staff including agronomists and quality control specialists along with 1 000 part time workers at the peak of the harvesting season, distributing some 1 220 million of FCFA in salaries. The sales levels achieved in this company have also contributed to the reducing of the known government fiscal budget problem.

In 2007, SOCAS produced 10 000 tonnes of tomato concentrates out of 2 700 hectares to meet the demand of its local as well as foreign clients. Its turnover in the same year amounted to about 12 billion of FCFA making this company the leader of tomato paste production in the Sub Saharan Africa region. To meet this performance, the company has relied partly on the company facilities for crop production, harvesting, packaging, cooling, refrigerated storage, and transport. This accounted for about 11 billion of FCFA in investments including its state of the art irrigation and labouratory facilities, thereby allowing the company to process about 100 000 tonnes of fresh tomato or 18 000 tonnes of tomato concentrate and cover the needs of its local consumers..

The company has also diversified its farm and industrial operations into the production of packaging materials (cans, etc)in a state of the art packaging manufacturing plant established in Savoigne. In addition to producing some canned vegetables, onion seedlings, aromatic plants, it now exports about 600 tonnes of fresh green beans and 300 tonnes of dried tomatoes.

that the overall state of preparedness and development by small farmers is too limited for them to become reliable suppliers of high quality agricultural products. They believe that the required investment in time, effort and money required to upgrade the technical, financial and managerial skills of smallholders to help them become responsible and reliable suppliers is simply too great to make contract purchases attractive.

This situation is explained in a recent analysis carried out to cover Senegal's non-traditional export products, whereby the main exporter (GDS), has chosen not to engage in contract farming to meet its French parent company's corporate policy. In this regard, it is important to recall here the emergence of more stringent and constantly changing EU requirements for food quality and food safety – such as Global Gap, traceability and maximum pesticide residue levels that have gradually become a barrier to trade for many exporters including many developing country small producers.

The general low supply capacity of the smallholders is seen by some as another factor which may have induce companies to vertically integrate the production stage of its operation by establishing its own production farms. The GDS agribusiness model (Box 2) contrasts with that of SOCAS in terms of generation of better pro-poor long-term outcomes. The GDS model is viewed as more poverty-perpetuating than poverty-eradicating in the longer run, and should be discouraged. Accordingly, it should be a policy of APIX to assess proposed large-scale agribusiness projects of this kind and see how best to retain investors who are interested in investing responsibly in agricultural ventures – including smallholders – while creating lasting agribusiness legacies in Senegal.

Drawing from available data, experience has shown that foreign investors willing to invest in the Senegal agriculture and food industry are more inclined to maintain a much greater control over the companies' stakes than engage in agribusiness joint-ventures with local partners. As mentioned earlier, only 57 percent of foreign investors are registered with APIX, meaning they rated the joint-venture as an effective business strategy for penetrating Senegal's agriculture, compared to 68 percent of investors registered with GIPC in Ghana. The relatively low rate of

BOX 2
Building responsible agribusiness model – the case of GDS in Senegal

In 2001, a French subsidiary company, Grands Domaines du Senegal (GDS) with food production and distribution affiliates in a number of countries in Europe, Africa, and Latin America began producing cherries tomatoes in Senegal. By the end of the 2006-2007 seasons, GDS accounted for some 99 percent of fresh tomato exports from Senegal. The company produces cherry tomatoes near Saint-Louis, in the Senegal river delta where land and water are abundant.

GDS deliberately cut past smallholders in an effort to vertically integrate the supply chain. The company's production of cherries tomatoes relied completely on company facilities for crop production, harvesting, packaging, cooling, refrigerated storage, and transport.

GDS established a conditioning station in the Senegal river delta for handling and processing fresh vegetables; it also invested in high-technology production practices that include mechanized and computerized drip irrigation, along with the application of fertilizer and pest control products through the drip system. These technologies along with the required inputs such as improved seeds, fertilizers, and photo-sanitary products were imported from the European Union.

agribusiness restructuring obtained in Senegal relative to Ghana suggests that there exists a higher propensity for foreign investors to trust a business partner in Ghana than in Senegal. There is a need to promote good corporate governance practices in the industry in order to create a more trust-building environment in Senegal.

Increased risks of seeing Senegal small farmers lose available fertile land areas

Like Senegal's other identified, diminishing natural resources such as forests and fisheries, arable land too is becoming a scarce productive input as a result of a growing land demand pressure. With 3.3 person per hectare, Senegal appears as one the most land-scarce countries relative to possible FDI competing destinations such as Côte d'Ivoire (1.1 person/hectare), Ghana (1.9 person/hectare) and Nigeria (1.2 person per hectare).

As the following graphs suggest (Figure 7), there is no free land areas readily available in Senegal without some sort of claim or ownership attached, suggesting the need to seriously tackle the long- standing land use and rights issues with measures gauged to remove this well-known FDI related constraint. More than 3 804 900 hectares of land were available for cultivation in Senegal in 2002 (Figures 7 and 8), of which 246 000 were irrigable, underscoring the very tight situation at hand for allocating the irrigable lands left among various future uses or destinations. This matter is to become a much more acute issue for potential investors in general and small farmers in particular with the implementation of the GOANA, which foresees the unlocking of much of the cultivated land areas already identified. Indeed, there is a long list of FDI companies whose possible production expansion operations may require securing appropriate land for key future foreign investments. In this connection, some 30 percent of the firms or institutions surveyed in a recent global study assessing the business climate in Senegal perceived access to land among the most difficult obstacles that investors face in the country[20].

These results suggest more competition ahead to access the increasingly scarce land space. Weak enforcement of laws and regulations, and poor administration by the rural councils as currently experienced in the country all combine to create

[20] African competitiveness report, see Senegal country report, world economic forum, at www.weforum.org.

SENEGAL

FIGURE 7
Total land areas available in Senegal, 2008

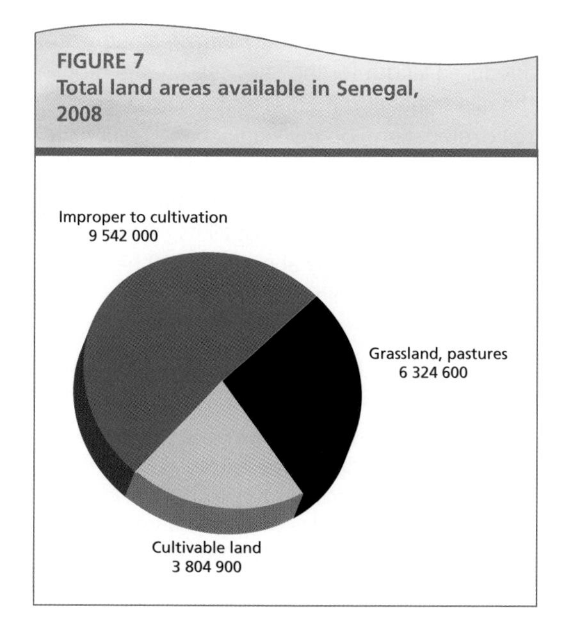

FIGURE 8
Allocation of available cultivable land areas in Senegal, in hectares, 2002

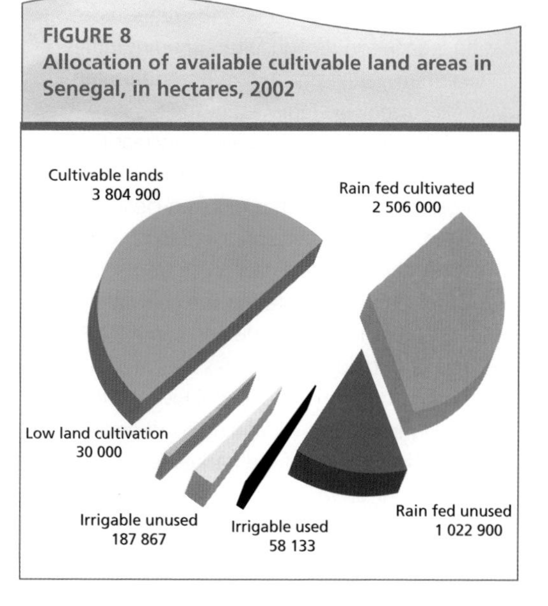

FIGURE 9
Allocation of irrigable land areas in Senegal, in hectares, 2007–2008

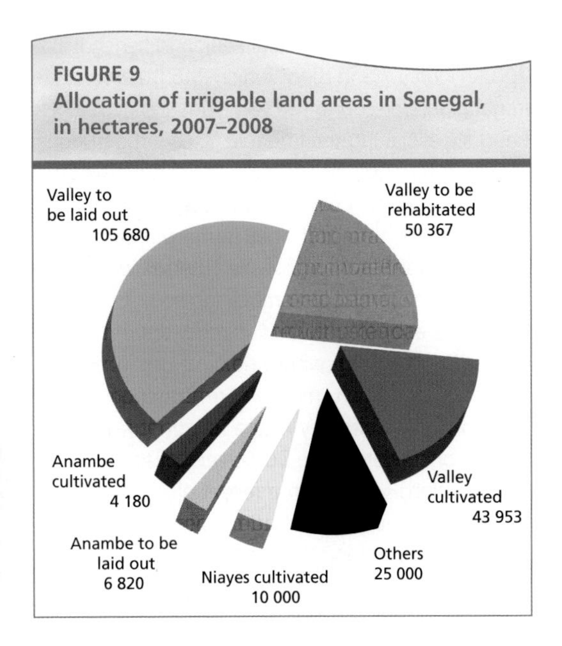

FIGURE 10
Available irrigable lands in Senegal, in hectares, 2008

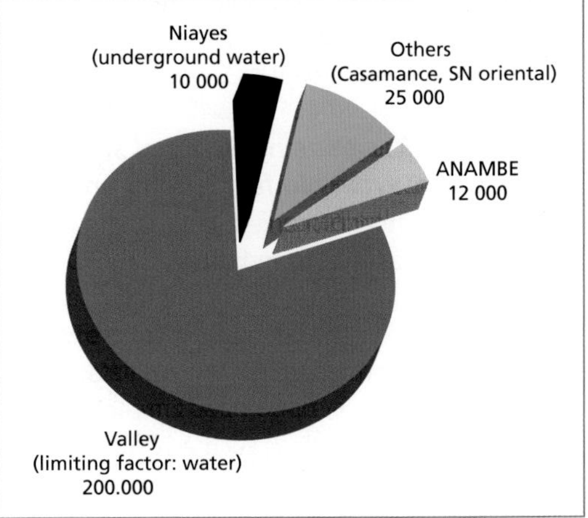

Niayes
(underground water)
10 000

Others
(Casamance, SN oriental)
25 000

ANAMBE
12 000

Valley
(limiting factor: water)
200.000

favourable environment for more corruption in the handling of land use and rights issues, pointing to more risks ahead for the poor to hold onto the fertile lands they farm.

5. Conclusions and recommendations

Senegal should adopt economic policies that promote responsible agribusiness investments built on sustainable processes to create more lasting legacies in its agricultural communities

Like many other agriculture-based farming ventures currently established in the Senegal River Valley (SRV) area, with the view of increasing agricultural outputs, the production of fresh tomatoes in this region has required clearing an estimated 2 653 hectares on average of forest cover over the past ten years. With also a total of 3 267 hectares allocated to the cultivation of fresh tomatoes in 2007/2008 and some 52 862 hectares of additional land areas allocated to other key food products in the Senegal River Valley as shown (Figures 9-10), expanding food production in existing key Senegal food baskets such as the Senegal River Valley, the Niayes region, the Calamanco and the Nambe will not persist without creating further environmental problems.

Indeed, Senegal is set to face more environmental problems, driven by its high population growth rate and the accompanying persistent degradation of its most valuable natural resources. In many rural and coastal areas, for example, soil fertility and productivity is stagnant or declining due to a combination of overuse and poor management of fertilizers and other inputs in regions such as the Senegal River Valley or the Niayes region. Rain precipitation patterns have dropped significantly over the past ten years (down 35 percent in quantity terms), along with a reduced length for the rainy season and drop in the rain frequency. There is also a high degree of soil degradation[21] observed in Senegal, especially for rice cultivation.

In should also be remembered that the country has exhausted almost all its natural forest resources. Indeed, only about 600 000 hectares of forest cover remain[22] to accommodate possible

[21] Soils are believed poor in nutrients such as P_2O_5, K_20, M_gO, S, Z_n, C_u.

[22] The protected forest area which is made up of all the protected areas including protected forests, perimeters for reforestation and restoration, national parks, biosphere reservations, full integral reservations, and special national reservations. This area covers 624 000 hectares, which represents 31.7 percent of the national territory.

BOX 3

A prospective FDI rice project in Senegal River Valley - some potential benefits and related risks

Senegal is set to implement a Saudi-Senegalese rice development project that could reduce its dependency on the world rice markets by 2015, should the investors involved successfully overcome a few remaining project hurdles. This thrust will involve private investors from Senegal (Societe agro-industrielle du Senegal (SAIS), S.A.) and Saudi Arabia with a goal to produce 1 000 000 tonnes of paddy rice locally of which 70 percent will be exported to Saudi Arabia. This rice project is set to take place on the left side of the Senegal river stretching over an area of some 30 km wide and 800 km long. The project will require land plots carefully selected from four departments covering Dagana, Podor, Matam and Bakel where rice is known to be grown from June to December. If implemented, the project will require unlocking a total gross land area of 119 583 hectares by the year 2015/2016, double rice cropping rice activities from June to December but also over the hot off-season.

Rice demand in Senegal is currently estimated to reach 900 000 tonnes per annum making this product the most consumed staple food item in Senegal. The supply of rice in the country is traditionally met by some 700 000 tonnes of rice imported from the world markets and another 175 000 tonnes provided domestically, of which 70 percent are outputs from the Senegal River valley.

Based on available data, the project stands to bring about a number of benefits to all involved. These include more than FCFA 100 billion in equivalent foreign currencies per annum by the fifth year of the project execution according to some preliminary estimates. It is estimated to cost about FCFA 581 billion (under a no state subsidies scenario) over 5 years with a financial plan structured as follows: 10percent - Senegalese in the form of land equity swap; 90 percent - Saudi in the form of cash.

The project outcomes build upon an internal rate of return of about 39.7 percent, an actualized benefit/cost ratio of 2.26, a payback time of 5 years. It is said to require some 66 435 hectares of cultivable land space to be secured absolutely in the Senegal river valley. Its investors will inject FCFA 54 176 billion on average per year in terms of direct incomes, distributed over the first five years of the project operations while strengthening the staffing and building partnerships with various service providers (producers, phosphates de matam, concerned rural communities, local consulting firms and contractors). The project promoters have also plans to make use of more intensified cropping methods spending some FCFA 21 933 billion and 54 585 billion on phosphates 18-46 and urea respectively over the same timeframe.

As planned, the project stands to bring about some new tangible income-generating benefits as well as real risks (social, environmental, etc.). This includes seeing some increased pressure on the most fertile lands available to the small holders established in the target regions.

climate change effects in the country. APIX should pay more attention to attracting agribusinesses that will invest responsibly in agriculture, paying due attention to the best RRR[23] practices available in order to maintain the country's fragile ecosystem.

With the levels of production discussed earlier, it is worth noting that SOCAS is one of Senegalese companies in the best position to meet opportunities emerging in the growing global food markets including West Africa. In

[23] RePlanting, ReUsing or ReCycling of forest based resources

2007, the volume of tomato imported worldwide was estimated at 5 855 000 tonnes or US$6.9 billion. That is about US$1 184 per tonnes in terms of unit value, which compared favourably to the levels of farm-gate tomato costs. To take advantage of these potentially profitable opportunities that lie ahead, Senegal must address some of the policy related deficiencies reported in several recent business global climate and competitiveness assessments. These areas include addressing, among others, the country's competition policy – covering sectors such as sugar, vegetable oil and tomato – to make them more transparent and competitive. There is indeed no need to buffer against competition among domestic investors which have failed to pay adequate attention to issues as important as corporate governance, innovation, R&D.

As already pointed out in some studies, there is a poorly established culture of good corporate governance in Senegal's agribusiness sector, including the tomato industry, which is said to be performing well on the grounds of the tangible economic and social benefits provided. In this sector, there is a poorly established culture of sharing key agribusiness information with the public and this ought to be confronted by the leadership of agribusiness companies such as SOCAS. Not surprisingly, there is no information shared for example of the economic bottom lines of this company on its own website; this requires changing.

APIX should promote a culture of timely information disclosure in the Senegal agribusiness sector to enhance assessment of the key sustainability indicators of interest in an objective manner. These should include:

- *Governance and management.* These factors comprise the importance of sound business principles, transparency, values and ethics in governing a company. For example, the agribusiness companies under review including SOCAS should, among other considerations, set in place basic industry governance structures.
- *Stakeholder engagement.* This factor addresses a company's engagement with its stakeholders regarding sustainable development issues. For example, companies may provide information on their sustainable development – environmental, social and economic – performance, principles and polices through meetings and communication with stakeholders. This can be done, for example, by regularly producing a public report shared on the company website, if any.
- *Environmental focus including environmental process improvement and environmental products/services development.* This factor addresses the companies' use of natural resources in the production of their goods and services. It also concerns the importance of the agribusiness companies embedding enough environmental principles in their products or service development. For example, the agribusinesses reviewed should indicate on their websites, among other communication methods, how they design new products specifically with a view to improving their environmental, social or economic impacts. In the case of the tomato industry, SOCAS should indicate how it involves its key suppliers, especially the smallholders with the best production practices available when reviewing and designing new tomato products and related services.
- *Socio-economic development with attention to local economic growth.* This factor addresses the agribusiness companies' commitments to the capture of economic benefits within the communities where they are operating, as well as contributing to the economy. In the case of the tomato industry, for example, SOCAS should be transparent enough to indicate on its website how it supports community development and capacity strengthening to generate wealth. These considerations extend to providing fair wages and benefits to contracting agricultural labour. The agribusiness companies under review including SOCAS should also show how they generate wealth and how these resources are distributed along the food value chain they serve.

- *Community development.* This factor addresses the companies' commitments to the social development of the community (beyond economic development). In the tomato industry for example, SOCAS should indicate on its website, among others, investments made in basic need projects to help the tomato small farms in becoming self-sustaining enterprises beyond the company's involvement.
- *Human resource management.* This factor addresses the agribusiness companies' commitments to providing a safe, high quality work environment for their employees – including management, staff and agricultural labour under contract. For example, the Senegal agribusiness companies including SOCAS should show on their web sites how they respect regulations covering working hours and payments for overtime and how fair the wage system in force in their respective industries is compared to the national average. In particular, agribusiness groups of longstanding presence such as Grands Moulins (SOCAS) should show the efforts made over the past 100 years to integrate the tomato farmer groups into the companies' shareholding systems.

Drawing from its areas of competitive advantage, Senegal has also passed several reform measures in recent years, to scale up its FDI inward efforts but without much success as yet. This policy has started paying off only recently, making Senegal the third fastest growing country behind Burkina Faso and Ghana in terms FDI inflows and the fourth largest FDI stock builder in West Africa behind Burkina Faso and Guinea. Despite this performance, the country is still lagging behind Ghana and Côte d'Ivoire, accounting only for 1.9 percent of FDI inflows and 1 percent of FDI stocks in the West African economies for 2005–2008. The benefits of these outcomes appear unevenly distributed when seen from a sectoral perspective: within the agriculture subsector, crop farming accounted for around 35 percent of registered agricultural projects or investment resources mobilized over the same period.

Foreign direct investment traditionally originated from France but this is changing in favour of newly established Asian investors such as the Arab states, Malaysia and China who are all expressing renewed interest in available land acquisition or leasing opportunities to farm energy and food crops in the country. The disparity in FDI distribution just mentioned remains also true at the geographical level. Thus, the irrigated lands of the Senegal River Valley remain the most appealing investment areas to agribusiness oriented investors.

Various bottlenecks, including land tenure insecurity and technical inefficiencies, have also continued to hamper the noticeable efforts being made in Senegal to drive more FDI into its agriculture sector. Despite the plethora of land-related institutions currently operating in the country, the government has not yet succeeded in simplifying the process of acquiring or leasing land in a transparent manner. Contrary to the perceived belief that land and water resources are abundant and readily available for engaging in any profitable agribusiness activity in Senegal, the remaining resources in these areas, especially in the Senegal River Valley, are shrinking fast. This underscores how sensitive land issues will become in the future, especially for smallholders whose livelihoods are highly dependent on rural income generating opportunities. There are no more large tracts of land now available for sales or leasing without some sort of claims or opposition attaching to them, suggesting the need to swiftly address the land tenure insecurity that is still prevalent in the country. Providing sustained financial support and a strong political will to map out clearly designated foreign investment areas earmarked for most promising investment projects (PAMOCA, free zone schemes) should be considered as a priority in helping to set up working land banks – which would contribute significantly towards raising the levels of FDI flowing into Senegal's agricultural sector.

The impacts of the FDI have also been favourably felt in many areas across the Senegalese economy in general including its food sector; in particular in the case of the tomato industry and the free zone scheme, which have been discussed extensively in this chapter, with the aim of illustrating selected FDI related economic, social and environmental effects. Foreign direct investment inflows to Senegal have helped ease various structural imbalances, including the state of food insecurity in the country, while, however, raising new areas of concern: the risks of seeing more and more small land users dispossessed of the land vital to their livelihoods, and the gradual exposing to more climate change effects. Though encouraging, the FDI results obtained so far are well below expectations in many other areas, viz. persistent agricultural yield gaps due to insufficient technology transfer, and the overall technical inefficiencies affecting the food sector.

The review suggests joint-venturing as the preferred mode of entry on the part of foreign investors to penetrate the agribusiness sector (57 percent of the registered FDI projects with APIX). Findings suggest the tomato sector as one of most profitable agribusiness activities, were good public policy, coupled with good corporate governance, would make a big difference in reducing poverty and food insecurity. Furthermore, the review of the SOCAS business model also suggests that there is still room for foreign investors to invest responsibly in profitable wealth-creating opportunities in Senegal while keeping in check their expected economic, social and ethical commitments. There exists a win-win solution to maximize companies' economic bottom lines while at the same time reducing the risks for all involved, including the smallholders. The findings also suggest the need to go beyond established or traditional agribusiness models (outgrowers and contract farming schemes), experienced so far, to include more inventive arrangements such as farmer ownership models whereby smallholders would be given a chance to use the lands they farm in equity swapping arrangements. This would help them to keep a greater share of the value additions created along the various food supply chains in which they are involved.

Finally, the benefits derived from FDI can be maximized in countries like Senegal and the risks attached can be minimized, but this would require taking appropriate institutional and policy measures in a timely and transparent fashion to match local working conditions with opportunities arising in the global energy, food, financial and carbon markets.

ZAMBIA:

Investment in agricultural land and inclusive business models[1]

1. Introduction

This chapter falls into two parts. First, it describes the policy framework and recent trends in large-scale agricultural investments in Zambia. Then, it examines two agricultural investment case studies: Kaleya Smallholders Company Ltd (Kascol) and Mpongwe Development Company Ltd (MDC). The latter went into voluntary liquidation in 2006 and its farms were bought up by ETC BioEnergy and another investor. At the time of the study, ETC Bioenergy sold the farms to another company, Zambeef.[2] Though liquidated in 2006, MDC continued to exist until 20 July 2011 when it was finally deregistered by the Registrar of companies. Neither investment project belongs to the recent wave of agricultural investments that has attracted much international attention over the past few years. Indeed, the two projects started in the 1970s and early 1980s as joint ventures between the Government of Zambia and the Commonwealth Development Corporation (CDC), and have been privatized in recent years. The involvement of the CDC reflected the development orientation of both projects at their inception. Given this circumstance and given the implementation time behind these two experiences, the case studies provide valuable insights on the longer-term development outcomes of best practice agricultural investments. These insights may be a useful contribution to today's international debates about agricultural investment.

Despite their similar historical roots, the two case studies are rather different. Kascol is an agribusiness company operating in Mazabuka district, in Zambia's Southern Province. It is a single-product company that produces sugar cane on a farm situated about 8 km south of Mazabuka, the main town in Mazabuka District. The sugar cane is sold to Zambia Sugar Company, which mills the cane into sugar for the local and export markets. According to the latest annual report of the Zambia Sugar Company, sugar exports to the European Union make up 62 percent of total sales, while the rest is sold in the local market (ZSC, 2011). Interviews with Zambia Sugar Company management indicated that sugar is also being exported to a number of countries inside Africa. Kascol started operating in 1980 and holds about 4 314.9 hectares of land, of which 2 265.3 hectares are fully developed and under cultivation. Kascol's approach to business is a combination of own-production and contract farming on land held by the company. Land is held as a 99-year lease, and Kascol subleases about 1 000 hectares of this land to about 160 outgrowers on the basis of 14-year renewable contracts. The model also involves equity participation and board representation for smallholder outgrower farmers: an organization of the outgrowers holds 13 percent of the equity in the company, and a district-level, sugar-cane grower association holds an additional 25 percent equity.

The ETC BioEnergy company (formerly Mpongwe Development Company, MDC), runs plantations for a total of 46 874 hectares in Mpongwe District, in Zambia's Copperbelt Province. The total landholding consists of

[1] This chapter is based on an original research report produced for FAO by Fison Mujenja, RuralNet Associates Ltd

[2] In this chapter, name references are made to both the Mpongwe Development Company and ETC BioEnergy, with a preference for the Mpongwe Development Company whenever reference is being made to events before 2007, and ETC BioEnergy when referring to events after 2007. The name Mpongwe Development Company is used in cases where events cut across the two periods.

three farm blocks, each with a separate 99 year lease. Farm number 4451 (Nampamba, 22 921 hectares) and farm number 4450 (Chambatata, 12 490 hectares) are used for crop production, while farm number 5388 (Kampemba, 11 463 hectares) is used for ranching. Of the total land area, 10 661 hectares are currently developed, of which 3 000 hectares are under irrigation. Various crops are produced, the mix sometimes changing from year to year. In the last production season, this mix consisted of wheat, maize, soybeans, rice, mixed (dried) beans, barley and jatropha. The "traditional" crops (wheat, maize and soybeans) handled by ETC BioEnergy, however, have been grown for almost as long as the farm has existed. Of the traditional crop, wheat is exclusively grown as an irrigated crop while soybeans and maize are mostly rain fed. In recent years, the company has also produced winter maize, which is irrigated.

The map in Chart 1 shows the location of the districts where the two projects are situated. The map also shows the overall, comparative population growth rate for the districts.

1.1 Research methodology

The research methodology involved interviews with company management and staff, villagers and other stakeholders. At the MDC, the research team conducted face-to-face, semi-structured interviews with company management; the Mpongwe District Council; a former member of Parliament who presided over land negotiations in the 1970s; residents of Mpongwe, and a focus group discussion with residents of the area of one of the chiefs who owned and controlled the land acquired by the Mpongwe Development Company (designated "Chief N" in this chapter). Similarly, at Kascol, the research team held face-to-face, semi-structured interviews with company management, and focus group discussions with two groups of outgrowers (both men and

CHART 1
Mpongwe and Mazabuka Districts, where the projects are located

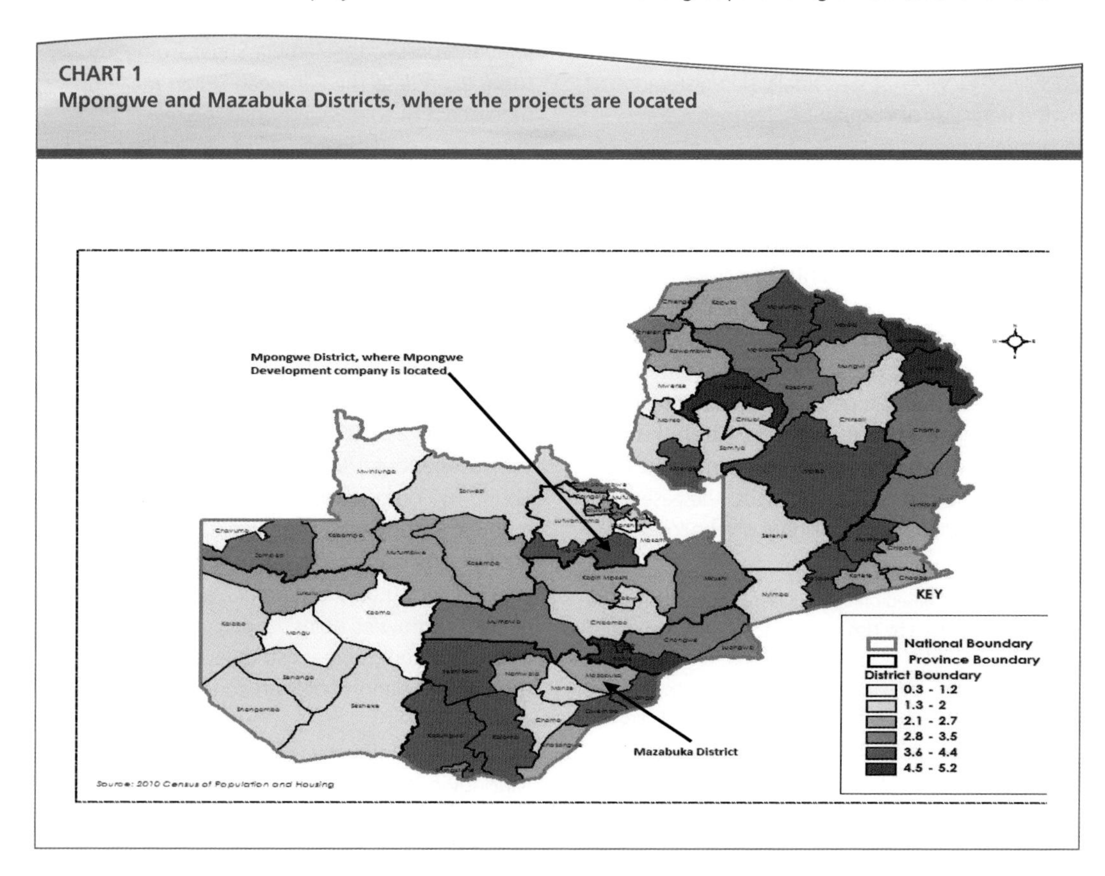

Source: Adapted from 2010 Census of Population and Housing Preliminary Report

women). An interview was also held with a descendant of one of the original owners of the farms bought by Kascol in the early 1980s. The team also interviewed a management representative at the Zambia Sugar Company, and another at Krookes Brothers, a sugar-cane estate. In addition to interviews, observations were also made about the farms and farm infrastructure.

1.2 Study limitations

Limited access to data resulted in gaps in the information presented, which consequently affected the thoroughness of the study. Both case studies involved private corporations that do not ordinarily disclose company information to the general public: ETC BioEnergy, for instance, restricted the choice of the researchers as to whom to interview in the company. The length of time during which both companies have existed made it impossible to carry out detailed analyses of performance and outcomes over the long term. Thus, the researchers were obliged to focus on the present (while being aware of the company's history to the extent possible). The tight timeframe within which the fieldwork was conducted also limited the number of interviews conducted.

This chapter contains a further four sections. Section 2 provides the national context and discusses the policy framework and recent trends in large-scale agricultural investments. Section 3 discusses the design and implementation of the investment projects. Section 4 analyses the socio-economic outcomes for the two projects and Section 5 covers the conclusions.

2. National context

2.1 Policy framework

This section deals with the policy framework for large-scale agricultural investment. It begins by describing matters related to land and then moves on to a brief consideration of agricultural policy.

Land policy and practice

Zambia's total land mass is approximately 75.2 million hectares, of which 12 percent (or 9 million hectares) is suitable for arable use (GRZ, 2002). Of the 9 million hectares of suitable land, about 1.7 million hectares are under cultivation (GRZ, 2009). The latter figure represents the total land under crop production and takes into account both subsistence and commercial farming.

Under Zambian law (Lands Act no. 20 of 1996, chapter 184 of the Laws of Zambia), all land in Zambia is vested in the President, and is held by him in perpetuity for and on behalf of the people of Zambia. For historical reasons, land in Zambia is generally divided into two categories: land in customary areas, which is referred to as customary land in this chapter, and state (or Crown) land. Customary land is land that was defined and reserved for indigenous peoples by the colonial masters under the Zambia (State Land and Reserves) Orders of 1928-1964, and under the Zambia (Trust Land) Orders of 1947-1964. About 94 percent of Zambia's land is said to be in customary areas. It should be noted, however, that continued reference to customary areas in official documents – including the Lands Act – is misleading. Many of the areas which were delineated and designated "customary" under the Zambia (State Land and Reserves) Orders of 1928-1964 are no longer customary in the sense in which the term must have been originally used. As noted later in the chapter, this has created further confusion by equating customary areas with customary tenure.

The remaining 6 percent is state land. State land was originally reserved for the exclusive use of European settlers (Roth *et al,* undated). Again, continued use of this term in reference to the present land situation leads to confusion and the tendency by some to exclusively equate the size of the land under leasehold tenure with the size of state land. This point is further elaborated later in the chapter. Customary areas tend to be areas with low agricultural potential due to poor soils, poor infrastructure, or both. State land, on the other hand, tends to be served with better transport and communications infrastructure, and has "attracted virtually all the skills and investment necessary for the development of the country's resources", according to Banda (2011).

That said, there are two major systems of land tenure in Zambia which, to some extent, parallel the two land categories just described: customary and leasehold. Land that is held under customary tenure is controlled by traditional rulers, locally known as chiefs, using local customary laws, as long as such laws are not in conflict with statutory law (e.g. the Lands Act of 1996). The chiefs are the custodians of customary law and they subsequently control land in their chiefdoms. Customary law is unwritten and is area-specific; that is, it depends on the unwritten traditions and customs of each chiefdom. It is estimated that some 94 percent of Zambia's land is held under customary tenure (GRZ, 2002). However, as noted above, this is not correct in that it appears to equate customary tenure with the size of customary areas. Land in customary areas can be held under leasehold title and this happens from time to time. Thus, land held under customary tenure must have reduced over the years as some of it is now held under leasehold tenure.

Customary land tenure is what most Zambians, especially in rural areas, are acquainted with. By virtue of belonging to a chiefdom, a person has the right to use and occupy land in that chiefdom, free of charge. However, the area chief has the right to withdraw land from anyone he deems to be violating the customs and traditions of the chiefdom. This is one of the sources of tenure insecurity under customary land tenure. Evidence from the field suggests that this is a major concern for some villagers in Chief N's area in Mpongwe district. The villagers complained, in a focus group discussion, that the traditional ruler was taking land from them and giving it to people coming into the area.

Leasehold tenure is regulated by statutory laws, which allow for a renewable maximum leasehold of up to 99 years. Some official sources put land under leasehold tenure at 6 percent of the total land mass (e.g. PRSP, 2002). However, as in the case for customary tenure, this may not be correct, equating as it does, leasehold tenure with the size of state land. As indicated above, usage of the term state land is a carryover from the colonial period and was defined in the Zambia (state and reserve land) Orders of 1928–1964. Zambia became independent in 1964, and there has not been any redefinition of the boundaries of state land since that date. State land was, and has remained, about 6 percent of the total land mass. The current practice is that land under leasehold tenure can come from either state land or from reserve and trust lands, which together form customary land. It is unlikely, therefore, that land under leasehold tenure corresponds to the 6 percent state land size.

Land under leasehold tenure is ordinarily used for residential, commercial, industrial, and agricultural purposes. Both nationals and foreigners can obtain leaseholds. In the case of foreigners, the Land Act of 1996 limits acquisition to the following circumstances: that the non-Zambian is a permanent resident in the Republic of Zambia, and a legally recognized investor; or that the non-Zambian is a company registered under the Companies Act, and less than 25 percent of the issued shares are owned by non-Zambians.

The Ministry of Lands, on behalf of the President, has the legal authority to alienate land. Land alienation applies to both state and customary land. In the case of customary land, land alienation by the President means converting land under customary land tenure into leasehold tenure. Box 1 below gives some details on how this is done. In some respects, it is desirable to convert land under customary tenure into leasehold tenure because customary tenure has very little protection under the law and is often subject to abuse by chiefs; sometimes even by the state.

The practical implication of customary tenure is that no matter how long villagers may have occupied or claimed ownership to a piece of land, they will have no officially registered title to it until the President alienates such land to them. Such alienation effectively converts the tenure status of the land from customary tenure to leasehold tenure. From that point on, the villager is required to pay ground rent to the state (the Ministry of Lands), becoming, in effect, a tenant to the state. Tenure change also means that the area chief loses control over the alienated land.

BOX 1
Land alienation in Zambia

The Ministry of Lands is the main ministry mandated to carry out the functions of land administration. Because the Ministry of Lands has no district-level structures, local authorities are appointed as agents to process applications and select suitable candidates on behalf of the Commissioner of Lands. Recommendations made by local authorities to the Ministry of Lands may be accepted or rejected by the Commissioner of Lands.

Alienation of State Land consists of the following:

a. Land Identification. Identification of land in any city, municipality, or district is the responsibility of the local or provincial planning authority concerned. Once land has been identified, the planning authority shall carry out its planning for various uses within the provisions of the Town and Country Planning Act and relevant regulations. Once the planning authority has planned and approved the area, the layout plans are forwarded to the Commissioner of Lands for examination of land availability.
b. Allocation of Land. Once land has been numbered and surveyed, the local authorities may advertise the stands in the news media or any transparent medium, inviting developers to apply to the Commissioner of Lands through the local authorities, using a prescribed form. On receipt of the applications, the local authorities will select the most suitable applicants for the stands and make recommendations in writing to the Commissioner of Lands, giving reasons supporting the recommendations. This recommendation letter will be accompanied by the full set of Council minutes. The Commissioner of Lands will consider the recommendations and may approve or disapprove them. The Commissioner of Lands will not approve a recommendation if it is apparent that doing so would cause injustice to others or if a recommendation is contrary to national interest or public policy.

Alienation of Customary Land involves a different process. Any person who holds land under customary tenure may convert it into a leasehold tenure not exceeding 99 years on application, in the manner prescribed. A person who has a right to the use and occupation of land under customary tenure, or has been using and occupying land for a period of not less than five years, may apply to the chief of the area where the land is situated. The chief shall consider the application and shall give or refuse consent. Where the chief refuses consent, s/he shall communicate such refusal to the applicant and the Commissioner of Lands, stating the reasons for such refusal in a prescribed form.

Source: Adapted from Ministry of Lands (http://www.ministryoflands.gov.zm/index.php?option=com_content&view=article&id=60&Itemid=87, accessed on 20 December 2011)

The cumbersome procedures of converting land from customary tenure to leasehold tenure, the subsequent obligation to pay ground rent to the state, and the fact that the concept of title deeds is extraneous to local practice are some of the reasons why most villagers do not obtain title deeds on the land which, in their view, is theirs by virtue of their (or their ancestors) having lived there even before Zambia was a nation. This tenure context creates a breeding ground for tensions in cases where the government takes land away from local groups and allocates it to an outside investor.

The situation is aggravated by the fact that while the government is able to convert land from customary to statutory tenure, there are currently no legal mechanisms for conversion from statutory tenure to customary tenure.

Ordinarily, this should not be a problem. However, in the case of Zambia (where the majority of its people cannot afford to hold land on statutory tenure), it poses a challenge and potential source of tension. Most Zambians use land for subsistence farming and the low productivity that characterizes subsistence farming implies that the subsistent farmer is only able to produce enough for own consumption and is severely cash constrained. Furthermore, since holding land on statutory tenure requires payment of ground rent to the state, which most poor subsistence farmers cannot afford, conversion of huge tracts of customary land to statutory tenure by the few who can afford to do so practically deprives the majority poor of a means of livelihood. When this happens, tension can potentially build up.

Both Mazabuka District, where Kascol is situated, and Mpongwe District, where the MDC is located, are rural areas with over 90 percent of the land being under customary tenure. No land in Mpongwe District and, in all probability, in Mazabuka District constitutes what was formerly designated Crown land (now called state land), implying that land that is currently under leasehold tenure in these areas was converted from customary tenure in customary areas – which consists of reserve land and trust land which the colonial government had reserved for indigenous use. The terms 'state land' and 'customary area' are here being used in the original sense, that is, state land means that portion of land originally reserved for European settlers and customary land is as that portion of land originally reserved for indigenous people. As discussed, land in customary areas is held by chiefs on behalf of their subjects, though ultimately this land is vested in the President for and on behalf of the people of Zambia. Land under customary tenure can be turned into leasehold tenure by following procedures laid out in the Lands Act. One important element of those procedures is the consent of the chief himself, although in practice chiefs do not always act in the best interests of their subjects.

The role of chiefs is illustrated by the acquisition of land by MDC. In the 1970s, land currently held by the MDC was under the control of chiefs, among them two chiefs designated in this chapter as Chief N and Chief L. It was a forested area used for hunting and gathering of wild products by villagers who lived near the area. In 1976, the Government of Zambia, through the Ministry of Agriculture, asked Chief N for land which they could use for commercial agriculture. Chief N referred the government officers to the family who held the piece of land in which the government was interested under customary tenure. At the time, a member of this family was area member of Parliament and he surrendered the land to the Ministry of Agriculture which, in fact, was facilitating land acquisition by the MDC

The procedure for the acquisition of customary land is largely the same today (see Box 1). Apart from titled land, land acquisition must involve local chiefs who, in consultation with their subjects, can either give or decline to give land for investment purposes. In practice, the government has taken measures to proactively facilitate investors' access to customary land. Under the Investment Act of 1995, an investment centre was established to facilitate investments in both agricultural and non-agricultural sectors. The centre was later amalgamated with various other government bodies, and is now known as the Zambia Development Agency (ZDA).

One of the roles of the Zambia Development Agency is to facilitate land acquisitions by investors, including through the creation of 'farming blocks'. Chiefs have been encouraged to give land to investors in the name of economic development. Thousands of hectares have since been given to investors by chiefs and have either been converted to leasehold tenure or are in the process of conversion.

In contrast to the land acquired by the Mpongwe Development Company in the 1970s, Kascol land was already under leasehold tenure when it was acquired by the company. The land was thus already earmarked for commercial agricultural use and part of it was already under agricultural use when it was acquired by Kascol. Similarly, in addition to the "virgin" land that MDC originally acquired in Nampamba, subsequent expansions of company operations involved acquiring titled land (Nchanga Farms) from an existing commercial establishment. The purchase of land which is already titled is largely a private arrangement between the buyer and seller. However, even in private transactions, the government often facilitates the acquisition of land by "big" investors, and this was the case with both Kascol and the MDC. In the Kascol case, the government persuaded three farmers who owned the land to sell it to Kascol. Although the original farmers may have been happy with the money they got (as indicated by a grandson of one of the original farmers), it is not clear to what extent the transaction was one between a willing-seller-willing-buyer, given that the government was heavily involved.

Agricultural policy

The overall goal of Zambia's agricultural policy is to promote a self-sustaining, export-led agricultural sector, which ensures increased household income and food security. Agriculture is seen as having the greatest potential in the fight to reduce poverty through its contribution to economic growth, and its inclusive nature. The emphasis on export-orientation implies a shift from the traditional focus on maize to other high value, exportable crops. Historically, Zambia's agricultural policy has favoured maize production. The shift to a more diversified agricultural product base is seen as something beneficial for the sector, as the emphasis on a single crop led to the neglect of infrastructure and service support to other equally rewarding agricultural activities in the sector, and ultimately, to the detriment of the sector as a whole.

The agricultural policy has continued on its path towards market liberalization – begun in the early 1990s – which involved the removal of consumer subsidies on maize and maize products and that of price controls. The policy encourages exports and imports of agricultural commodities and inputs and thus, in the past 15 years or so, the policy thrust has centred around (a) consolidating the liberalization of agricultural marketing; (b) strengthening the liberalization of the trade and pricing policy; and (c) streamlining the land tenure system (PRSP, 2002). Current policy intentions are to: (i) develop and implement policies and programmes that support crop diversification, livestock and fisheries production, increased productivity in crops and livestock, sustainable land and water management, including forestry, agro-forestry, climate change adaptation and mitigation and other environmentally friendly agricultural systems; (ii) facilitate equitable access to land for agricultural purposes; (iii) adhere to predictable, rule-based market and trade policies and strengthen public-private coordination and dialogue; (iv) facilitate the private sector to scale-up investments in production, input and output markets, processing and value addition in crops, livestock and fisheries; extension linkages focusing on Public Private Partnerships (GRZ, 2011).

In spite of these policy pronouncements and intentions, there still seems to be a bias towards maize production, probably understandably so, because maize is the country's staple grain. In 2001/2002, Zambia experienced a severe maize shortage and was offered genetically modified (GMO) maize donations by the WFP, which it rejected out of environmental and health concerns. This move appears to have pushed the government back to maize subsidies through the introduction of the targeted Fertilizer Support Programme (PSP), renamed Farmer Input Support Programme (FISP) in 2009. The maize bumper harvests that have been experienced over the last three consecutive farming seasons may be attributed partly to the FISP. The government also introduced its own Food Reserve Agency which buys maize from small-scale farmers. This was clearly a reaction to the slow pace at which the private sector was filling the vacuum left by the abolishing of government controlled marketing companies at the height of liberalization policies.

Apart from these isolated success stories in agriculture, the country does not yet boast a vibrant agriculture sector. The fact that less than 2 million hectares are under cultivation means that Zambia's agricultural potential has still to be realized and government is making efforts to encourage agricultural investments in these areas. To that end, investment legislation includes a number of general safeguards for investors: free repatriation of net profits and debt payments; safeguards on investment protection (including full compensation based on market value for expropriations); and facilitation services provided by the Zambia Development Agency (e.g. in obtaining water, electric power, transport, and communication services and facilities required for their investments, in regularizing investor immigration status, or in acquiring other licences necessary to operate a business in any particular sector). Tax incentives are also provided, including:

- Implements, machinery and plants used for farming, manufacturing or tourism qualify for a wear-and-tear allowance of 50 percent of the cost per year in the first two years;
- Duty-free importation of most capital equipment for the mining and agriculture sectors;
- Corporation tax at 10 percent on income from farming;
- Farm works allowance of 100 percent of expenditure on stumping, clearing, prevention of soil erosion, boreholes, aerial and geophysical surveys and water conservation;
- Development allowance of 10 percent of the cost of capital expenditure on growing of coffee, banana plants, citrus fruits or similar plants;
- Farm improvement allowance – capital expenditure incurred on farm improvement is allowable for the year when the expenditure is incurred;
- Dividends paid out of farming profits are exempt for the first five years the distributing company commences business;
- For rural enterprises, tax chargeable reduced by one-seventh for the first 5 years; and
- for business enterprises operating in a priority sector under the Zambia Development Agency Act 2006, a 0 percent tax rate for the first 5 years , a rate reduced by 50 percent from years 6 to 8, and a rate reduced by 25 percent from years 9 to 10.

2.2 Recent trends in large-scale agricultural investments

Zambia's economy has traditionally been dependent on mining, especially copper production. However, agriculture has often been given emphasis by successive governments at various points in the history of the country. In the 1970s, state enterprises dominated the Zambian economy and private sector investments played a minimal role. This trend was initiated by the 1968 economic reforms and institutional changes that favoured increasing state control of the economy. The government of then President, Kenneth Kaunda, encouraged Zambians to "go back to the land" and it actively participated in agricultural production and marketing through state-owned farms and the then National Agricultural Marketing Board (NAMBOARD). In this context, the government became the major investor in large-scale agricultural projects.

Both Kascol and the MDC were a consequence of this policy. The major objective for agricultural investments was to "increase agricultural production to achieve self-sufficiency in staple foods, both nationally and regionally where possible, and to provide raw materials for agro-industries" (GRZ, 1979). Because of the need to achieve self-sufficiency in staple foods, the major focus of large-scale agricultural investments was on cereals and livestock production. The MDC and a related company, which Mpongwe acquired after government divestiture – Munkumpu Ipumbu Crop Farm and Kampemba Ranch – are examples of such focus. On the other hand, Kascol is an example of a large-scale agricultural investment aimed at providing raw materials for agro-industries.

The trend of public investments in agricultural land was reversed in the 1990s when Zambia shifted from a characteristically command economy to a market economy. Agricultural policy reforms were undertaken, the main thrust of which was the liberalization of the agricultural sector and the promotion of private sector participation in production and marketing of agricultural inputs and outputs. Enterprises that were fully or partially state-owned were privatized and private investment was encouraged.

The MDC, which was jointly owned by the Government of Zambia and the Commonwealth Development Corporation (CDC), was effectively sold to the CDC by way of increasing its shareholding in the company from 50 percent to 70 percent. The government reduced its shareholding to 30 percent, and this equity was meant to be transferred to the Privatization Trust

Fund (PTF) for subsequent public floatation. This did not happen and by 2005, the CDC had 100 percent ownership of the company. In a surprising turn of events, the MDC went into voluntary liquidation in 2006 and its assets were sold to other companies, the main one being ETC BioEnergy.

The government's equity stake in Kascol was partly through the Development Bank of Zambia, and partly through the Zambia Sugar Company; when the latter was sold to Tate and Lyle in 1995, Kascol was effectively privatized.

The trend of private investments in agriculture has continued in recent years. In the ten-year period starting from 2000 to 2009, total pledged investments in agriculture have been on an upswing, reaching US$315 027 378 from US$8 343 207 according to data from the Zambia Development Agency (see Figure 1).[3]

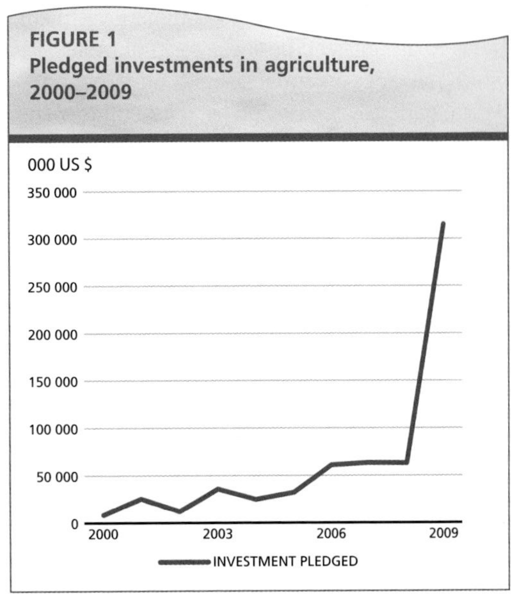

FIGURE 1
Pledged investments in agriculture, 2000–2009

Source: Zambia Development Agency

These figures represent both start-up companies and investments in existing companies, and indications are that the majority of the investments are purchases of existing farms. For instance, the data includes the purchase of MDC assets by ETC BioEnergy in 2007 by an investor of Indian origin, with a pledged investment of US$59 648 687. This was the second largest pledge in the agricultural sector in ten years. It also includes the purchase of Munkumpu Farms, once part of Mpongwe Development Company, by Somawhe Estates Ltd, with a pledged investment of US$14 060 000. Somawhe is owned by a Danish investor.

In 2011, ETC BioEnergy sold its farms and associated assets to Zambeef Products Plc at US$47 390 000. Zambeef Products Plc is a Zambian agribusiness company involved in the production, processing, distribution and retailing of beef, chickens, pork, eggs, milk, dairy products, flour and bread, edible oil and stockfeed through its own retailing network throughout Zambia and West Africa. Could this signify a new trend – where foreign-owned companies are bought by locally-owned companies? It is perhaps too early to make a case out of this one isolated incidence.

What is worth noting, however, is that there are exceptions to the apparent trend of new investments being focused on purchases and expansions of existing farms. In a few cases, completely new farms have been started, although this phenomenon has taken place mostly in the biofuels subsector. Investments were started in earnest a few years ago, driven by the rising prices of fossil fuels. Since 2010, however, investments in biofuel crops, especially jatropha, have almost ground to a halt. The fall in fossil fuel prices that has undoubtedly made investments in jatropha unattractive, is probably one of the reasons why ETC BioEnergy sold Mpongwe farms in 2011. ETC BioEnergy planted 500 hectares of jatropha on an estate that was originally being used to grow coffee, and had plans to expand the area under jatropha to 12 000, which, however, did not materialize.

The preference for investing in existing farms may be explained by various factors, among them the high cost of land clearing for virgin land. It is estimated that it costs about US$900 to clear one hectare of land. At this rate, one would need an investment of close to one million dollars to clear land which is just slightly over a thousand hectares. The other possible reason is that

[3] These figures represent the investments pledged when obtaining investment licences. The amounts actually invested may be different.

investors may be unwilling to commit investment funds to an untested business concept. Actis, which manages an agribusiness fund targeting Africa, generally focuses on established firms and avoids greenfield start-ups.

There could be a third reason: most of the new investments are being made in areas that are generally accessible by road and rail. These are the areas that have traditionally attracted investments in commercial agriculture and even though these areas may have unused land, most such land is owned by commercial entities, even though only small proportions of that land are being used. This gives room for expansion and modernization by injecting money into an already existing establishment. Graham Rae, managing director of Zambezi Ranching & Cropping Ltd is reported to have said: "When we first moved to Zambia, only 100 hectares of Zambezi Ranching & Cropping Ltd were cultivated but we've changed that .... We are now cropping 4 000 hectares with room for expansion" (Armitage, 2011). Zambezi Ranching & Cropping Ltd is one of the recent investments in commercial agriculture. Most commercial farmers in Zambia hold undeveloped/underutilized farm land which offers room for expansion should there be an injection of capital (and perceived product demand, of course). Kascol and Mpongwe farms also have undeveloped land and have thus significant room for expansion.

Another trend appears to be emerging: the government has recognized that one of the key constraints to the growth of commercial farming in the country is poor infrastructure and has come up with a land development programme that identifies and demarcates land and provides basic infrastructures and facilities, such as trunk roads, bridges, electricity, dams, schools and health centres. A number of farm blocks have been identified (Table 1) for that purpose. These farm blocks are essentially virgin land and if the programme succeeds, investments in new farms is likely to increase. Most of the land in the farming blocks is being taken up by Zambian farmers. It remains to be seen to what extent these areas will develop into fully-fledged commercial farms.

Additionally, a large number of Zambians in urban areas have woken up to the prospect of being found landless should the current trend of land purchases by commercial entities continue, and have obtained tracts of land in villages from their chiefs. Currently, most such land is yet to be put on title. The challenge lies, of course, in obtaining the capital to develop such land.

TABLE 1
Farm blocks earmarked for commercial agriculture

Farm block	Province	District	Size (ha)
Nasanga	Central	Serenje	155 000
Kalumwange	Western	Kaoma	100 000
Luena	Luapula	Kawambwa	100 000
Manshya	Northern	Mpika	147 000
Mikelenge/Luma	North-Western	Solwezi	100 000
Musakashi (SADA)	Copper-belt	Mufulira	100 000
Muku	Lusaka	Kafue	100 000
Simango	Southern	Livingstone	100 000
Mwase-Phangwe	Eastern	Lundazi	100 000

3. Design and implementation of the investment projects

3.1 Origin and overview of the businesses

The MDC and Kascol represent two distinct organizational and transactional configurations. The organizational configuration of Kascol combines elements of contract farming, tenancy farming and joint ownership. This configuration has evolved over time, initially starting as contract-tenancy farming and culminating into smallholder outgrower equity ownership. The MDC, on the other hand, is a business with a rural development dimension. It is centred on large plantations and does not involve collaboration with smallholders, or equity ownership by low-income groups. The main development contribution is seen in employment generation in a rural setting, where poverty levels are around 80 percent; in payment of public revenues; in contributing to the food security of its employees and of urban dwellers in the nearby towns on the Copperbelt Province; and, more generally, in opening up the area to investment in agriculture by demonstrating success and by persuading the government to improve infrastructures.

The Kascol and MDC investment projects were each conceived of as both a business opportunity through which shareholders would obtain a reasonable return on their investments, and as a development initiative that would help propel the poor in the respective districts out of poverty. The creation of Kascol was a response to a business opportunity arising from the increased demand for sugar and the derived demand for more cane to feed the sugar-processing factory at the Zambia Sugar Company. The original shareholders that partnered up with the Zambian government, particularly Barclays Bank (a purely private, profit-seeking entity), saw an opportunity for maximizing shareholder wealth. The Commonwealth Development Corporation (CDC) and the Development Bank of Zambia, which were also involved as shareholders, had explicit development objectives. This led to the inclusion of the outgrower scheme in Kascol. MDC was set up to exploit a business opportunity arising from the growing demand for agricultural produce in the country and the region. However, to the initial owners of the company (the Government of Zambia and CDC), Mpongwe Development Company was not only a business opportunity but also a development opportunity, as can be gauged from its name.

The Kascol and MDC investment projects reflect a vision of agricultural modernization through large-scale agricultural enterprises. They were a response to two policy measures: the first was to increase agricultural production to achieve self-sufficiency in staple foods, both nationally and regionally, and provide raw materials for agro-industries; and the second was to create employment and income opportunities in the rural areas in order to counter rural-urban migration (GRZ, Third National Development Plan , 1979). The design of the two investment projects reflects these elements. Both projects are in the agricultural sector and both are rural-based. The central government played a direct role in the formation of both Kascol and the MDC. Firstly, the government was a shareholder in both companies. Secondly, it was involved in the land purchase negotiations in the case of Kascol, where land was purchased from existing farmers. The government also facilitated the allocation of land by Mpongwe chiefs to the MDC. As discussed, both Kascol and the MDC were more recently privatized, reflecting a shift in national policy towards liberalization and privatization. In the case of Kascol, privatization enabled associations of local farmers to acquire equity stakes in the company – as will be discussed below.

3.2 The impact of privatization

The privatization drive started in 1992 and was conducted within the wider context of economic liberalization and against a backdrop of apparent economic decline that started in the 1970s, barely a decade after gaining political independence from the United Kingdom. The main objective of economic liberalization was to arrest this economic decline and privatization was offered as the kingpin of that process. According to Cheelo and Munalula (2005), citing the World Bank, the immediate objectives of privatization were:

1. To scale down the government's direct involvement in the operations of enterprises;
2. To reduce the administrative load associated with this direct involvement;
3. To minimize state bureaucracy in enterprise operations;
4. To reduce the costs of capital expenditure and subsidies from public funds;
5. To promote competition and improve efficiency of enterprise operations;
6. To encourage wide ownership of shares;
7. To promote the growth of capital markets;
8. To stimulate both local and foreign investment;
9. To promote new capital investment.

In that regard, many state-owned enterprises were privatized, including Kaleya Smallholders and Mpongwe Development Company. As stated above, one of the immediate objectives of privatization was to improve efficiency of enterprise operations. This particular goal has generated a lot of interest and in Zambia attempts have been made to measure the impact of privatization on firm performance. One of the most rigorous studies is that by Cheelo and Munalula (2005) who, using panel data, took an econometric approach to measuring the impact of privatization on firm performance. They came up with the following conclusions:

1. There were significant differences in the performance of privatized firms between their pre- and post-privatization periods in terms of improvements in operating efficiency, capital investment (investment in land and building, and investment in plant and machinery).
2. The influence of liberalization was arguably more important in determining turnover and profitability performance than change of ownership (i.e. privatization).
3. Privatization had a negative impact on firm employment levels, at least in the short term.

How do Kascol and the MDC measure up to these conclusions? The impact of privatization on Kascol are not sharply defined because the company was, for all practical purposes, in private hands even before the privatization programme was embarked on. In fact, it was not even on the list of state-owned enterprises that were to be privatized. As stated elsewhere in this chapter, the government had no direct equity stake in Kascol – it was a state enterprise to the extent that two government-controlled entities – the Zambia Sugar Company and the Development Bank of Zambia – had shares in it; 50 percent in total. Thus, when the Zambia Sugar Company was privatized, the effect on the operational management of Kascol was minimal. The situation was, however, different for the MDC. Originally a joint venture between the Government of Zambia (GoZ) and CDC, the company was 60 percent owned by the Government of Zambia just before privatization (Kaunga, undated). After privatization, there was a significant injection of capital into the company and a major expansion programme got underway, about 5 000 hectares were to be cleared, together with major investments in capital equipment.

Indications are that the MDC was one of the most viable agricultural investments in the CDC portfolio. Anecdotal evidence suggests that by 1996-1997, the MDC was more profitable than the then Zambia Consolidated Copper Mines (ZCCM). In terms of employment, there is no evidence to suggest that there was an immediate drop in employment levels following privatization, perhaps because of the expansion programme that the company had embarked on soon after privatization which required an increase in manpower levels. In 1996, a number of young university graduates were recruited (including the author), mostly in middle management positions. In 1998, the MDC was merged with a newly created milling company – Mpongwe Milling – and another CDC-owned farm – Munkumpu-Ipumbu Farm. The merger did not negatively affect employment levels. In the same year, however, it became clear that tough times lay ahead as CDC itself was facing imminent privatization and the financial return threshold for remaining in the CDC portfolio was to be raised.

Following the change in strategy within CDC, a number of assets in the agricultural business were sold off between 2000 to 2003 (Tyler, undated), including Nanga Farms and York Farm in Zambia, but the MDC was unaffected, implying that it was not immediately considered for disposal though plans to do so were still in the offing. The changes within CDC had a ripple effect on the MDC and there was much discontent among employees, some of who thought the company was no longer being well-managed. In 2006, the MDC went into voluntary liquidation and its assets were sold off. Two new companies, ETC BioEnergy and Somawhe Estates Ltd, were now the new owners of the farms. Even though ETC BioEnergy pledged to inject capital in the company, employment levels were significantly reduced.

With regard to the impact of privatization on firm performance, it is admittedly difficult to separate the impact of privatization from that of market liberalization. To say that privatization had an impact on company performance is equivalent to attributing performance to the form of ownership and control. While a causal relationship might exist, no study of the Zambian experience has successfully measured that relationship. That said, a World Bank Post-Privatization Study on observed that "performance by companies purchased through pre-emptive rights sales (usually to foreign investors holding minority shares and a management contract) was unaffected by privatization" (Serlemitsos & Fusco, 2003, p. 6). This, apparently, was the case for both the Mpongwe Development Company and the Kaleya Smallholder Company.

There is a significant possibility that liberalization in general had a greater impact on company performance than privatization on the Mpongwe Development Company, at least in the short term. For instance, an orientation towards the export market had a positive effect on company performance (Serlemitsos & Fusco, 2003). On the other hand, liberalization, by opening up the economy to increased competition, negatively affected the performance of many of the privatized companies that were totally dependent on the local market.

3.3 The economic inclusion of local, low-income people in the investment projects

This section discusses the economic inclusion of low-income people in the investment projects. The discussion is centred on the concept of inclusive business models.

According to the UNDP (2010), an inclusive business model includes "people with low incomes on the demand side as clients and customers, and/or on the supply side as employees, producers and business owners at various points in the value chain" (UNDP, 2010). The goal of an inclusive business is neither philanthropic nor pure Corporate Social Responsibility (CSR), but pursuance of a business opportunity in a low-income market in such a way as to meaningfully provide tangible benefits to the low-income sections of society, while making sufficient returns to justify the investment. It is helpful to assess the degree and quality of inclusion of low-income groups in a business by considering four factors: ownership (that is, ownership of the business and control over key assets like land or processing facilities), voice (that is, participation in the management of the enterprise), risk (the sharing of production, marketing and other risks), and reward (the distribution of the costs and benefits generated by the project (Vermeulen and Cotula, 2010).

Ownership

The ownership structures of both Kascol and the MDC present similarities and differences. Both companies started with similar owners. Kascol was originally owned by the Government of Zambia (through Zambia Sugar Company, a then state-owned enterprise which owned 25 percent shares in Kascol; and through the Development Bank of Zambia, a development finance institution established in the early 1970s by an Act of Parliament), CDC and Barclays Bank. The MDC was owned by the Government of Zambia and CDC, each having a 50 percent share in the early stages of company's development. As discussed, the substantial government involvement reflects the prevailing government policy at the time the companies were established.

The CDC's involvement shows that both companies were, to some extent, formed to contribute to national economic growth and poverty reduction. The Commonwealth Development Corporation (CDC) is the development finance institution (DFI) of the Government of the United Kingdom, and is currently established as a company owned by the Department for International Development (DFiD). The CDC has had an economic development agenda for poor countries since its establishment in 1948. One of its objectives at the time of investments, it would seem, was to ultimately divest itself of its investments in the companies once they had matured enough to serve their development role without the CDC's continued support. For instance, in a 1995 press release by the Zambia Privatization Agency announcing the acquisition of Munkumpu Farm and Kampemba Ranch by the CDC, this objective was explicitly stated (ZPA, undated). The CDC eventually divested from both Kascol and the MDC.

Fieldwork for this study suggests that the understanding by Kascol outgrowers was that CDC shares would ultimately be sold or perhaps even donated to them. This was perhaps a misinterpretation by outgrowers of CDC's intention to divest from the company once fully established. When CDC eventually divested from Kascol, it sold its shares at market price. The outgrowers could only afford to buy a relatively small percentage of shares (13 percent) through a bank loan. The significance of this, however, is that outgrowers engaged in the outgrower scheme have an equity stake in Kascol. This is an important difference compared to MDC, where people with low incomes have no equity stake in the company.

Kascol outgrowers own their shares through a Trust – the Kaleya Smallholders Trust. This Trust is part of a consortium known as View Point Investment Holdings. In addition to the trust, the consortium also includes two companies – Nzimbe Ltd and Kascol Consultants. Collectively, the three members of the consortium hold 50 percent of the shares in Kascol, which were sold by CDC and Barclays Bank. The Development Bank of Zambia has maintained its 25 percent equity stake.

The remaining 25 percent of Kascol shares are held by the Mazabuka Cane Growers Association, an association of cane growers who supply cane to Zambia Sugar Company Plc. The Mazabuka Cane Growers Association assists cane growers in Mazabuka District to improve cane production and productivity. It acquired its equity stake in Kascol through a donation from the Zambia Sugar Company Plc, which previously owned this equity stake. The donation was probably intended to ensure close collaboration with cane suppliers and thus assure continuity of supply of cane to the sugar factory. The effect of the 25 percent stake held by the Mazabuka Cane Growers Association and of the 13 percent stake held by the Kaleya Smallholders Trust is that bodies representing local farmers own a substantial share of Kascol. The demand for cane by the Zambia Sugar factory is big enough to take in all the cane supplied by the cane growers in Mazabuka and as such, relations among suppliers (for instance, between Kascol outgrowers and other cane growers in the district) are virtually non-competitive. However, having a single buyer of cane may work to the disadvantage of the cane suppliers (going by Michael Porter's oft-cited model of competitive forces).

As noted above, another important difference between Kascol and the MDC is that the former but not the latter involves an outgrower scheme. Of the 4 314.9 hectares of land held by Kascol through a long-term lease, about 1 000 hectares are subleased to some 160 outgrowers (contract farmers) on the basis of 14-year rental contracts. Each outgrower has, on average, 6.5 hectares of land. Only about half (50 percent) of the land leased to Kascol is under cultivation. Part of the land is rocky and therefore not suitable for cultivation. More importantly, Kascol sources suggest that irrigation water availability represents the main limit on how much land can be brought under cultivation. Processing facilities (a sugar processing plant located within a 10 km radius of the farm) are owned to Zambia Sugar Company, to which Kascol sells its entire produce.

Voice

Participation in the management of an enterprise is related to the amount of control and influence that an individual or group has on the strategic and/or operational decisions of the enterprise. The amount of control that outgrowers and other low-income people have in an enterprise is related to the role they play in the value chain – whether they are shareholders, suppliers/producers, employees or customers, and how critical the company perceives their role to be in its survival.

At Kascol, low-income groups participate in the value creation process as shareholders, suppliers/producers and employees. They are represented at the board level, where the chairperson of Kaleya Smallholder Farmers Association is a member. The board is the highest decision-making organ of Kascol. The Kaleya Smallholder Farmers Association is a producer association which seeks to assist farmers in issues of production and outgrower welfare. Sources from among the outgrowers indicated that their representation on the board was not as effective as they would have liked, particularly in matters regarding the sharing of rewards from cane production. The outgrowers are of the view that the 55 percent of the outgrower gross sales which goes to the company represents more than its fair share of the proceeds. Since the Mazabuka Cane Growers Association is also represented on the Kascol board, the interests of outgrowers would be expected to be taken care of in an effective way. But this may not necessarily be the case, as the other members of the Mazabuka Cane Growers Association are independent commercial farmers who deal directly with the Zambia Sugar Company (the buyer). It is possible that the board representative from the Mazabuka Cane Growers Association is more inclined towards the interests of bigger commercial cane farmers than those of smallholder outgrowers. The smallholder outgrowers do not deal directly with the Zambia Sugar Company because the transactions costs would be higher on the part of Zambia Sugar Company if it opted for that approach, and also because the smallholder outgrowers produce on contract with Kascol.

Apart from board representation, smallholder outgrowers are not ordinarily involved in the day-to-day running of Kascol, except during the tendering process, when the company is deciding to procure major inputs for use on smallholder outgrower farms. To facilitate collaboration between smallholder outgrowers and company management on a more regular basis than would be warranted by the involvement of the board member, a Smallholder Relations Officer, who is a fulltime employee of Kascol, is engaged for that purpose. He is the link between smallholders and the company.

In contrast, low-income people at the MDC are not represented on the company's board and they barely participate in determining the direction of the company. Their role is confined to that of employees, and virtually all of them hold non-management positions.

Thus, it would appear that the low-income people have a greater voice in Kascol, where they participate as producers, shareholders and employees, than in MDC where they only participate as employees. Even though the low-income groups at Kascol are not directly involved in the day-to-day management of the company, they exert an influence in the choice of senior managers due to their representation at board level, taking into account, however, the limitations noted above on the extent of the effectiveness of this representation. In addition, Kascol is heavily dependent on outgrowers, as these produce close to 50 percent of the sugar cane that the company sells to the Zambia Sugar Company. This circumstance would be expected to increase the leverage of the outgrowers.

Risk

Risk relates to possibility of loss of assets or income-earning potential. Clearly, this is in turn related to the contribution each individual makes towards the value creation process and the value of the rewards derived from participating in value-creation. Again, the contribution in the value-creation process is dependent on the role each person plays in the value-chain – that is, whether a shareholder, supplier/producer, employee or customer. In Kascol, low-income groups participate as producers, shareholders and employees. Therefore, the risk they bear is higher than in the case of MDC, where they only participate as employees. In Kascol, smallholder outgrowers face the risk of loss of assets and income-earning potential. In MDC, low-income groups face the risk of loss of income earned through wage employment. The severity of loss (and the probability of it occurring), influences the decision whether or not to put in place a risk management system. At Kascol, where the severity of loss seems higher, smallholder outgrowers, in conjunction with Kascol management, have put in place a risk management system: crop insurance. Kascol has taken out a single crop insurance policy for the whole sugar estate (which includes the fields of smallholder outgrowers), in its name. The smallholder outgrowers pay for the insurance by allowing Kascol to deduct a small percentage of money from their cane sale proceeds. In the MDC, there is no such insurance designed specifically to protect the assets of the low-income categories, except the legal requirement of facilitating the remittance of statutory contributions to employee pension schemes, which Kascol also does. According to the Pension Scheme Regulation (Amendment) Act number 27 of 2005, a pension scheme means "any scheme or arrangement, other than a contract for life insurance, whether established by a written law for the time being in force or by any other instrument, under which persons are entitled to benefits in the form of payments, determined by age, length of service, amount of earnings or otherwise and payment primarily upon retirement, or upon death, termination of service, or upon the occurrence of such other event as may be specified in such written law or other instrument." The pension schemes are thus safeguards against loss of income arising from retirement, job loss or death (in which case the beneficiary will be the surviving relative(s)). They do not safeguard against reduction in wages. The MDC has also taken out crop insurance policies.

Rewards

In an equitable system, rewards, like risks, are related to the individual's contribution in the value creation process. In practice, the value of economic rewards depends on various factors, including the forces of demand and supply for the factors of production contributed and the products produced. The Kascol investment project is an interesting case in that low-income groups participate as shareholders, producers and employees. The annual incomes of outgrowers are generally higher than those of their counterparts who are engaged as employees. On average, an outgrower obtains a net income of up to ZMK15 000 000 (US$3 167.40) per year from a good harvest of cane, while the average annual wage income of a unionized employee at Kascol is currently around ZMK3 657 120 (US$772.24).[4] In addition, through their equity participation, the outgrowers are entitled to a dividend whenever the company declares one. So far, the dividends have been used to pay back the loan obtained from a commercial bank to buy shares from CDC and Barclays Bank. The loan is likely to be cleared within three years.

It would also appear that outgrowers are wealthier than employees. The researchers' observations were that outgrowers had more assets (some even had cars) than those who were working for a salary. This excludes those in management positions. Apart from tangible economic rewards, psychological rewards also seem important and these depend on an individual's perceptions and values. Those individuals who value independence would rather be outgrowers producing for the company, than employees. Currently, no outgrower is also a Kascol employee. Some of the outgrowers were former employees who chose to become outgrowers. Those who remain in employment are, presumably, individuals who prefer the certainty and regularity of wage income – or people who cannot afford to become outgrowers.

3.4 Constraints and success factors

The Kascol and MDC investment projects have faced some factors that have restricted their

[4] The research team could not obtain salary scales for non-unionized permanent workers who, in both companies, are regarded as management staff. Those interviewed could not disclose this information.

operations and others that have accounted for their successes. Some of these factors are now described, beginning with constraining factors and then moving on to success factors.

Low levels of financial returns

While investments in the agribusiness may be profitable, returns on investment are generally low compared to other sectors of the economy. Low levels of financial returns in the agribusiness sector are what motivated CDC's divestiture from the MDC and many other agribusinesses in Africa. The case of low returns in the agribusiness sector has long been recognized by investors. Tyler (undated) , quoting from the 1972 CDC annual report, states that *"many agricultural projects, particularly involving smallholders... have had to be ruled out in the past because ... the overall rate of return is well below that necessary to cover the service of the capital invested"*. This makes it a lot harder for the sector to attract private sector investors. Tyler (ibid), quoting from the 2000 CDC annual report quotes the Chairman of CDC as having said: *"It was with considerable reluctance that the board concluded that many of our agribusiness investments, with which CDC has been proudly associated throughout its history, are unlikely to meet our minimum financial return requirements. We have therefore substantially written down the values attributed to them, to reflect a 'for sale' rather than 'going concern' status."* Gauging from two of the biggest and most successful agribusinesses in Zambia – Zambia Sugar and Zambeef – which, incidentally, are associated with the present case studies, the average return on investment for agribusinesses in Zambia could be around 10 percent (Table 2). The average return on net assets for ZAMBEEF, the new owners of Mpongwe Farms, was 10.3 percent for the years 2010 and 2011. Similarly, the average return for Zambia Sugar PLC – the single buyer of Kascol of cane – was 10 percent for the years 2010 and 2009. Note the contrast in the average return for Arcades Development, a newly established shopping mall in Lusaka, whose average for 2010 and 2009 was 27 percent.

That said however, the individual, yearly returns are very disparate from those of other businesses. The returns, for instance, compare favourably with the returns on the CDC portfolio for the years 2007 and 2006 which posted returns on net assets of 14 percent and 12 percent, respectively. It would appear, though, that returns from agribusinesses are subject to wide variations, probably a reflection of the sensitivity of agribusiness, especially agricultural commodities, to both local and global economic conditions.

TABLE 2
Return on net assets for selected companies

Company	Return on net assets		
	2010	2009	Average
		%	
Zambia Sugar Plc	13.1	7.5	10.3
Zambeef Products Plc	14	6	10
Arcades Development Plc	45	8	27

Source: Computed from company financial statements

There is, however, a likelihood of a sustained rise in returns in agribusiness given the rising trend in global food prices. Increased demand for food is expected to increase even as world population increases and countries like Zambia that still have arable land may reap high returns from investments in agriculture.

High operating costs

As noted elsewhere in this chapter, Zambia has vast tracts of arable land which are not being utilized (except for gathering wild fruits and other forest products), but poor infrastructures in most parts of the country where this land is found makes agriculture investments an expensive venture. Poor transport and communications infrastructures, the absence of commercial services such as banking and suppliers of inputs, materials and operating requirements significantly increase operating costs. Additionally, investors who choose to invest in such places are often forced to make additional investments in assets that are not directly related to their core business. The MDC, for example, was obliged to invest in road maintenance and telecommunications equipment and such investments have a depressing effect on returns. Further, in the

1990s, the MDC had high vehicle maintenance costs due to the bad road leading to the nearest town, Luanshya.

Imbalance of power relations in the cane supply chain

The case of Kascol is that of many suppliers and a single buyer. There is only one buyer of cane in Mazabuka. This has created an imbalance of power relations in the supply chain. The suppliers of cane are dependent on Zambia Sugar PLC, which can therefore dictate the price of cane supplied by cane growers. Outgrowers at Kascol complained that they are told the price at which their cane will be sold and have no say in the setting of that price.

In spite of these constraints, there are many factors that induce investors to continue investing in both the Kascol and Mpongwe Farms. The following are some of the key ones.

Favourable investment climate

The favourable investment environment created since the early 1990s, when Zambia liberalized its economy, has made continued investment in the case study businesses an attractive option. The section on Recent Trends in Large Scale Agriculture Investments outlines some of the factors that have attracted investors to the Zambian Agriculture sector. Following liberalization of the economy, the MDC was able to set its own price for maize and soybeans and to export its commodities and earn income in foreign currency at a time when the base lending rates in Zambian currency were over 50 percent. Its international customers included Glencore International plc, and Otterbea; in 1996, the company sold all of its 3 788 tonnes of soya crop to Otterbea of South Africa.

Comparatively low transaction costs in the outgrower scheme

The oft-stated downside of contract farming systems involving a large number of small outgrowers are the high transaction costs. Kascol, however, has managed to keep these costs at manageable levels because 1) the outgrowers are geographically concentrated on one farm which makes it easier to provide them with extension services and to supervise their farming activities; 2) the outgrowers use Kascol land, and this increases compliance levels in contractual matters as the cost of eviction from company land is high on the part of the outgrower should they abrogate the contract; and 3) outgrowers do not have alternative buyers of cane and this prevents from side-selling their cane.

4. Socio-economic outcomes

While the socio-economic outcomes of the two projects are difficult to determine due to the problem of attribution and to constrained data availability, there is evidence that the companies have had some impact, both positive and negative, on the poor and their environment. The following section takes into account the situation prior to the investments and then contrasts this with the socio-economic outcomes of the investments.

4.1 The situation prior to the investments

It is impossible to produce a proper assessment of the development context in the project areas at the time of project inception. Too much time has passed since then, and data on key socio-economic indicators is in scarce supply. However, it is possible to make some general observations.

In the mid-1970s, when the negotiations for the acquisitions of land for the agricultural investments were taking place, Zambia was a young nation. It had only been an independent state for ten years. One of the major challenges faced was human capital. The country had very few schools: by 1976, it had 2 743 primary schools (most of these just went up to the fourth grade) and 121 secondary schools. College education was scarce. In 1976, Zambia had 13 teacher training colleges, 14 technical and vocational training colleges and 1 university which had opened ten years earlier, in 1966. Mpongwe District (then part of Ndola Rural District) at that time had about 9 primary schools while Mazabuka District had about 39 schools (GRZ, EdAssist Database, 2002). However, compared to Mpongwe and many other rural districts, Mazabuka District (where Kascol is located) was much better off. Mazabuka's location along the main (and the country's first) railway line, attracted missionaries and white settlers (mainly as commercial farmers) and, as such, the Kascol project catchment area benefited from earlier investments in human resource projects undertaken by the government, missionaries and commercial establishments (e.g. the Zambia Sugar Company). By 1980, a number of schools existed in the project catchment area at both primary and secondary levels. These included Mazabuka Basic School, Saint Columbus Primary School, Kaonga Primary School, Saint Edmonds Secondary School, and Mazabuka Girls High School. Vocational training centres, however, were very few, and perhaps only the Zambia Institute for Animal Husbandry was found within the project area. A farmer-training centre that catered to the needs of the entire Mazabuka district existed about 60 km away, but the distance meant that it was of limited use to the farmers in the project area.

The vast majority of the rural population in the project catchment areas probably made a living through subsistence farming and herding. To date, about 90 percent of Zambians who live in rural areas derive their livelihoods from agriculture (CSO, 2003). Only 6.3 percent of the rural population is in paid employment, implying that 93.7 percent are most likely engaged in independent agricultural activities, whether for subsistence or for commercial production.

While formal employment opportunities existed for work in government offices (both central and local) and in parastatals that were being established by the government in the first decade after independence, few rural people, who made up the majority of residents in the catchment areas of the two case studies, could get such jobs, due to lack of education and training. Most, therefore, could only work as 'labourers', a term used to refer to unskilled labour in Zambia. In Mazabuka District, where Kascol is located, some farming enterprises, notably the Zambia Sugar Company, had just been established. These enterprises provided whatever type of employment the local residents could pick up (cane cutting, office cleaning, etc.).

FIGURE 2
Wage employment in Zambia, 1972–1980

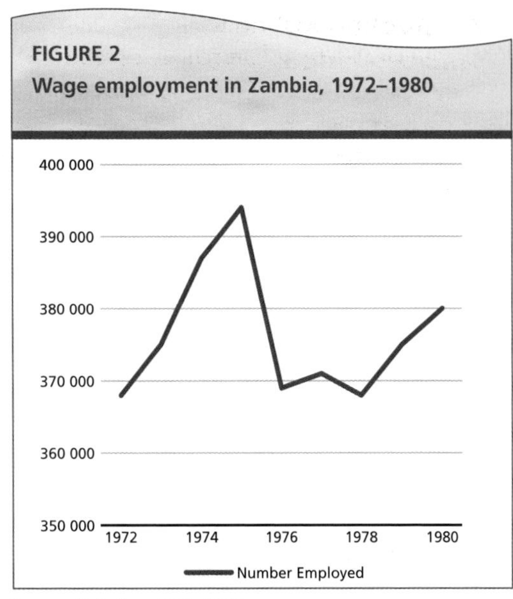

Source: Data from CSO, 1986

Management and other jobs requiring technical expertise were mostly in the hands of expatriates. Mpongwe District had even fewer, if any, opportunities for wage employment.

The employment situation was particularly bad in the second decade after independence, because although Zambia recorded a growth in wage employment due to growth in the economy in the first ten years after independence (1964–1973), from the mid-1970s growth in wage employment grew less than growth in the labour force (CSO, 1986) (see Figure 2). This period marked the beginning of Zambia's economic decline, largely due to the decline in copper prices on the world market, increased oil prices and policy mishaps.

4.2. Direct livelihood contributions

According to Scoones (1998), "a livelihood comprises the capabilities, assets (including both material and social resources) and activities required for a means of living". The same author adds further that "a livelihood is sustainable when it can cope with and recover from stresses and shocks, maintain or enhance its capabilities and assets, while not undermining the natural resource base". The terms 'livelihoods' and 'employment' are inexorably linked and according to the Zambia Labour Force Survey (1986) a "person is employed if he performs some work for pay, profit or family gain", and this includes subsistence farming. Thus, employment can include a wage for the employed person or independent production of a product for direct consumption or market sales (Scoones, 1998).

Wage employment

One of the policy objectives at the time Kascol and MDC were being set up was to create employment and income opportunities in the rural areas; both companies were seen as an operationalization of this policy. Today, Kascol is one of the largest employers in Mazabuka District and Mpongwe is the biggest employer in Mpongwe District. When the project started in the 1980s, Kascol had over 300 permanent staff – both management and non-management. By 1999, the number had dropped to 78 and this has been maintained to the present. The drop in the number of employees came with changes in ownership structure. The company underwent a restructuring process which led to over half of the employees being laid off and salaries being reduced for those who remained. Apart from those who work directly on the farm, Kascol also runs a clinic which employs four staff. Besides permanent staff, Kascol employs between 250 and 350 seasonal workers, mostly cane cutters. These seasonal workers are employed for up to eight months in a year and, as such, receive an income for two-thirds of the year. The ratio of workers to area cultivated in hectares is about 0.38, or 38 workers per 100 hectares.

Between 2004 and 2007, MDC employed an annual average of 457 full time workers and 1 082 seasonal workers – a combined average of 1 539 workers per annum. This figure translates to roughly 38 percent of all salaried employees in the district in the 2000s.[5] The figure in terms of workers per cultivated land in hectares is 0.14 or 14 workers per 100 hectares, significantly lower than that of Kascol. The difference is most likely a reflection of the difference in the levels of mechanization in the two companies: the MDC is more highly mechanized than Kascol. At the time of interviews (July 2011), the company had 520 permanent staff and 1 200 seasonal workers.

These figures are certainly low by international standards, though still significant within the Zambian context. A recent survey of Zambian businesses sponsored by the World Bank and others (Clarke *et al*, 2010) found that "even large Zambian enterprises are small by international standards. Close to half have between 51 and 70 employees – just above the notional cut off size of 50 employees for medium-sized enterprises". Thus, a company like MDC, which employs over 500 persons in a year, is, by Zambian standards, a large employer. This is especially true if such a company is in a rural area – as in the case of the two businesses studied.

That said, the level of job creation appears small compared to the size of the rural labour force. In 2000, Zambia had a total labour force of 3 165 151 persons, of which 2 151 776 were in rural areas (CSO, 2003). The rural labour force was 64 percent of the total labour force, implying that Zambia's labour force was (and still is) predominantly rural. Of the rural labour force, only 6.3 percent (about 133 205 persons) were in wage employment.

In 2005 (the year for which comparable national data on wages is available), unionized/ general workers at Kascol were getting a minimum of ZMK704 711 (US$206.28) per month, while seasonal workers were getting a minimum of ZMK15 816 (US$4.63) per day.[6] In the same year, the average wage of an unskilled, unionized worker at MDC was ZMK300 000 (US$87.81). For instance, Brian, who in 2005 was working as a centre pivot irrigation equipment operator at the MDC, was earning ZMK370 000 (US$108.30) per month, while Peter who worked as a security guard was earning ZMK300 000 (US$87.81) per month. These figures are higher than the national average figure of ZMK293 621

[5] This figure has been estimated by taking into account a population of 64 371 in 2000 for Mpongwe District (CSO, 2000) and a wage employment rate of 6.3 percent in rural Zambia, according to 2000 national census.

[6] Exchange rate of ZMK3 416.34 to a 1 US$ (Bank of Zambia, average 2005 rate).

(US$85.95) per month for all salaried workers in 2005, and much higher than the national average figure of ZMK105 426 (US$30.86) for those employed in the agricultural sector (CSO, 2005), though farm labourers would be expected to have lower pay. Currently, unionized workers at Kascol get a minimum of ZMK15 238 (US$3.22) per day, which is about ZMK304 760 (US$64.35), less than half what they used to get in 2005.[7] Management explained that this major reduction in salaries was due to a move from an enterprise-based salary scale to an industry-wide scale. An enterprise-based salary scale is negotiated at an enterprise level, while an industry-wide salary scale is negotiated and set at an industry level. Similar salaries are paid at the MDC (now Zambeef Mpongwe Farms), where unionized workers get a minimum of ZMK419 000 (US$88.48) per month. This figure is the legal minimum wage obtaining in the country. Based on the above figures, it would seem that Kascol is paying less than the legal minimum wage for staff in non-management positions. As noted elsewhere in this chapter, those interviewed could not disclose the salaries of management staff.

Who gets the jobs

For both Kascol and MDC, management jobs are predominantly held by individuals who come from outside the surrounding communities, while low-skills jobs are held by locals. In the case of seasonal labour, the demand for labour may be too high for the surrounding communities to meet and the two companies have often obtained such labour from outside their respective districts. At Kascol, most of the seasonal workers (cane cutters) are migrant labourers from other parts of the country, the Western Province in particular. This has been the trend from project inception, as locals (Mazabuka residents) who are traditionally cattle keepers, have often viewed cane cutting as an unattractive employment option. With the depletion of cattle stocks due to disease outbreaks, however, a number of locals have also started seeking seasonal employment as cane cutters. Similarly at MDC, during coffee picking years, the company used to hire coffee pickers from other districts. In 2011, the company had 121 locals in permanent employment (out of the 518) and most of the seasonal workers are locals (Mpongwe community members).

From a gender perspective, a parallel study on gender in large-scale agricultural investment, also supported by FAO and steered by IIED, suggests that Kascol does not have a deliberate policy for affirmative action or promoting gender equity in decision-making positions, though some efforts are being made to encourage women applicants once a vacancy arises (Wonani, in press). Thus, out of 78 permanent employees, only 8 (10 percent) are female. Table 3 below shows the positions held by female staff in the Kascol. The situation is slightly better with regard to outgrowers, as 27 percent of the smallholder outgrowers are female, though it is possible that the larger proportion of female smallholder outgrowers is linked to inheritances arising from the demise of the original male smallholder outgrowers.

Livelihood opportunities other than direct employment

As discussed, Kascol has contracted 160 outgrowers who supply the company. These outgrowers have annual sales of up to ZMK60 000 000 (about US$12 669.6) each. After deducting the cost of fertilizers and chemicals (average of ZMK26 000 000, US$5 490.29), water (ZMK12 000 000, or US$2 533.98, on average), transport and other,

TABLE 3
Numbers and positions of female staff in Kascol

Position	Number of staff
Management Accounting Officer	1
Agriculture Management Trainee	1
Human Resources Assistant	1
Environmental Health technologist	1
Nurse	1
Secretary	2
Zone Leader	1

Source: Wonani, 2012

[7] Exchange rate of ZMK4 735.74 to a 1 US$ (Bank of Zambia, average 2010 rate).

the outgrower would remain with something like ZMK15 000 000 (US$3 167.40). Income from the sale of sugar cane is variable and when the harvest is poor, or when there price is unfavourable, incomes can be much lower than stated here. In a situation where an outgrower fails to get enough income to sustain him/her for the rest of year, Kascol gives an "advance" (that is, a loan) of ZMK1 000 000 (US$211) to the farmer. The price of sugar cane is set by the Zambia Sugar Company, which takes into account several factors including the demand for sugar on the local and international market.

Outgrowers also receive dividends from their equity stake in the company, although as discussed, this revenue stream has so far been used to repay a bank loan. As already mentioned, outgrowers tend to have higher incomes, better living conditions and greater self-satisfaction than wage labourers employed by Kascol. Besides income from sugar sales, each outgrower has about 0.5 hectares of farmland used for residence and for staple crop (maize) production for household consumption. Kascol's mode of operation also impacts the livelihoods of local suppliers who provide agricultural inputs and operating materials directly to Kascol, and those who supply to the Zambia Sugar Company factory which mills Kascol cane into sugar. There are more men (73 percent) than women (27 percent) among the 160 outgrowers. Older farmers appear to dominate; the few youths that are involved are heirs who have taken over the estate from parents who have either died or are too old to farm.

In the case of the MDC, the main direct livelihood contribution is through wage employment. Other avenues through which the MDC may have positively impacted on the livelihoods of local communities appear limited. Virtually all its inputs, other than unskilled labour, are sourced from outside Mpongwe District, and this implies that the low-income members of the surrounding communities do not participate in the value chain as suppliers of inputs. Additionally, all its products are sold to corporate customers outside Mpongwe District, so that the low-income members of the surrounding communities do not participate in the value chain as consumers of affordable products.

The comparison between the Kascol and MDC models suggests that a model in which low-income groups participate more in value creation (as shareholders and suppliers, as well as labourers) offers more potential for local livelihoods than models that are mainly centred on wage employment.

Productivity, technology transfer and skills development

Both businesses offer training, but only to those members of the communities that are directly involved in the production activities of the businesses. Kascol provides training to outgrowers in areas such as cane production, farm management and good citizenship. Workshops and seminars are also held on matters relating to the health of outgrowers (e.g. HIV/AIDS). The MDC/ETC BioEnergy only provides training to its workers. However, the company believes that some transfer of expertise takes place by employing local farmers as seasonal workers. It was not possible for this study to verify this claim. However, earlier research by the study author in the same areas found that smallholders who obtain occasional employment with the company tend to have a higher productivity than those who do not (FinScope MSME Study, 2008).

4.3 Impact on food security

The impact of the cases studied on food security is difficult to quantify in the absence of data on production and price trends. However, since two of the key elements of food security are the availability of, and access to, food, it can be argued that both projects have had significant impacts on the food security situation of employees, outgrowers and urban dwellers. As noted elsewhere in this chapter, both projects have enabled sections of the low income groups to earn cash incomes. These incomes are used to buy food – and this constitutes access, which is an element of food security. The MDC's contribution to food security in the nearby towns on the Copperbelt Province is significant. Though data on the MDC's maize production is unavailable, we can gauge its contribution by considering aggregate production data at the district level. In 1996/97, for instance, Ndola Rural District, where the MDC is located,[8] contributed about 80 percent of the total maize produced on the Copperbelt (Table 4). In other words, one district produced virtually all the maize crop in the province.

Maize production estimates for 2004/2005 makes the contribution of MDC even clearer as the data is disaggregated by farmer category. The large-scale farmer category in Mpongwe District's contribution to total maize production on the Copperbelt was 45 percent and this was the highest production level in the region (see Table 5).

4.4. Public revenues and public infrastructure development

Kascol pays land fees (ground rent) to the Ministry of Lands. On average, the company pays ZMK130 million (about US$27 450.83 at an exchange rate of US$1 to ZMK4 735.74). The annual tax to the government by Kascol is about ZMK500 million (about US$105 580.12). The annual contribution to Water Board by Kascol is ZMK 68 million (US$ 14 358.90). Payments for ground rent, water rights and tax go the central government. The local authority receives about ZMK40 million (about US$8 446.41) for rates and billboards. Councils charge a fee for advertisements and other information displays by organizations on billboards in their districts. A breakdown of public revenues provided by Kascol in 2010 is presented in Figure 3.

ETC BioEnergy, on the other hand, has an agreement with the Government of the Republic of Zambia – the Investment Protection and Protocol Agreement – whereby the company enjoys a tax holiday. This means that for the first three years of investment in Zambia (starting in 2007), ETC BioEnergy has been exempted from paying tax. ETC has, however, been paying ground rent and the latest figures are as shown in Table 6.

ETC BioEnergy also pays annual water rights and the 2010 figure was ZMK43 442 000 (US$9 173.22). Figure 4 shows that the biggest proportion of ETC BioEnergy's payments is for ground rent. The volume of water utilised by ETC is 102 000 m^3 / day. The cost of the water rights is calculated on the basis of water volumes as follows:

- Up to 500m^3 per day: ZMK5 000 (US$1.06) per day;
- For every m^3 above 500m^3/day: ZMK2 (US$0.0004) per m^3;
- Charge for registration: ZMK2 000 (US$0.42).

In comparison, Kascol, although the smaller of the two projects, contributes more to public

[8] Since 2000, following a change in district names and boundaries, the MDC lies in Mpongwe District, as discussed throughout the chapter.

TABLE 4
Maize production estimates, Copperbelt Province, 1996/97

District	Expected production (mt)	% Production
Chililabombwe	2 876.76	5
Chingola	1 338.12	3
Kalulushi	1 725.21	3
Kitwe	496.08	1
Luanshya	2 006.28	4
Mufulira	2 249.28	4
Ndola Rural	42 410.34	80
Total	53 102.07	100

Data Source: CSO Crop Forecast Surveys

TABLE 5
Maize production estimates, Copperbelt Province, 2004/2005

District	Farmer category	Expected production (mt)	% Production
Chingola	Large-Scale	9 708.89	9
Chingola	Small-Scale & Medium	1 081.04	1
Kalulushi	Large-Scale	770.35	1
Kalulushi	Small-Scale & Medium	1 524.24	1
Kitwe	Large-Scale	453.86	0
Kitwe	Small-Scale & Medium	1 743.99	2
Luanshya	Large-Scale	68.4	0
Luanshya	Small-Scale & Medium	6 741.56	6
Masaiti	Large-Scale	2 765.93	2
Masaiti	Small-Scale & Medium	8 094.12	7
Mpongwe	Large-Scale	51 761.8	45
Mpongwe	Small-Scale & Medium	14 224.89	12
Mufulira	Large-Scale	307.8	0
Mufulira	Small-Scale & Medium	1 534.44	1
Ndola	Large-Scale	510.2	0
Ndola Urban	Small-Scale & Medium	6 379.9	6
Total		**113 943.25**	**100**

Data Source: CSO Crop Forecast Surveys

FIGURE 3
Kascol's contribution to public revenues (2010)

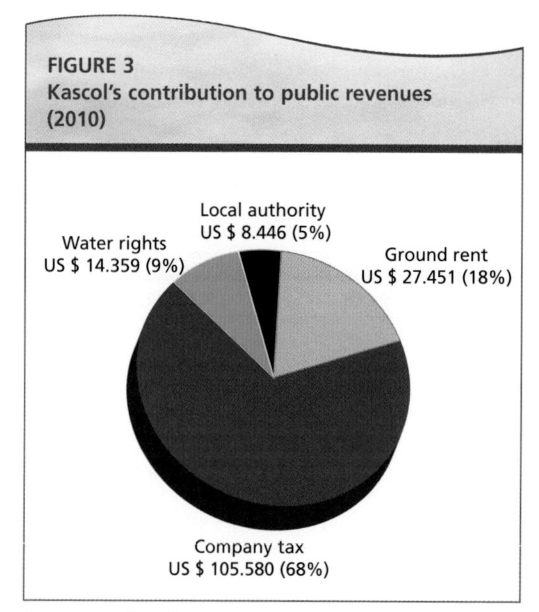

Data source: Kascol

FIGURE 4
ETC bioenergy's contribution to public revenues (2010)

Water rights
US $ 9.173 (16%)

Ground Rent
US $ 47.184 (84%)

Data source: Kascol

TABLE 6
Annual ground rent payments for ETC Bioenergy (2010)

Name of farm	Ground rate (in ZMK)	US$
Chambatata (4450)	61 044 860	12 890.25
Nampamba (4451)	105 044 860	22 181.30
Kampamba (5388)	57 359 860	12 112.12
Total	223 449 580	47 183.67

Data source: ETC BioEnergy

TABLE 7
A comparison of Kascol and ETC contributions to public revenue (2010)

	Ground rent	Corp. tax	Water rights	Local authority	Total public revenue (US$)	Ratios	
						Public revenue to developed land	Public revenue to total land
Kascol	27 450.83	105 580.1	143 58.9	8 446.41	155 836.26	68.79/ha	36.12/ha
ETC	47 183.67		9 172.22		56 355.89	5.29/ha	1.20/ha

Data source: Kascol and ETC Bioenergy

revenue than ETC BioEnergy, on the basis of the 2010 figures. In 2010, Kascol contributed a total of US$155 836.26, compared to US$5 355.89 contributed by ETC BioEnergy. For every hectare developed, Kascol contributed US$68.79 to public revenues, compare to ETC BioEnergy's US$5.92 (see Table 7). The difference is due to ETC's tax holiday – the tax incentive given to investors by the Government of Zambia. ETC Bioenergy qualified for this incentive because it came in as a new investor in 2007 and bought assets from the MDC, which went into voluntary liquidation in 2006. It is possible that the MDC, prior to liquidation, was contributing more to public revenues than ETC BioEnergy.

Both companies have provided infrastructure and social amenities for their operations and employees. Kascol provides housing (316 units) for its employees. It also has a clinic with four staff members. The clinic serves the company employees and outgrowers. The company also supports a basic school within the estate. Transport is provided for taking schoolchildren to schools outside the estate. Kascol has also built nine water boreholes, which supply water to houses for its staff and to the residences of outgrowers. Roads and irrigation facilities have been constructed within the estate. These various facilities are not accessible to third parties not related to the project.

ETC BioEnergy maintains an access road owned by the local authority. Apart from this, the company has not invested in public infrastructure projects, but it provides housing for its employees. ETC BioEnergy has five company clinics; a doctor visits the clinics every two weeks. The number used to be seven under the MDC but two have since closed. Senior and middle managers have an insurance scheme with Company Clinic, a privately owned surgery in Kitwe; this scheme is not available to other employees. The company also has a school with 330 pupils and 15 teachers. The number of schools was two during the period under the MDC but one was sold together with Munkumpu Farms. Services are provided free of charge to company employees.

The company also provides housing to its permanent staff and the staff residential areas have the basic social amenities (such as markets, shops, and social clubs).

4.5 Social and environmental impacts

Though agriculture is a beneficial and desirable activity, it often brings with it social and environmental costs that are not often taken into account when assessing its full impact. This is at least true in the Zambian context. Thus when Kascol and MDC were being set up in the 1980s, environmental and social impact assessments were not undertaken, nor have they been undertaken at any other time in the lives of the projects. One reason for this is that at the time the projects were being set up, and to some extent even now, Zambia lacked the capacity to carry out environmental assessments at a scale and with a scope able to include the majority of agricultural projects. At the time of project inception, the Zambia Environmental Management Agency (ZEMA), which regulates environmental issues, had not yet been established. The Environment Council of Zambia regulations cover issues like environmental impact assessments, air pollution, waste management, pesticides and toxic substances (PTS), water pollution, hazardous waste, and ozone depleting substances (ODS).

Presently, both companies claim to adhere to ZEMA standards. However, both projects generate toxic emissions into the air, soil and water. Additionally, land clearing on a scale practised by the two projects has the potential to negatively impact the ecosystem. With regard to emissions, both projects are heavy users of chemical fertilizers and some of this fertilizer certainly finds itself in areas beyond the farms through such means as running water, seepage into underground water systems and evaporation. Mpongwe Development Company relies heavily on chemical weed killers. These chemicals are applied through aerial means. This method of application creates the possibility for the chemicals to be blown far beyond the farm with consequent damage to the atmosphere. A related problem is the safe disposal of used containers for pesticides and other toxic substances. Poorly disposed, used containers are a real danger to the surrounding communities who pick up these containers and start reusing them for such things as water and food storage. The villagers are often not aware of the dangers inherent to the use of such containers.

Kascol contributes to polluting the atmosphere through the burning of sugar cane plantations in readiness for harvest. This is done annually and is a source of particulate matter (smog) and of oxides of nitrogen (NOx). Further, some environmental concerns may not directly be addressed by ZEMA. A study by German et al (2010) revealed that, for example, industrial-scale biofuel plantations negatively impact the environment through deforestation. In this regard, it is reasonable to conclude that the jatropha plantation developed by ETC BioEnergy contributed to deforestation. From the resource point of view, the abstraction of water for irrigation purposes is certainly having some effect on the water resource. Both companies are heavy users of irrigation water and though the impact could not be determined, the abstraction of water to irrigate 2 000 hectares must have a telling effect on the resource.

On a positive side, Mpongwe Farms practise zero tillage which is an aspect of conservation farming. Kascol also only tills the land once in seven years. Mpongwe does not burn the fields. These practices are more environmentally friendly than the slash-and-burn method which is practised by the majority of subsistent farms in the country.

With regard to the social impact, it is impossible to assess the livelihood impacts of the loss of land at the time the investment projects started. In both cases, too much time has passed to permit meaningful assessment. It is, however, worth noting that no involuntary resettlement was involved, as the land was acquired from existing commercial operations or through deals negotiated with local landholders (see above). The area which makes up the Mpongwe farms was unoccupied when it was being acquired, even though the nearby villages claimed ownership of it and it was communally used for gathering forest resources. As noted earlier, permission had to be sought from the family who held the piece of land in order to have land allocated to Mpongwe Farms in the Nampamba area.

However, although at the time of allocation of land to Mpongwe farms the villagers saw no serious problems, growing land scarcity in the project catchment areas is now being felt by local people. This growing land scarcity is being driven by demographic changes: the population of Mpongwe District has been growing at the fastest rate in the Copperbelt Province for the past three decades (CSO, 2011). Growing land scarcity is compounded by the existence of big commercial farms that have been established in the area, e.g. Dar Farms, which is said to comprise about 76 000 hectares, in addition to the area taken up by the MDC.[9]

The severity of the problem of land scarcity may be appreciated when one considers that apart from the small numbers of people that are in paid employment, the majority of Mpongwe residents derive their livelihoods from the cultivation of crops for own consumption. For these people, land is their most important asset. Growing land scarcity erodes local access to a most important livelihood source for rural dwellers. This situation has resulted in tensions at the local level and with the company. The MDC's land wrangles with the so-called 'squatters' on their farm exemplifies this tension. Squatters started encroaching on the MDC land around 2003. Initially, these were mostly individuals retiring from the MDC and companies in urban areas. Later some local villagers joined in. When the company noted the presence of squatters around the farmland, it engaged independent surveyors to verify the farm boundaries. It was ascertained that the squatters had in fact encroached on company land. Squatters resisted attempts to remove them, and sued the company. The case was resolved in favour of the MDC in the Supreme Court. But getting all the squatters to move out of the farm has not been easy even after the court ruling. As a compromise, some of the squatters have been allowed to stay on the edge of the farm, while others have moved out. The focus group discussion held with N villagers revealed that growing land scarcity is fostering hostile attitudes among villagers towards large-scale land investments in their area. A former member of Parliament at the time of the MDC land transaction, now a villager in the area of Chief N, regretted that he had given land to the MDC back in the 1970s and wished it could be repossessed. The company is aware of the growing dissatisfaction among villagers due to land pressure and is considering introducing an outgrower scheme whereby land would be subleased to local farmers for these to produce for the company on contract.

For some residents of the area, the blame for these tensions lies also with their chief who, they claim, has continued to give out land to investors in exchange for gifts, with little regard to the needs of his subjects. They argue that he does not want them to get leasehold title: *"It is not easy to get title deeds. The chief refuses to give us (application) forms because he fears that he would lose control of the land"*, said the members of the focus group discussion at Kantatamo Market. The loss of control that the villagers were referring to is the chief's inability to give land to investors once the villagers get leasehold title. Without leasehold title, the chief can easily take away land from the villagers and allocate it to investors.

The social impact of the problem of land scarcity may also be gauged from a statement made by one villager, a former member of Parliament for Mpongwe who claims to have given the land to MDC in the 1970s: *"If I heard that Nampamba (Farm) was closed, I would be very happy because we would get back the land"*. Yet, it is nearly impossible for the villagers to get back land that is held on leasehold title.

Constrained access to land may throw many into poverty, particularly if agricultural investments do not generate sufficient employment for local people. The youths are especially vulnerable, as their land access is particularly limited and they may face growing food needs as they establish new households. In the area, the youths who are setting up their independent households can only get pieces of land from that held by their parents,

[9] Land scarcity is location-specific. There are areas in Mpongwe District where land is still available, but these areas are far from the road and, as such, they are unattractive to both the investor and the villager.

assuming their parents have enough. It is difficult to find suitable "virgin" land that can be brought into production.

On the other hand, the MDC has land lying idle - only about 22.7 percent of the farm is being utilized. Growing land scarcity and idle plantation land are the two factors that underpin the encroachment by squatters on company land. Legal battles have been fought over the matter and, from the legal point of view, ETC has emerged a winner. Socially, however, the company has had to contend with growing hostility exhibited in such behaviour as deliberate damage to company property from some community members. A senior manager intimated that the majority of the people held in prison cells in Luanshya town probably had something to do with the disturbances at the MDC. As mentioned previously, the company has tried to reach a compromise by allowing squatters to use company land for free, on the understanding that they can be moved out when the company needs the land. Another solution under consideration is to turn the squatters into tenant or contract farmers.

Besides land scarcity, other types of adverse impacts on local communities are also possible, but they could not be documented by this research. For example, reference made by Kascol staff to irrigation water constraints as a limit to the land area that can be cultivated suggests that there may be a problem of water scarcity. More research is needed to establish whether this is indeed the case and, if so, what impacts it has on people around the farm.

5. Conclusions and recommendations

This chapter has examined two experiences of large-scale agricultural investments in Zambia. Both experiences have been implemented for decades – the MDC since the 1970s, Kascol since the early 1980s. While this circumstance made it impossible for this study to assess the socio-economic impacts of the projects on local livelihoods at the time of their establishment (for example, with regard to impacts linked to loss of land), the sufficient implementation time has enabled us to learn lessons on how agricultural investments may work in the longer term.

The important caveat to this consideration is that, for much of their duration and until their recent privatization, both projects had a strong development component beyond commercial returns. This characteristic makes the project significantly different to the many, more recent investments that are being carried out in many parts of Africa as part of the ongoing global land rush. Recent changes in employment conditions, including wage levels, following the privatization of the two companies, illustrate the difference between investment mainly or solely driven by commercial returns and projects with an explicit development objective.

In both experiences, enabling factors have played an important role in making the ventures possible. A key enabling factor in the initial stages of the Kascol project was the configuration of expertise and contributions provided by the different shareholders. The Zambia Sugar Company and the Commonwealth Development Corporation brought production and management expertise, while the two banks brought financial resources to the new company. The CDC, in particular, managed the initial set-up and provided management expertise. The involvement of the CDC may also have played a role in lowering the risk profile of the project, thereby making it more appealing to private investors and lowering the rate of return needed to make the project a commercially interesting proposition.

The early association with the CDC was an important enabling factor also for the MDC. The CDC injected cash into the company and provided management expertise. The abundant rains and fertile soils of Mpongwe District have also contributed to its success. The company sits on an aquifer which has proved invaluable as a source of irrigation water. Partly due to the high productivity of the company, the road infrastructure has been developed and this has presumably reduced the company's transport costs. Current company management believes the company (that is, ETC BioEnergy) is making a sufficient return on investment to satisfy the shareholders. Even though the MDC went into liquidation in 2006, it was neither a loss maker nor was it facing liquidity problems – it was liquidated because it could not meet the higher targets of return on investments demanded by its owners, the CDC. Presumably, the company was liquidated instead of being sold as a going concern because its owners believed there was more value in selling a 'dead cow than a live one' – they may have obtained more by selling individual assets than by selling the company as a going concern.

Both companies are large employers in the Zambian context, but overall job numbers have remained small compared to the rural labour force. Both projects have experienced downward pressures on wages. And while both projects have created jobs in skilled positions, the rural poor are most likely to take up low-paying jobs due to their generally low educational status. These circumstances raise real questions as to the best ways of reducing poverty in rural areas. Investment projects that maximize positive economic linkages with local rural areas through multiple avenues appear to have the highest potential for impact on poverty.

More specifically, investments that include low-income groups as producers in the supply chain and as shareholders in decision-making and profit-sharing are more promising than models that only purport to involve the poor as wage labourers. At Kascol, outgrowers involved in the outgrower scheme appear to have higher incomes, better living conditions and higher levels of overall self-satisfaction than their counterparts who work as wage labourers in the same company. Similarly, joint ownership of the company, whereby local groups have an equity stake in the business, provides the poor with additional income opportunities and with avenues to oversee the management of the business. That said, the higher returns associated with participation in a business as a supplier or a shareholder are also associated with higher business risks. But mechanisms can be developed to manage some of these risks, particularly with regard to crop insurance.

On the other hand, growing scarcity of valuable land in parts of the country is resulting in tensions around agricultural investments involving large plantations. This is particularly relevant in those parts of the country that appear attractive to outside investors, for instance due to water availability or soil fertility. This is the case of the Mpongwe area, where commercial pressures on the land are on the increase. The circumstance is exacerbated by the different ways in which formal legislation and local people view land ownership in customary areas. Even in projects that have been established for a very long time, this situation can result in conflict over land, including encroachment and litigation.

Neither project is using the land allocated to it to its full potential. Where land is becoming scarcer and investors hold big tracts of land that they are not using, it may make economic sense to rent land to local farmers as part of contract farming arrangements. Renting unused land to local farmers can also help stem hostile attitudes by poorer groups towards large-scale operators. As mentioned previously, the MDC is contemplating this option as a possible solution to encroachments on its land by villagers that have run out of land for farming as a result of demographic growth and increasing commercial pressures.

These case studies have demonstrated that large-scale investments in agriculture have beneficial impacts on the poor, largely on three fronts: first, through contributions to the nation's gross domestic product (GDP); second, through direct provision of income to the low-income groups; and third, through access to social services. However, large-scale investment in agriculture can also negatively impact on the poor by limiting access to their most important asset – land.

Recommendations

1. Since poverty is predominantly rural, agricultural investments have a significant potential in providing the rural poor with jobs and income growth. However, agricultural investment in rural areas is expensive, and financial returns are generally low. Additionally, where investments have taken place, the rural poor often take up low-paying jobs due

to their lack of education and skills. In addressing the daunting problem of rural poverty, the government should put in place policies that promote inclusive investments in agriculture. Investments in rural infrastructure should be the top priority. The government should also double efforts in the provision of education to the rural population: adult education, and vocational training should be extended to rural areas.

2. Consistent with previous studies on privatization, e.g. Cheelo and Munalula (2005), employment levels and/or conditions of service generally fall after privatization. The decline in employment is likely to persist when there is no significant growth in the size of a firm. Improvements in efficiency that accompany privatization should be followed by action to support firm growth in order to generate new jobs. Firm growth will most likely require that investors actually inject into the company the amounts pledged during the bidding process. Effective monitoring and enforcement of the contractual relationship by government authorities play an important role in this.

3. The country as a whole has abundant arable land, but access to arable land by poor people is becoming increasingly difficult in some places. Land policy should strike a balance between the need for agricultural investments and the interest of poor farmers. Before any land is allocated to an investor, the government should take into account both the present and future land needs of local people. The impact of large-scale investments on demographic changes should also be taken into account in estimating future land needs. Ideally, the discipline of town and country planning should be developed to the extent where it takes into account the peculiar conditions that obtain in rural areas. With increasing population and the need for rural investments, it is becoming clear that land use planning should be extended to rural areas. The current system where chiefs are heavily involved in land allocation is not working well in some cases. The chiefs are demonstrating a growing unreliability in matters of protecting the needs of their subjects.

4. There is a tendency for investors to take up large tracts of land which they do not fully utilize, and by so doing, they create artificial shortages of land in some cases. Investors should only obtain land that they can utilize within a reasonable span of time. Depriving the poor of access to land is socially and economically irresponsible. The government should proactively protect the rights of the poor by ensuring that investors do not hold on to land they do not utilize for many years. Mechanisms should be put in place to allocate to investors only such land as they can utilize within a reasonable span of time, and to withdraw land from investors who do not comply with agreed development plans.

5. A key enabling factor in the initial stages and in the ultimate successes of the Kascol and Mpongwe investment projects was the association with the Commonwealth Development Corporation (CDC). The CDC provided investment funds and managerial expertise to both companies in return for comparatively low levels of financial returns. Although not set up as a development aid agency, the CDC was in practice a vehicle for delivering development aid to poor countries. It invested in the least developed countries and in sectors that would ordinarily be shunned by private investors; it undertook investment projects where financial returns were too low, too distant or too risky to attract capital flows from private investors. The Government of the United Kingdom did not require it to make a profit beyond that needed to service its debts (Tyler, undated). Strategically targeted development aid can play an important role in promoting commercially viable and socially inclusive models of agricultural investments.

6. Investment projects that include low-income groups as producers in the supply chain and as shareholders in decision-making and profit-sharing are more promising than models that only involve the poor as wage labourers. In countries where governments are genuinely committed to poverty reduction, this observation creates a powerful argument for effective policy interventions aimed at promoting equitable inclusion of the low-income as producers and equity owners. Given the benefits of collective action, collective equity ownership through a trust is perhaps more advantageous to the low-income groups than having individual poor people each owning few shares.

7. The outgrower scheme at Kaleya Smallholder Company has managed to keep at manageable levels the transactions costs that arise from dealing with a large number of outgrowers by avoiding the geographical dispersion of outgrowers. It has done this partly by keeping all the outgrowers on the company estate. While the overall number of Kascol outgrowers is limited, this experience shows that collaborative models are possible and can be commercially viable. This experience provides insights on practical ways to include smallholders in investment processes, both as suppliers and shareholders, and it is hoped that this chapter may help feed lessons learned into international policy debates about agricultural investments in the global South.

8. One of the challenges faced by this study was data availability. This was undoubtedly related to the choice of companies for the case studies. The companies under study are private entities not listed on the stock exchange. As such, they do not make their financial data public. A quantitative study would be of great benefit if it involved publicly listed companies. However, in Zambia very few agribusiness companies are listed on the stock exchange and make their data publicly available. A difficulty related to collecting data from private companies is assurance of confidentiality. Management of private companies want to be sure that data made available to the researcher will be kept private, and they do not want their identity to be made public. Notwithstanding cost implications, this may be achieved much more easily when a bigger sample of companies is involved in a study rather than one or two case studies. The confidentiality challenge is not a big issue when obtaining data from rural small-scale farmers. It would be interesting for future studies to carry out quantitative analyses of externalities of large-scale agricultural production. The economic cost to communities and the larger economy of holding unused land by large-scale agricultural entities could also be investigated. Such studies could also be extended to individuals who hold huge tracts of unused land. It has become fashionable for many to obtain huge tracts of land which then stay idle for a long time; this creates artificial shortages of land.

Often, the financial returns of large-scale agricultural investments are not known. Future studies could consider investigating the financial returns of agribusiness in Africa. Such studies could include the cost of developing a large-scale farming operation. Comparative analyses of the performance of agribusinesses and firms in other sectors could also be done. Other studies could target sources of equity capital for agribusinesses in agriculture.

References

Banda, C. T. (2011). *Institutional, administrative, and management aspects of land tenure in Zambia.* Retrieved October 2011, from http://dlc.dlib.indiana.edu/dlc/bitstream/handle/10535/7622/Institutional,%20administrative,%20and%20management%20aspects%20of%20land%20tenure%20in%20Zambia.pdf

Central Stastical Office. (1990). *1990 Census of Popilation and Housing.* Lusaka.

Cheelo, C., & Munalula, T. (2005). *The Impact of Privatization on Firm Performance in Zambia.* UNZA.

Clarke, G. R., Munro, J., Pearson, R. V., Shah, M. K., & Sheppard, M. (2010). *The profile and productivity of Zambian Businesses.* GRZ, Zambia Business Forum, FinMark Trust and the World Bank.

Cotula, L., & Vermeulen, S. (2010). *Making the most of agricultural investment: A survey of business models that provide opportunities for smallholders.* London/Rome/Bern: IIED/FAO/IFAD/SDC.

Cotula, L., & Vermeulen, S. (2010). *Making the most of agricultural investment: A survey of business models that provide opportunities for smallholders.* London/Rome/Bern: IIED/FAO/IFAD/SDC.

CSO. (2003). *2000 Census of Population and Housing.* Lusaka.

CSO. (2004). *2004 Living Conditions Monitoring Survey.* Lusaka: Central Statistical Office.

CSO. (2011). *2010 Census of Population and Housing Preliminary Results.* Lusaka: Central Statistical Office.

CSO. (1986). *Country Profile.* Lusaka: Central Statistical Office.

CSO. (1986). *Labor Force Survey.* Lusaka: Central Statistical Office.

George, G., & Bock, A. J. (2011). The Business Model in Practice and its Implications for Entrepreneurship Research. *Entrepreneurship Theory and Practice , 35* (1), 83-111.

German, L., Schoneveld, G., Skutch, M., Andriani, R., Obidzinski, K., Pacheco, P., et al. (2010, December). *The local social and environmental impacts of biofuel feedstock expansion: A synthesis of case studies from Asia, Africa and Latin America.* Retrieved October 2011, from Cifor InforBriefs.

GRZ. (2011). *Agriculture Statistical Bulletin.* Retrieved October 2011, from http://www.countrystat.org

GRZ. (2009). *Economic Report.* Lusaka: GRZ, Ministry of Finance and National Planning.

GRZ. (2002). EdAssist Database. Lusaka: GRZ, Ministrty of Education.

GRZ. (2002). *Poverty Reduction Strategy Paper.* Lusaka: GRZ.

GRZ. (2011). *Sixth National Development Plan.* Lusaka: GRZ.

GRZ. (1979). *Third National Development Plan .* Lusaka: National Commission for Development Planning.

Kaunga, E. C. (undated). Retrieved February 2012, from Privatisation: The Zambian Experience: http://unpan1.un.org/intradoc/groups/public/documents/aapam/unpan028227.pdf

Leonard, R., & Cotula, L. (Eds.). (2010). *Alternatives to land acquisitions: Agricultural investment and collaborative business models, IIED/SDC/IFAD/CTV, London/Bern/Rome/Maputo.*

Roth, M., Khan, A. M., & Zulu, M. C. (n.d.). *Legal Framework and Administration of Land Policy in Zambia.* Retrieved October 2011, from http://www.aec.msu.edu/fs2/zambia/resources/Chapter1.pdf

Scoones, I. (1998). Sustainable Rural Livelihoods: A Framework For Analysis. *IDS Working Paper 72* .

Serlemitsos, J., & Fusco, H. (2003). *Zambia Post-Privatization Study.* Washington DC: The World Bank.

Tyler, G. (Undated). *Background paper for the Competitive Commercial Agriculture in Sub–Saharan Africa Study.* Retrieved 2012, from http://siteresources.worldbank.org/INTAFRICA/Resources/257994-1215457178567/CCAA_Colonial.pdf

UNDP. (2010). *Business Solutions to Poverty– How inclusive business models create opportunities for all in Emerging Europe and Central Asia.* UNDP Regional Bureau for Europe and the Commonwealth of Independent States.

Wonani, C. (in press). *Gender dynamics of large-scale investment including inclusive business models in agricultural land: the case of Kaleya smallholder company limited (Kascol).*

ZPA. (Undated). *Bound copies of press releases.* Lusaka.

ZPA. (Undated). *Zambia Privatization Agency Bound Copies of Press Releases.* Unpublished.

ZSC. (2011). *Zambia Sugar PLC Annual Report 2011.* Mazabuka: Zambia Sugar PLC.

PART FIVE

SYNTHESIS OF FINDINGS

The findings of the studies presented in this publication confirm the increasing trend of foreign investment into the agricultural sector of the surveyed countries. Resource-seeking investment accounts for a substantial share of the increase, although it was not possible to quantify it more precisely due to the lack of recent disaggregated data. In spite of the increasing trend, the share of total FDI that is directed to agriculture remains low with respect to other sectors. In almost all the surveyed countries it was under 5 percent of total FDI inflows and in the majority of them it was even below 2 percent.

The case studies provide indications on the various types of impacts (economic, social and environmental) of foreign agricultural investment at national and local levels in the host country, although it is difficult to draw general conclusions due to the limitations inherent in the case study approach. There are substantial differences in the analysed contexts and models. The observed impacts were very diverse and depended on a range of factors as discussed below. Also, the time frame of agricultural investment is very long and the full impacts may only materialize after decades.

1. National impacts of agricultural FDI

At the national level, the studies found some evidence that FDI contributed positively to increase in agricultural production and yields. In Ghana, investment by a single transnational company was expected to contribute significantly to the increase of the nation's total palm oil output. In Uganda, companies such as Tilda (U) Ltd contributed to the production of rice, which nearly doubled over the last decade due to the introduction of a new rice variety (Nerica 4).

Some case studies suggest that FDI favours the diversification of crops. In Senegal FDI contributed positively to the production of some high quality fresh fruits. In the cases where the investment targeted export markets, evidence of higher export earnings was found for some countries. For example, in Ghana, exports earnings of non-traditional agricultural commodities (e.g. fruits and vegetables) increased four-fold between 2000 and 2009 due to FDI. A positive effect was also observed in Senegal, where for example, the volume of mango exports grew rapidly. These findings are consistent with those of another study by FAO which found that, in Egypt and Morocco, agricultural production projects to which foreign investment contributed had considerable positive impacts on agricultural exports. In Morocco, for example, foreign companies have contributed to the development of early tomato exports to Western Europe, a high-value market. The studies suggest that FDI contributed to an increase in value added. In some cases, foreign investments led to the adoption of higher standards. A variety of standards was adopted. Some were product-related standards, while other standards addressed the production and/or processing process (e.g. use of the ISO 9000 series and GlobalGAP standards in Uganda). Certification to widely-recognized voluntary standards can increase market access and value-added for farmers and export businesses.

Employment creation, in particular in rural areas, where most of the poor live, is often presented as an expected advantage of international investment in agriculture. The study observed this effect in Ghana, where employment creation by FDI in agriculture was estimated to exceed 180 000 jobs in the period 2001-2008 and Uganda where an estimated 3 000 jobs were provided by eleven transnational corporations in the agricultural sector in 2009. These findings complement those of another study by FAO which found that in Sudan over 6 500 jobs were created over the period 2000-2008. Beyond the observed effects, some surveyed projects expect to have high employment creation effects in the future.

The country studies show mixed results on technology transfer. There are some positive examples of adoption of new production

This chapter was prepared by Pascal Liu, Trade and Markets Division, FAO

technology, such as in the tomato export industry in Senegal or the adoption of improved crop varieties, such as the introduction of a new rice variety in Uganda. In some investment projects involving outgrower schemes and contract farming, small farmers have acquired new skills through either formal training organized by the project's promoters or by working on the nucleus farm (learning by doing). However, the studies suggest that the actual transfer of technology is seldom up to the level announced by the investors.

Some investments led to the development of new infrastructure or improvement in existing ones (e.g. roads, storage facilities, cold stores), either directly by the investor or indirectly, when the government built the infrastructure as part of the investment contract. Yet, in a few cases access to the new infrastructure was restricted to the investor's operations only and local people could not use it.

The studies found evidence of negative environmental impacts, mainly due to the intensification of production generated by the investment which puts higher pressure on natural resources. The intensive use of land and water may result in the degradation and depletion of these resources. There is some local evidence of reduction in forest cover and biodiversity as a result of the investor's activities. These adverse effects are often due to the lack of proper environmental impact assessment (EIA) prior to the investment and the absence of effective environmental management system (EMS) during its implementation. In some cases the EIA was conducted in a superficial manner and could not assess the full risks of the project. Nevertheless, some investment projects led to the adoption of environment-friendly technology (e.g. drip irrigation; adoption of organic farming methods in the case of ITFC in Ghana).

Finally, the data suggest that returns to investment tend to be higher where the investor builds on existing ventures in a gradual approach, as opposed to new ventures which are the most risky type of investment. Greenfield investments to establish very large farms in unknown areas and relatively new industries (such as biofuels) are probably too risky to be recommended as a strategy for agricultural development.

2. Local impacts of agricultural FDI

At the local level, the studies suggest that one of the main short-term benefits of FDI is the generation of employment, although there are limitations. First, the new jobs created by an investment project may not all be sustainable. In some case studies it was observed that projects are labour intensive during the initial phase but become increasingly mechanized later on, thus reducing future labour requirements. Similarly, a change in the type of crops cultivated on the farm may reduce the job number, as some crops are less labour intensive. Second, the new jobs are not always taken up by local people; labourers may come from other areas or even from abroad. Third, the net employment creation effect may be limited if the new jobs replace former ones or self-employment. Beside the purely quantitative aspects, the quality of the new jobs is important. For example, replacing independent small-scale farmers with low-skilled and poorly-paid worker jobs may threaten the resilience and sustainability of the local food system.

Evidence of other benefits to the local economy included higher prices obtained by farmers selling to the nucleus farm, as well as income generation when the nucleus farm sub-contracted some services (e.g. soil preparation, weeding) to local residents. In addition, positive spillover effects were observed in cases where small farmers who worked for the investment project as wage earners reinvested the earnings to increase the productivity in their own farm.

While the studies found substantial evidence on the local impacts of agricultural FDI, it is difficult to draw general conclusions, as the effects depend on a number of factors among which the business model and local context are critical.

3. The significance of business models

The studies examined a variety of business models with various levels of involvement of local farmers. The models ranged from the most classical and simple one where local farmers are at best involved as waged workers (acquisition of land by a foreign company for establishing a large plantation) through more inclusive but still classical ones (outgrower schemes, contract farming) to more innovative models where outgrowers are also shareholders of the business venture.

The case studies suggest that positive effects on local communities are unlikely to arise when the investment involves large-scale land acquisition, especially when the land was previously utilized in some ways (including informally). Then, the evidence suggests that the disadvantages far outweigh the benefits. The only economic benefits found were job creation (although there were some limitations as explained above).

On the other hand, there is ample evidence of the risks of large-scale land acquisition in countries where governance is weak. Many developing countries do not have in place all the necessary legal or procedural mechanisms to protect local rights, be they formal or informal in nature. Local interests, livelihood patterns and welfare are seldom taken into account adequately when contracts are signed with outside investors. Land deals are too often characterized by lack of transparency, creating opportunities for corruption. In situations where the government and customary authorities have the power to allocate collective land, the absence of clearly defined land rights, the lack of transparency in transactions for land allocation and arbitrary decisions may harm local communities and hamper economic development. The gap between legality – whereby the government may formally own much if not all the land – and legitimacy – whereby local people feel the land they have used for generations is theirs - exposes local groups to the risk of dispossession and investors to that of local contestation. The fact that many land deals are being negotiated behind closed doors and without local consultation compounds these problems.

The negative social impacts found include the displacement of local smallholders (often with inadequate or no compensation at all), the loss of grazing land for pastoralists, the loss of income for local communities, and in general, negative impacts on livelihoods due to reduced access to resources, which may lead to social fragmentation. In the case of an Economic Land Concession project in Cambodia, villagers suddenly lost access to the nearby forest when the investing company fenced it off. In a few days they lost the possibility to collect timber and non-wood products such as herbs and berries, which had provided an important income for their households.

Not surprisingly, these negative effects generate opposition by the local community and may lead to hostile action against the investor. In the case of the above project in Cambodia the villagers protested vehemently against the investor and the local authorities had to step in. In the case of ETC Bioenergy in Ghana, local people damaged the company's property. In the case of the Solar Harvest Ltd project in Ghana, the fire on the company's Jatropha plantation might have been the result of a widely shared perception that the company appropriated community land against the interests of local people.

The studies suggest that large-scale acquisition of land can raise problems even when the transaction is transparent, complies with the established regulations and relates to land that is not utilized (formally or informally) by anyone at the moment of the transaction. In the case of the ETC Bioenergy project in Zambia the purchase was completely transparent and strictly followed the regulatory procedures. The land acquired by ETC had belonged to commercial companies for many years, therefore no local person was dispossessed of their land and the villagers were not against the transaction. However, the large size of the transaction has contributed to putting upward pressure on land prices in a region where land is becoming increasingly scarce due to demographic factors. ETC holds some 47 000 hectares but utilizes less than a quarter of this area, the remainder being left idle. Villagers that have run out of land for farming as a result of demographic growth and increasing commercial pressures have encroached on the company's idle land. The company has fought back with law suits and won, but this has generated the hostility of some community members. Deliberate damage to the company's property has been reported.

These considerations suggest that even from the perspective of the investor, large-scale land acquisition may not be the most profitable model. In spite of the possible discounts on prices (or rents) and tax rebates granted by the host country, transaction costs are generally high. This is because land is such an important asset for rural people that their interests are likely to diverge from those of the investor sooner or later. Consequently, it may be more effective for the investor not to compete with local farmers but on the contrary provide what they do not have, i.e. financial capital, modern technology, management expertise, marketing skills and business know-how. The findings suggest that investment projects that do not involve the local community actively at an early stage tend to be ill-designed and are likely to fail.

4. Inclusive business models

In view of the above observations on the impacts of large-scale land acquisitions there is a case for the promotion of different business models that involve smallholder farmers while letting them keep the ownership of their land. In these models, local farmers and other members of the local community are active partners. There is a variety of such models with varying degrees of participation, including the classical contract farming arrangement, management contracts, outgrower schemes and more innovative models such as joint ventures between the foreign investors and a local farmer cooperative. One of the main objectives of the studies was to gather evidence on the impacts of these models on the local community and its economic development. The case studies found evidence of positive effects on the local economy. These effects include value addition for outgrowers and higher incomes. In the case of the Kaleya Smallholders Company (Kascol) in Zambia, local farmers involved in the outgrower scheme appear to have higher incomes and better living conditions than their counterparts who are wage workers in the same company. A similar observation was made for farmers participating in the outgrower scheme of the Integrated Tamale Fruit Company (ITFC). They reported that their income was much higher than the average local farm income of US$300 per year and growing. Their target income after the repayment of the loan to ITFC is US$2 000, which is more than six times the average income. The study of Kascol in Zambia found that on average an outgrower obtained a net income that was four times as high as the wage income of a unionized employee in the same company. In addition, the outgrowers had more assets than the employees.

Different types of business models were analysed with varying levels of inclusiveness. The most inclusive ones gave shares in the company to the association of local farmers participating in the outgrower scheme. This was the case of Kascol and Mali Biocarburant S.A. (MBSA). Holding shares enables the farmer association to have a say in the management of the company and in decisions on production and marketing. This increases local farmers' sense of ownership of the project. At the same time, it also increases the economic risks they bear. It may also create unrealistic expectations that may be disappointed in the initial phase of the business. In the first years of project implementation the shareholders of MBSA and Kascol earned no or little dividends as explained below.

Further, the surveyed inclusive business models successfully introduced and disseminated new technology and know-how. The technology is low-cost and adapted to small-scale farmers (e.g. use of irrigation pipes for the mango seedlings in Ghana), which makes its dissemination to a wider group of farmers easier. New methods of work were introduced to enhance productivity. Some of them promote active collaboration between farmers (e.g. work in groups), thus strengthening the local community. The inclusive business models emphasize continuous improvement and practical training. Farmers are not viewed as a passive recipient of training, but rather as a source of feedback for the improvement of the course and as potential trainers. Inclusive business models tend to give farmers an active role to train their peers. For example, the MBSA project has organized farmer field schools to train a large number of farmers to Jatropha production.

Some of the inclusive models have helped farmers obtain certification against standards that are demanded in export markets. For example, the ITFC project in Ghana helped its outgrowers become certified to the European Union's organic agriculture standard and the GlobalGAP standard. Through the organic certification the farmers have gained access to the European Union's market, the largest market for organic foods in the world, where they can obtain higher prices for their products. Similarly, the GlobalGAP certification gives farmers access to several large-scale retail chains in developed markets. Thus, successful inclusive projects can help small-scale farmers in developing countries access international value chains and increase the value of their products. The access is not direct in the initial stage, in the sense that the farmers export through the investing company. However, through this experience the farmer organization learns the requirements of export markets and gains marketing skills. This gradually builds its capacity to export directly in the longer term.

The studies suggest that in inclusive business models the investor (alone or with partner organizations) provides a number of public goods and services to the local community. These include housing, health and education. In the Tamale case study, for example, the project along with development NGOs built housing facilities for teachers, renovated the school and provided text books, nutritionally-balanced school meals and clean drinking water. It also provided training on health issues, prevention of bush fire, protection of biodiversity and local tree species and responsible harvesting of medicinal herbs. In the Kascol case study the company provided training on health. It should be noted that the provision of public goods is not a specific characteristic of such models. There are several examples of large plantations and nucleus estates that provide health, water or schooling benefits to local people. The difference may lie in the fact that in inclusive projects local people have more say in the selection of services to be provided in priority and in their management.

In addition to these tangible benefits, inclusive business models provide other types of advantages that may be even more important to long-term economic and social development. These include strengthening rural people's self-esteem; outgrowers expressed higher levels of self-satisfaction, motivation and feeling of ownership than waged workers. The surveyed inclusive projects supported the organization of the local community, in particular farmers in the form of associations or cooperatives. They strengthened existing farmer associations and, when no such organization existed, they facilitated efforts by local farmers to create one. For example, the ITFC project and partner organizations (including the Ministry of Food and Agriculture) facilitated the establishment of the Organic Mango Outgrowers Association (OMOA) in Tamale. The association has some 1 200 members. In this way, inclusive business models contribute to the development of institutional capital. These are important assets for the expansion of the business venture and the growth of the local economy. The lack of effective farmer organizations is one of the main constraints to agricultural development in sub-Saharan Africa.

However, the studies suggest that while inclusive business models can potentially benefit local economic development to a larger extent than large-scale land acquisition, the model *per se* is not a guarantee that the expected benefits will arise. More importantly, the expected positive impacts are unlikely to arise in the short term. Some commentators have tended to portray inclusive business models as the ideal solution to the agricultural investment dilemma. However, the studies reveal a more mixed picture. They could only find limited evidence of the expected benefits. The time factor is critical. Some investment projects have less than five years of implementation. In view of the long time frame for returns on agricultural investments, it is obviously too early for their full effects to be observed. Time will provide more evidence on the impacts on local economic development.

Furthermore, inclusive business models face strong challenges and high costs in their initial phase. They are unlikely to be profitable during the first years and this has an impact on the participating local farmers. For example, MBSA has not paid dividends to farmer shareholders because it has not made any profit yet, while in the case of Kascol the dividends are used to pay back a bank loan that outgrowers took to purchase their shares in the company. In the case of ITFC it will take 14 years before growers can obtain the full income from their mango orchards (although some limited income has materialized and increased gradually). This long time frame is the result of two factors: the long cycle of tree crops that require several years before entering into full production and the need for outgrowers to repay the loans to the company during the first years. The fact that the venture is not profitable in the starting phase is not a surprise: investing in developing country agriculture requires considerable of time, capital and expertise. The recovery period is long by definition.

In the case of inclusive business models, these characteristics are compounded by higher transaction costs due to the large number of local participants. Some outgrower schemes may involve hundreds of small-scale farmers. Their education level is generally lower than that of businessmen, not to mention the language barrier. Even in the presence of a common language, communication problems can affect the development of the project. These problems commonly arise between the investor and the local community, but also within the community, in particular between the local leaders (who are supposed to represent the interests of the community in the project but might have conflicts of interests) and the rest of the community.

The studies suggest that even in the case of inclusive business models there is generally one individual (or at best a handful of individuals) who drives the project due to his/her specific knowledge, background, experience and connections. Arguably there is a need for strong leaders and they have an important driving role to play especially at the inception of the project, but they may later become an obstacle to the greater participation of the community and the redistribution of profits. The capture of the benefits by the local elite is still possible, although its probability may be lower than in the classical form of foreign investment.

In order to overcome these various types of constraints, inclusive investment models need substantial external support (public and private) initially to ensure that the expected benefits materialize. In the case of Kascol, the presence of the Commonwealth Development Corporation (CDC) as shareholder was critical. The CDC provided capital with a demand on the return to investment that was below the level demanded by commercial lenders. It also designed and managed the establishment of the project and brought much-needed expertise in management. Another large shareholder, the Zambia Sugar Company provided expertise in agricultural production and local knowledge. In the case of the ITFC, the mango production project received support from a variety of private and public sector institutions, including a number of Dutch NGOs, Ghana's Ministry of Agriculture. the African Development Foundation, the United Nations Development Programme and the World Bank. Similarly, MBSA in Mali has been supported by international donors and NGOs. The Royal Tropical Institute (KIT) from the Netherlands is one of its majority shareholders.

It must be borne in mind that these partnerships bring together players (agribusiness, local farmers) with very different negotiating power, and that sustained support to farmers groups is key to making them work.

The presence of enabling factors is critical in determining the success of a business model. The existence of farmer organizations that are effective and genuinely represent the local farmers is certainly an important success factor. The internal dynamics, decision making mechanisms and workings of the farmer organization will determine its success in negotiating a good deal with the investor and engage in sustainable growth. Farmer organizations that are based on effective participation of members, democratic decision making and redistribution of profits are likely to be more successful in the long-term even though the nature of the democratic decision making may create some delays in the initial phase. But this is the price to be paid for building consensus and trust that will prove an invaluable asset in the medium and long terms.

The presence in the locality of farmer leaders that have technical knowledge, experience of the crop, a vision for development and access to decision makers is another significant enabling factor, as illustrated by the MBSA case study. These leaders with a vision are pioneers and drivers that can induce profound changes in the local community. Their role is determinant in the initial phase of the project. However, democratic mechanisms for the appointment and renewal of leaders should be in place. Strong leaders who stay in control of the project for too long might eventually become an obstacle to development in the longer term, as they can act as a barrier between the foreign investor and the other members of the community.

The contractual arrangements that govern the relationship between the investor and the local farmers are critical. However, the farmers seldom have the technical and legal knowledge necessary for the negotiation of fair contract terms and conditions. The difference in bargaining power often creates an unbalanced deal with a contract that favours the investor disproportionately over the local community and the host country. The local authorities can only provide limited help, as they often are in a similar situation of insufficient expertise. In some of the case studies the intervention of independent third parties (e.g. development agencies or the African Development Bank in the case of Mali) was useful to provide the missing knowledge and increase the bargaining power of the local community.

5. Other determinants of impacts

The studies suggest that while the type of business model is an essential determinant, it is not sufficient per se to ensure positive outcomes. Other factors have a strong influence on the impacts of agricultural FDI on the local community, its economic development and the wider economy of the host country. These factors are discussed below.

5.1 Good governance

The existence of a good governance system in the host country appears to be a key determinant, if not the most important one. The quality and adequacy of laws and regulations, their effective enforcement and the existence of grievance and redress mechanisms are extremely important factors. The land tenure system, laws and regulations and clear property rights should create conditions that ensure secure access to land for investors and local people. Also essential is the existence of adequate regulations on investment, agriculture, water, the use of natural resources and other sectors related to agricultural production and their effective implementation. Good governance, the rule of law, accountability, transparency, peace, stability, the absence of corruption and participation are conducive to more sustainable investment projects.

Conversely, investment projects that fail or have negative impacts on the local community and the environment are generally the result of governance failures. The existence of effective national institutions that have the capacity to effectively review investment proposals, improve their design, involve local stakeholders and enforce regulations is essential. The capacity of the host government to monitor and enforce investment contracts is an important factor, including the capability of local governmental institutions to intervene to correct failures and develop mechanisms for mediation and conflict resolution.

5.2 Local context

The social and economic conditions in the area where the investment is made are important determinants. The presence of adequate infrastructure and an educated workforce increases the rate of success. Communities that have a good level of organization, solidarity, collaboration and where members participate actively in decision making and have a relatively high level of education and technical knowledge are likely to negotiate better deals. As a result, the agreement will be more balanced, the likelihood of opposition lower and the project will have a higher probability of success.

In the local context, the capacity of civil society organizations (CSOs), in particular farmer organizations plays an important role. A well-functioning local group of farmers can be a strong asset for a foreign investor. It will complement the technology, capital and management expertise of the investor with local assets (especially natural assets such as land and water), skills and knowledge. This may create a win-win partnership.

5.3 Involvement of local stakeholders

The active involvement of local civil society organizations in the project, in particular local farmer organizations is a critical factor. This point has been discussed in details in the above section on inclusive business models.

5.4 Formulation and negotiation process

The process through which the investment project is negotiated, designed and planned is essential. Processes that are transparent, inclusive, participatory, democratic and documented tend to lead to more successful and sustainable investment outcomes, even if these characteristics mean that delays will be likely at the initial stage.

5.5 Contents of investment contract

The terms and conditions of the investment contract will determine the relationship between the project partners, the sharing of responsibilities, decision making, benefits and risks. Investment contracts are often too general and vague. There is a need for well-specified and enforceable terms. In particular, the contract should specify the benefits that the investor will bring to the local community (e.g. number of jobs created, type of infrastructure built and training provided).

5.6 Profile of the investor

The profile of the investing company, its management and technical skills, its experience in the production of the crop and its priority objectives (e.g. speculation, long-term development, long time horizon for financial return) will have an important impact on the outcome of the investment. In the Kascol case, the CDC's willingness to accept lower than average financial returns on its capital in the initial phase contributed to the success of the project. The ability of local project managers to maintain good communication with the local community and forge partnerships with its members is critical.

5.7 Support from third parties

The presence of impartial and effective external support from third parties is an important enabling factor, especially in the case of inclusive business models as detailed above. Good intentions are not sufficient; supporting organizations should have the relevant experience, skills and knowledge if they are to play an effective supportive role.

5.8 Type of production system and crops

The impacts on the local economy will also depend on the production system and crops selected by the investor. Production systems that rely on a large quantity of imported synthetic inputs and equipment are unlikely to create backward linkages with the local economy. Conversely, other systems make a large use of local inputs. This is the case of agro-ecological farming and organic agriculture. The type of crops selected by the project is also important. Crops such as coffee, fruits and vegetables are more conducive to the involvement of smallholder farmers than industrial crops.

PART SIX

CONCLUSIONS AND RECOMMENDATIONS

The available data confirm that there has been a marked rise in FDI into the agri-food sector of developing countries since 2007. Although agricultural FDI flows contracted after their peak of 2009, their level in the 2010-2011 period was still higher than the average for 2003-2007. The flows are characterized by regional patterns whereby intra-regional flows are greater than inter-regional flows except for Africa. The share of FDI that goes to the agri-food sector almost doubled between the periods 2000-2005 and 2006-2008 but is still low compared to other economic sectors, accounting for less than 5 percent over the period 2006-2008. The bulk of agricultural FDI flows is directed to the food manufacturing sector, while primary agricultural production accounted for less than 10 percent over the period 2006-2008. More recent trends for foreign investment in the primary agricultural sector are difficult to track due to the lack of recent disaggregated data.

The case studies presented in this book provide some evidence on the various types of impacts (economic, social and environmental) of foreign agricultural investment at national and local levels in the host country. The observed impacts were very diverse and the studies shed light on the various factors that condition the success or failure of agricultural investments. Overall, their findings are consistent with those of other studies, although it is difficult to draw general conclusions due to the limitations inherent in the case study approach.

1. Large-scale land acquisition

Over the last four years many analysts, development agencies, NGOs and the media have focused on one specific category of primary agricultural investment, namely acquisition of agricultural land on a large scale. This emphasis is due to the numerous economic, political, social and environmental implications that land acquisition has, especially if it is done by foreigners or for them. Estimates of the area acquired by foreign firms vary substantially across sources due to methodological differences. The more reliable cross-checked figures are not as high as what the media headlines suggest. Nevertheless, they do show that foreign investment in agricultural land in developing countries has increased markedly over the past decade. More importantly, the lands acquired by foreign investors tend to be among the best ones, with good soil quality, high production potential, irrigation and proximity to infrastructure and markets. As a majority of foreign investment projects aim at export markets or the production of biofuels, they may pose a threat to food security in low-income food-deficit countries, especially if they replace food crops that were destined for the local market. The net effect on food security will also depend on the additional income generated by the project, its sustainability and how it is distributed in the local economy.

Large-scale acquisition of agricultural land can have other adverse impacts, especially in countries where there is a lack of good governance, rule of law, transparency and clear land tenure rights. These negative effects include the displacement of smallholder farmers, the loss of grazing land for pastoralists, the loss of incomes and livelihoods for rural people, the depletion of productive resources, and in general, negative impacts on local livelihoods due to reduced access to resources, which may lead to social fragmentation. There is also evidence of adverse environmental impacts, in particular the degradation of natural resources such as land, water, forests and biodiversity. The case studies show that when such impacts arise they generate opposition to the project by local people, which at times translate in occupation of part of the land or hostile action such as damage to the company's property. Opposition can force the investor to engage in costly and time-consuming litigation and lawsuits; it increases transaction costs and reduces the return to the investment. The negative effects are likely to be worse when the company only utilizes a small share of the land it has acquired in areas where land is high in demand.

This Chapter was prepared by Pascal Liu, Trade and Markets Division, FAO

While a number of studies document the negative impacts of large-scale land acquisition in developing countries, there is much less evidence of its benefits to the host country, especially in the short term and at local level. The main type of benefits appears to be the generation of employment, but there are questions as to the sustainability of the created jobs. In several projects the number of jobs has decreased over time and, in any case, was lower than what was initially announced by the investor. There is also the issue of the quality of the created employment and who benefits, as managerial positions tend to be occupied by expatriates or persons originating from other areas than that where the project is located. In some projects, even low-skilled worker jobs were mainly taken up by non-locals. Another expected advantage of FDI in developing countries is the transfer of technology. In the case of large-scale land acquisition the evidence is mixed. There is obviously a time dimension in the assessment of this effect as for other outcomes of investment. It may be that the investment was too recent for the transfer of technology to have occurred or to be observed.

In conclusion, the studies suggest that for investment involving large-scale land acquisitions in countries where land rights are unclear and insecure the disadvantages often outweigh the few benefits to the local community, especially in the short run. This outcome is even more likely when the acquired land was previously utilized by local people whether in a formal or informal manner. Consequently, acquisition of already-utilized land to establish new large farms should be avoided and other forms of investment should be considered. Even from the investor's perspective, business models that do not involve the transfer of land control are likely to be more profitable.

2. Inclusive business models

The studies suggest that investment projects which give local farmers an active role and leave them in control of their land tend to have positive effects on local economic and social development. Successful projects combine the strengths of the investor (capital, management expertise and technology) with those of local farmers (labour, land, traditional know-how and knowledge of the local conditions). This combination can provide the basis for win-win outcomes. Business models that leave farmers in control of their land give them incentive to invest in the improvement of the land. Since the bulk of agricultural investment comes from farmers themselves, these models are more likely to raise the level of agricultural investment in developing countries. Also, inclusive business models empower farmers by giving them a say in the implementation of the project or even its management. In some cases, farmers are shareholders and therefore they are joint owners of the business. These characteristics make inclusive business models more conducive to sustainable development than land acquisition.

However, their benefits do not arise immediately. The time factor is essential. By their nature inclusive models involve more stakeholders, hence building consensus on the project requires time and decision-making is slower. Transaction costs are high, especially in the initial phase. They should be viewed as a necessary investment that will enable higher returns in the longer term. However, most companies need relatively rapid returns to their investment and their time frame is not compatible with that of local economic development. There is a need for 'patient capital' provided by investors with a longer time horizon initially to ensure that the expected benefits materialize. Such investors are usually from the public sector (e.g. governments, development banks and sovereign-wealth funds) or the non-profit sector, but some private companies such as "impact investors" and "social investors" also have longer time frames and their number is increasing.

The high transaction costs inherent in inclusive models and their heterogeneous nature makes them very fragile in the beginning. There is a high risk that faced with high initial costs, slow progress and the absence of tangible benefits in the starting phase, both the investor and local players might become discouraged and abandon the project. This situation increases the possibility of misunderstanding, suspicion and distrust. Consequently, inclusive investment models require substantial support from an independent and competent third party which can play the role of honest broker and facilitate collaboration between the investor and the local community. The projects surveyed in this book received substantial support from a variety of institutions such as governmental agencies, foreign development agencies, NGOs and multilateral banks.

Support is also needed to strengthen the capacity of local farmer organizations so that they can become a more solid business partner for the foreign investor. The support organization can help them raise their bargaining power to create a more level playing field. Training the leaders of farmer groups will help them represent better their members in the negotiations, communicate more effectively with them, adopt effective management practices and promote democratic decision making in their organization.

There is a broad variety of inclusive business models and the studies suggest that none of them can be presented as the ideal solution to agricultural development in all contexts. There is no one-size-fits-all. Different situations will require different models. Local economic and social factors including the level of organization of the community, the strength of local institutions, the technical level of farmers and the effectiveness of their organizations will condition the type of model which is most likely to succeed. In cases where farmers are unable or reluctant to create an organization, contract farming may be the most appropriate model. Conversely, in communities where there is a strong tradition of collaboration and effective farmer organizations, an outgrower scheme giving farmers a share of the capital, or possibly a joint-venture between the investing company and a farmer cooperative, may be the most appropriate option. Other factors that condition the success of business models include the national legal and institutional framework, the specific terms and conditions of the investment contract and the experience, skills and motivations of the investor.

Obviously there is a tradeoff between benefits and risks for local farmers. The higher their participation in the venture and their share of its benefits, the more risks they bear in case of failure. Therefore, their level of involvement in the business should be commensurate to the strength of their organization. Weak farmer organizations should avoid being directly exposed to responsibilities and risks that they do not have the capability to manage. Nevertheless, risk reduction strategies and tools (e.g. crop insurance) can be devised. External assistance can help farmer organizations develop such mechanisms. It can also take on the responsibility for some of the risks in the initial phase of the project and gradually transfer the responsibility back to the farmer organization as it becomes stronger.

3. International guidance

Among the many factors that condition the impacts of foreign investment on the local economy, the domestic laws and institutions governing agricultural investment and land tenure are critical. However, in developing countries they are often inadequate to ensure sustainable agricultural development, especially in terms of enforcement. Developing country governments and local institutions need support in the form of policy advice, capacity building and technical assistance. Useful guidance can be obtained from some of the international agreements that have been adopted in recent years. In particular, after three years of international consultations involving governments, civil-society organizations and companies, the Committee on World Food Security (CFS) adopted in May 2012 the Voluntary Guidelines on the Responsible Governance of Tenure of Land, Fisheries and Forests in the context of National Food Security (VGGT)[2]. The VGGT serve as a reference and provide guidance to improve the governance of tenure of land, fisheries and forests with the overarching goal of achieving food security for all. Implementation guides relating to specific issues are being developed. One of them will deal with agricultural investment. Another important

[2] http://www.fao.org/nr/tenure/voluntary-guidelines/en/

internationally-agreed instrument is the FAO Voluntary Guidelines on the Right to Food[3].

In addition, the CFS is preparing to launch a consultation process for the development and broader ownership of principles for responsible agricultural investment that enhance food security and nutrition. It is expected that the principles resulting from the consultations will have international recognition and serve to guide agricultural investment. They will refer to and build on the VGGT. The consultations will take into account various existing instruments, including the voluntary principles for responsible agricultural investment that respect rights, livelihoods and resources (PRAI)[4]. The PRAI have been jointly formulated by FAO, IFAD, UNCTAD and the World Bank (the Interagency Working Group - IAWG) to serve as a possible reference framework for governments in the development of national policies, laws and regulations, or in the negotiation of international investment agreements and individual investment contracts. The PRAI are a set of very general principles that need to be translated into more operational guidance. To this end, the IAWG implements pilot projects with governments, investors and civil-society organizations in selected developing countries. The results of the projects will feed into the CFS consultations.

4. Recommendations

4.1 Further research on the impacts of agricultural investment

While the findings of the case studies indicate useful directions, one should refrain to draw general conclusions for several reasons. First of all, the case study approach has inherent limitations and cannot fully capture the wide variety of situations. Some observed changes may be due to other factors than the considered investment. Another reason is the issue of time frame. Most studies analyse recent investments, while the full effects may materialize many years after the investment has taken place. Over the long run, the outcomes of a project may change drastically. Finally, it is difficult to compare the results due to the differences in local contexts. Several research institutes and development agencies have conducted case studies on the impacts of FDI so far. However, comparing their results and drawing general conclusions is uneasy, as the studies use different analytical frameworks. There is a need for normalizing the approaches of the various research activities on the impacts of agricultural investment.

It is recommended that the organizations working in this area develop a common analytical framework that would be applied to all studies. They could develop a typology using the business model as the entry point, building on the available results of studies. The development of a common system could build on the Investment Development Indicators Framework, an analytical framework developed by UNCTAD for assessing the development impacts of investment. The Framework encompasses input-output analysis (backward and forward linkages) and a range of impact indicators in categories such as employment, economic value added and sustainable development. This instrument might be complemented as appropriate (e.g. at the level of the farm and/or of the local area) by the World Agricultural Watch's analytical tool. FAO and CIRAD collaborate on the establishment of the World Agricultural Watch to monitor structural changes at the farm/local area level and assess their effects on the three dimensions of sustainable development. More meta-analysis is necessary. The approach should consider the counterfactual. The common analytical tool should consider different geographical scales and time frames to capture the full effects of investment over space and time. It should examine the structural changes induced by the investment project over the short, medium and long term, at both macro- and microeconomic levels.

The case studies have not found sufficient evidence on the impacts on food security, although this is a fundamental issue in the current

[3] www.fao.org/righttofood/publi_01_en.htm

[4] www.responsibleagroinvestment.org

debates on resource-seeking foreign investment in agriculture. The lack of conclusive findings is probably linked to the fact that most of the surveyed investments were made recently, hence it was too early to evaluate their effects on food security. More gender-disaggregated analysis is also necessary to assess the differential effects of investment models on men and women.

The research has identified a large number of factors that determine the impacts of investment. These factors need to be clustered in broader categories, and their respective importance in different contexts should be assessed.

The financial returns of large-scale agricultural investments are seldom known. Future studies could consider analysing the financial returns of agribusiness in developing regions, especially Sub-Saharan Africa. Such studies could include the cost of developing a large-scale farming operation. Comparative analyses of the performance of agribusinesses and firms in other sectors could also be done.

The main topic of the studies was foreign agricultural investment. However, it is difficult to dissociate foreign and domestic investment, as they are often intertwined and complementary. The current analyses should be broadened to all forms and source of agricultural investment, including domestic investment, which is far greater than foreign investment. In the international debate on "land grabbing", concerns have focused on the role of foreign investors. This can be easily explained by the implications in terms of sovereignty, national food security and other politically-sensitive issues. Yet, in most developing countries large domestic investors acquire more land than foreign ones and there is no evidence that these acquisitions are more respectful of the rights and interests of local communities. The research and debate on large-scale land acquisition should therefore include domestic investors more systematically and give them similar attention to foreign ones.

More importantly, investments by farmers account for the bulk of agricultural investment and play an essential role in ensuring food security in developing countries. Small-scale farmers and their families are both a fundamental source of agricultural investment and the possible victims of food insecurity. Investment by small-scale and family farms should be the focus of the research on increasing agricultural investment in low-income food deficit countries.

4.2 Policies for promoting investment for sustainable agricultural development

In order to take advantage of the opportunities offered by FDI while minimizing its risks, developing-country governments should ensure that the policies, laws and regulations governing land tenure and agricultural investment are consistent and mutually supportive in order to avoid loopholes and contradictions. There is a need for a coherent and comprehensive policy on agricultural investment, bringing together scattered provisions from different policies and laws. National land tenure systems must be clear and accountable. Governments should follow the Voluntary Guidelines on the Responsible Governance of Tenure of Land, Fisheries and Forests in the context of National Food Security. The duration of land leases could be suited to the economics of investment projects, taking into account the area where the project is located, the size of the land to be leased, the business model and the economic activity. Governments could consider establishing a maximum area for land acquisitions. They should proactively protect the rights of the poor by ensuring that investors do not hold on to land they do not utilize for many years. Mechanisms should be put in place to allocate to investors only the land area that they can utilize within a reasonable span of time, and to withdraw land from investors who do not comply with agreed development plans.

Land allocation contracts could be published to ensure transparency and public scrutiny of the fairness of the deal. Coordination with related sectors (such as water management) should be enhanced so that the national political and institutional framework is conducive to sustainable agricultural development.

Furthermore, measures are needed to strengthen the institutional capacity to monitor and enforce existing laws and regulations. It is necessary to strengthen the mechanisms to promote accountability in decision making related to land allocation. Land allocations should be subject to the free, prior and informed consent of local landholders. This would require going beyond generic consultation requirements already included in laws regulating impact assessment studies. Precise requirements and criteria should be set for meaningful environmental and social impact assessments. Impact assessment should be conducted with wide participation from concerned stakeholders. The reports of the assessments should be scrutinized by competent institutions where local people could have access. Investment contracts with companies should clearly specify that any land acquisition requires the consent of local landholders.

More generally, improving governance, transparency, accountability and the rule of law in all sectors will increase the positive impacts of FDI at both local and national levels.

In addition to ensuring an enabling environment, governments should take active steps to enhance the participation of local landholders and farmers in the design and implementation of the investment projects. Economically-sound projects that give local actors an active role and a say in decision-making should be favoured. Their financial participation through the distribution of shares in the business can be considered in order to promote better sharing of the project's benefits. The shares could be jointly held in a collective trust to promote collaboration among local actors. Higher participation by local actors means higher benefits for them in case of success but also higher risks. Mechanisms should therefore be devised to reduce the risks, especially in the early stage of the project. Host governments could provide investors with incentives to involve local people, for example through discounts on the rent or taxes.

Government authorities should monitor and enforce the implementation of the contract by the investor. They should provide independent and effective mechanisms for channelling grievance and resolving dispute. They should support the development of organizations that genuinely represent local stakeholders, in particular farmer organizations, and strengthen their capacity.

More generally, governments should utilize the international guidance instruments indicated above.

Further, investments in rural infrastructure should be a priority. Governments should also invest more in the provision of education to rural communities, including vocational training and technical extension.

Finally, although this is outside the scope of this publication, it is important for governments to keep in mind that FDI only accounts for a small share of total agricultural investment and that national policies should give more emphasis to increasing domestic investment, in particular by farmers, as they account for the bulk of investment in agriculture.

4.3 Increasing the effectiveness of support

The studies suggest that good governance at both national and local levels is one of the most important conditions, if not the first one, for positive impacts of foreign investment on local development. The organizations that provide assistance to governments, farmer organizations and other stakeholders in developing countries should support efforts to strengthen the governance systems at national and local levels. Before designing support programmes they should analyse the needs of governments and local civil society organizations in terms of capacity building, policy advice and technical support. They should provide developing country governments with guidance, including practical assistance in analysing investment proposals and making informed decisions. To this end, they could set up a specialized technical assistance facility and training programmes for government officials. The facility could train local consultants who could then act as local resources. They could help governments formulate policies that orient FDI in directions that enhance smallholder family farm investments and livelihoods and national food security. They could support consultations between governments and major actors, in particular small-scale producers, for developing such policies. The capacity of government agencies to negotiate contracts with investors should be strengthened.

In addition, it is essential to strengthen the capacity of local communities, farmer groups and other civil society organizations to analyse and negotiate projects with investors and governments. There is often an asymmetry of information and power. A more level playing field must be created. Local communities must be informed of their rights. Legal assistance should be provided when an investment project is being considered. There is a need for training, capacity building, technical advice and assistance at all levels.

Development organizations should support inclusive negotiation processes at various levels, especially in the negotiations between the government and the investor, between the government and the local community, between the investor and the local community, and between the different stakeholders within the local community. They should assist developing country governments in formulating agricultural development strategies that focus on investment that nurtures sustainable food security and supports family farming. This formulation work should be done in collaboration with farmer groups and other organization genuinely representing the various interest groups of civil society.

Support programmes should identify schemes that reconcile the development objectives of host countries and local communities with the commercial objectives of investors. Strategically targeted development aid can play an important role in promoting commercially viable and socially inclusive models of agricultural investment. It is possible to bridge the gap by providing investors with incentives to design and implement projects that yield sustainable benefits to the local community as a whole.

Development organizations should raise the awareness of key investors (e.g. industry leaders and business champions) on the importance of a responsible approach to agricultural investment to their own business interests. Engaging them in the development and implementation of guidance tools such as principles for responsible agricultural investment is likely to promote adoption and a sense of ownership. They could become the best advocates of a responsible investment approach vis-à-vis other companies. This would accelerate the pace of dissemination and buy-in of the guidance tools.

With the support of development organizations, governments should devise tools to reduce the risks of foreign agricultural investment for local farmer organizations. The risk management strategy should be adapted to the level of development of the local organization.

Support programmes should promote and support multi-stakeholder partnerships for sustainable development. The roles of the public and private sectors are complementary and cannot always be substituted for each other, although their respective roles may differ by location depending, among other factors, on the level of economic and institutional development of the country and the nature of the agricultural development challenges at hand. A successful strategy for agricultural investment must be based on a partnership between governments, donors and farmers, in which each delivers in its area of responsibility.

4.4 A more proactive role for civil society organizations

Agricultural investment projects are more likely to benefit local economic and social development when local farmers and landholders play an active role. In order to be successful, inclusive business models require effective local organizations.

These organizations should be consulted and the detailed arrangements of the project should be negotiated with them, including the sharing of benefits, as discussed above. Further, local NGOs should actively engage in raising community awareness regarding civil rights and how to exercise those rights. They should advocate for better recognition of community rights by investing companies and local authorities.

Local NGOs should closely monitor potential conflicts between local communities and investing companies, keep records and inform the public. They can play the role of whistle-blower and draw the attention of the local and national authorities when a project is creating hardship to the local community or damaging natural resources. In order to represent the interests of their members effectively, these organizations should establish procedures to ensure that they function in a democratic and effective manner. Their management should be transparent and accountable to all members. Organizations aiming to be the voice of the local community must ensure that groups that tend to be underrepresented such as women, the youth, landless farmers and migrant workers are given a say in the decision-making.